INFORMATION OV

A complete list of titles in the IEEE PCS Professional Engineering Communication Series appears at the end of this book.

INFORMATION OVERLOAD

An International Challenge for Professional Engineers and Technical Communicators

Edited by

Judith B. Strother
Florida Institute of Technology

Jan M. Ulijn
Open University of The Netherlands

Zohra Fazal
Florida Institute of Technology

Sponsored by IEEE Professional Communication Society

Professional
Communication
Society

IEEE PRESS

A JOHN WILEY & SONS, INC., PUBLICATION

Library of Congress Cataloging-in-Publication Data:

Information overload : an international challenge to professional engineers and technical communicators / [edited by] Judith B. Strother, Jan M. Ulijn, Zohra Fazal.
    pages cm. – (IEEE PCS professional engineering communication series ; 2)
  Includes bibliographical references.
  ISBN 978-1-118-23013-8 (pbk.)
1.   Information resources management. 2. Time management. I. Strother, Judith B. II. Ulijn, J. M.
III. Fazal, Zohra.
T58.64.I5285 2012
658.4'038–dc23

2012004721

10 9 8 7 6 5 4 3 2 1

This book is dedicated to our parents—to the memory of Judy Strother's parents, Thelma and Bailey Banks, and her stepmother, Grace Alexander Banks; Jan Ulijn's parents, Wim and Riek Ulijn-van den Oever; and Zohra Fazal's father, Muradali Fazal, and to the honor of her mother, Nasseem Fazal. We will be eternally grateful for their many contributions toward making us who we are today.

CONTENTS

LIST OF PRACTICAL INSIGHTS FROM CORPORATIONS

A2Z Global LLC
Alvogen
Applied Global Technologies (AGT)
The Dutch Employers' Association (AWVN)
College of Aeronautics, Florida Institute of Technology (Florida Tech)
Harris Corporation
IBM
Laboratory for Quality Software (LaQuSo)
Limburg Media Group (LMG)
Xerox

LIST OF FIGURES

List of Practical Insights Figures

LIST OF TABLES

FOREWORD

As I am writing this foreword, I am keenly aware of the many e-mails I still have to answer, the phone calls that I still have to return, and the piles of papers on my desk that I still have to process. I cannot resist occasionally checking my e-mail, but I try not to reply to or be distracted by any message. I set my telephone directly to voicemail, so I can catch incoming messages later. A few times, however, I do interrupt the writing process: there are some e-mails that I really should respond to today, and I check my voicemail anyway to see if there are important messages. Soon, I find myself making calls after all, and I grow frustrated with the interruption that I initiated myself. I try multitasking, listening to voicemail messages while simultaneously reading text on my screen. I switch back and forth between the screen of my PC and that of my smartphone. No, this is not working well! Let us concentrate on the writing process!

To prepare for the writing of this foreword, I consulted various information sources, including Web of Science, Google, and Google Scholar. Finding information on the topic of information overload has turned out to be easy, but there is too much! Google gives me 3.9 million hits for "information overload" and Google Scholar gives me 61,000 articles. Where should I start? Given my other obligations, I only have about 10 hours to read up on the latest literature, so after 3 hours of searching and collecting information, I am left with only 7 hours to read the various articles I have collected. I need to read fast and smart; otherwise, I will not have the required information for my essay, as I am already past my deadline. I realize that my essay will add another 50 or 60 KB of information to the more than 2 billion GB of information that will be produced and disseminated this year [1].

I am, it seems, one of those individuals suffering from information overload, this disease of modern times that affects so many of us. Information overload has, in recent decades, become a phenomenon that is complained about in offices and homes, commented on in mass media, and studied by scholars. Books and articles warn us of the dangers of data smog and information glut and exhort us to get rid of information clutter and get control of our lives. Psychologists study how information overload leads to information anxiety, work stress, and information fatigue syndrome, and warn us about their perilous consequences. In management and organization studies, scholars consider how information overload negatively affects productivity and worker well-being and how the overload may be managed more effectively. In addition, in fields like information technology, library and information studies, and management information systems, researchers study how information systems may be developed and set up so as to help reduce information overload.

The concept of information overload was popularized in the 1970s by Alvin Toffler in his 1970 book *Future Shock* and in later publications [2]. It has since then been picked up as a topic of study in various academic fields, including library and information science, management, organization studies, psychology, and others. With the proliferation of new information technologies, information overload has been receiving increasing attention as a research topic. Although there are many conceptions and definitions of information overload, most conceptions revolve around two central ideas: they define information overload as corresponding to situations in which the amount of information supplied to an individual exceeds his or her information processing capabilities, and they emphasize the negative consequences of these situations on the functioning or well-being of such an individual. Information overload, in most conceptions, is a state of having too much information to digest, with resulting negative effects on performance or well-being. Moreover, in many conceptions, it is construed as a problem specific to the workplace. In other conceptions, it is defined more broadly, as a general condition of modern life that can manifest itself in different contexts and activities.

The academic study of information overload precedes the rise of the Internet and the emergence of e-mail and mobile and social media in the 1990s and 2000s. Classical notions of information overload, developed before the rise of these new media, tend to define a particular type of information overload, which I call *task-related information overload*. Task-related information overload concerns situations in which there is too much potentially relevant information to perform certain tasks or make certain decisions, relative to the information processing abilities and the amount of time and resources that an individual has available. As a result, the individual becomes overburdened and either performs tasks poorly or spends too much time and too many resources on the task, thereby causing problems in other areas.

Task-related information overload is simply a situation in which an individual performing a task has to wade through massive amounts of information to find and interpret those pieces of information that are most relevant to the task at hand, and finds that he or she does not have enough time and resources to perform the task well, given the information requirements of the task. A typical example would be a situation in which an employer is asked to write an internal report of the safety procedures in a factory and is confronted with dozens of internal documents that may contain information about the factory's safety procedures, with very little time to process them. Another example is a case in which an individual wants to learn about atherosclerosis, only to find that his or her PC gives 12,000,000+ entries on this topic to choose from and to realize that it is difficult to determine which entries contain reliable information that is relevant to the topic.

In the 1990s and 2000s, the emergence of new communication media like e-mail, mobile communications, and social media introduced a new kind of information overload. These media confront individuals with a continuous bombardment of messages that require their attention and interrupt their activities. The resulting type of information overload may be called *message overload* (or, alternatively, *communication overload*). In message overload, the problem is not so much the abundance of information available for a given task, but rather the abundance of received messages, each of which invites a new task to be performed. Unlike a mere piece of information, a message has a sender and one or more receivers. While some messages merely contain some information that the sender wants its receivers to read or see, others invite replies or define tasks that its receivers are expected or invited to perform. Messages can cause overload in two ways. First, if people receive too many messages, they fall behind in their responses to them and feel overloaded in the processing of messages and any further tasks that their proper processing requires. Second, messages often interrupt activity and divert attention and may invite multitasking activities, thereby contributing to a feeling of overload.

Information overload is sometimes associated with a third type of overload, which I call *media overload*. In media overload, the overload does not result from having too much information to perform a task or from being bombarded by messages that ask for attention and a response. Rather, overload results from having too much choice in media content and being lured by media messages that invite ever more consumption of media content. Media overload is the result of the abundance of content and channels in contemporary media, including television, radio, the Internet and mobile and social media, music, games, movies, lifestyle and infotainment sites, news sites, blogs, and twitter feeds. As a result of this abundance of content, media consumers may become distracted from everyday activities and become superficial channel surfers without a focus in their media consumption activities, lost in a sea of media content.

Information overload thus comes in at least three forms, which include task-related, message, and media overload. In the workplace, the first two are the most important, whereas in the home, the third is often the most important. Information overload in the workplace often correlates with *work overload*, which refers to situations in which the workload is greater than the amount that workers can take on given the time and resources that they have available. Work overload can exist independent of information overload, for example, in professions such as construction work or farming in which the processing of information is not a major part of the job. But in today's offices, work overload is strongly correlated with information overload. Today's office workers are bombarded with messages that define or prompt tasks, and these tasks often require the collection and processing of large amounts of information under time pressure. Both the number of messages to be processed and the information intensity of tasks to be performed are major contributors to work overload in office environments.

That information overload has negative consequences for the functioning and well-being of individuals is often considered to be an integral part of its definition. Information overload negatively affects performance because it engenders poor decision making, poor execution of tasks, loss of time because of interruptions in work activity, and diminished creativity. It also negatively affects well-being by contributing to stress, anxiety, fatigue, loss of motivation, and even depression. From an organizational viewpoint, information overload in workers may result in loss of productivity, poor strategy and implementation processes, and resulting economic losses [3].

Why has information load become such a pressing issue in our time? Some of its more immediate causes are found in changing practices of producing, transmitting, storing, and consuming information, which are themselves driven by recent developments in information and communication technology (ICT). Developments in ICT have enabled a variety of developments that contribute to information overload: increased production of new information; easy and cheap duplication, storage, and transmission of information; an increase in new information channels such as e-mail and instant messaging; and a democratization of the roles of information producer and provider, which allows anyone to easily generate content and disseminate it to others, which then results in an abundance of new content and has contributed to a low signal-to-noise ratio.

It is not just developments in ICT, however, that have yielded information overload. It is also transformations of workplaces and organizations, as well as larger changes in culture and society. Work has changed because of globalization and economic restructuring. In his well-known study of the information society, Manuel Castells argues that since the 1970s, a new, global capitalist economic system has been forged, which is highly competitive, flexible in its labor and production processes, and organized around ever changing networks of individuals and organizations rather than factories. These networks are held together by ICT. In this new economic structure, there is an imperative of constant communication and the constant production and absorption of information. Information and communication processes are moreover subject to competitive standards of efficiency and quality standards [4].

Everyday life is also changing in the information society that has emerged since the 1970s. Contemporary society is a highly liberalized consumer society, centered around individual choice and opportunity. People draw up their own plans, make their own lifestyle choices, create their own social networks, and choose the products and services they want. They do not necessarily trust authorities and experts; instead, they form their opinions after consulting multiple information sources. This highly individualized way of life requires constant management of information and communication processes. It results in what Heylighen has called *opportunity overload*, the paralysis that results from having too many options to choose from in life: too many ways to spend one's vacation, too many brands of detergents to choose from, too many ways of going from *A* to *B* [5]. The desire to manage opportunity overload is itself a cause of information overload, as it causes people to seek forever more information to guide their decision-making processes.

These factors suggest that information overload will not go away easily. Information overload seems to be an aspect of the modern condition, bound up with the way we have organized the modern economy and life. So what are our prospects for limiting information overload? Part of the solution may be found in better information technology. Information systems may be designed to do some of the work of processing and structuring information for us and to help us interpret and recognize relevant information. Better search engines and filtering systems may be developed, intelligent agents may help us with information queries, and the development of semantic web technologies may further facilitate information retrieval and organization. As philosopher Hubert Dreyfus has argued, however, intelligent processing of information requires that information systems are capable of discerning relevant from irrelevant information. This is a skill that computers are not particularly good at [6]. There are limits, therefore, to the extent to which information systems can help manage information overload for us.

Part of the solution must therefore lie in better information management by humans and organizations. Workers may be trained to master better techniques for sifting through information, and they may learn to be more selective and efficient communicators in the messages they send out. Organizations may be redesigned so as to support a more efficient transmission and utilization of information resources. Fields like management and organization studies, library and information science, communication studies, and psychology may help us in developing these improved tools for information management.

Computer Scientist David Levy has argued, however, that good information management is not enough to avoid information overload. As he argues, workers also need a space and time for thinking, reflection, and extensive reading. They need to be able to step back from the constant stream of information piled onto them, and devote time to creative thinking, musing, and careful consideration [7]. Organizations may want to consider creating such room for their workers, running against the trend of an ever-increasing production and consumption of information. Productivity and quality may eventually improve because of it.

The present book considers causes of information overload, its harmful consequences, and possible steps toward a reduction of information overload. Therefore, it makes an interdisciplinary contribution to the study of information overload that should be of value to professionals in a variety of fields.

PHILIP BREY

About the Author

Philip Brey, Ph.D., is currently a Full Professor of Philosophy of Technology and the Chair of the Department of Philosophy, University of Twente, Enschede, The Netherlands. He is also the Director of the 3TU Centre of Excellence for Ethics and Technology. His research interest is in the philosophy of technology, with special emphasis on ethical issues in technology and the philosophy of information and communication technology. He is particularly concerned with the implications of information technology for the quality of life. He has extensively published on the social and ethical impacts of information technology. His most recent publication is *The Good Life in a Technological Age*.

REFERENCES

J. Gantz and D Reinsel, *Extracting Value from Chaos*. IDC Corporation Report, 2011. Available at http://idcdocserv.com/1142. Accessed June 30, 2012.

A. Toffler, *Future Shock*, New York: Random House, 1970.

M. Eppler and J. Mengis, "The concept of information overload: A review of literature from organization science, accounting, marketing, MIS, and related disciplines," *The Information Society*, vol. 20, no. 5, pp. 1–20, 2004.

M. Castells. *The Rise of the Network Society. Information Age*, vol. 1, Oxford: Blackwell, 1996.

F. Heylighen. Complexity and information overload in society: Why increasing efficiency leads to decreasing control. ECCO working paper, 2002. Available at http://pespmc1.vub.ac.be/Papers/Info-Overload.pdf. Accessed June 30, 2012.

H. L. Dreyfus. *On the Internet*. London: Routledge, 2001, pp. 8–26.

D. Levy. "Information overload," in *The Handbook of Information and Computer Ethics*, K. Himma and H. Tavani, Eds, Hoboken, NJ: John Wiley and Sons, Inc., 2008.

PREFACE

This book is the culmination of four years of work that developed from an idea that germinated during the International Professional Communication Conference (IPCC) 2008 in Montreal, Canada. The conference chair, Dr. Kirk St. Amant, pointed out that the program committee had received a number of proposals that would have made great papers, but that did not fit the theme of that particular conference. A number of these proposals addressed a timely issue—information overload.

We all experience information overload, and it shows no signs of abetting. We wondered: What is it? How do we manage it? What does the research say about it? As we started looking into the literature, we realized the paucity of solid information and research on the topic of information overload. Our discussions quickly evolved into a book project, during which additional authors—from both the United States and abroad—were invited to contribute. In addition, we felt a strong need for corporate input, so we invited key corporations to share their strategies for coping with the challenge—for both their employees and their organizations. This book is the result of significant efforts from all involved.

We hope that, with its wide range of insights from this international group of scholars and practitioners, this book helps you, the reader, further understand the challenges and costs of information overload. In addition, we sincerely hope that the strategies presented herein will help you not only reduce the problem for those around you, but will also devise and/or improve your own system of dealing with the information deluge that is all too common in today's information economy.

ACKNOWLEDGMENTS

This book would not have come to fruition without the generous assistance and guidance from a number of colleagues on both sides of the Atlantic.

Special thanks go to Traci Nathans-Kelly, IEEE Series Editor, *Professional Engineering Communication*, IEEE Professional Communication Society, for her insights, her willingness to help with every phase of the manuscript preparation process, and her unflagging good humor. We would also like to thank Taisuke Soda, Senior Editor, and Mary Hatcher, Associate Editor, Wiley/IEEE Press as well as Sanchari Sil, Customer Service Manager, Thomson Digital, for their assistance with the publishing process. We are also grateful for the support of The Professional Communication Society of the IEEE and its leadership. In particular, during the early stages of conceptualizing and preparing the manuscript, Kirk St. Amant provided valuable insights, for which we are very appreciative.

For our *Practical Insights* boxes, a number of corporate executives and professionals generously gave of their time for interviews about their companies' challenges and best practices in the area of information overload. These executives include the following.

A2Z Global LLC	Theodora Landgren, Director
Alvogen	Dr. Svafa Grönfeldt, Chief Organizational Development Officer
AGT	Josh Lott, Software Engineer
AWVN	Dr. Arjen Verhoeff, Senior Advisor
College of Aeronautics, Florida Tech	Captain Nicholas Kasdaglis, U.S. Air Force, Retired, Aerospace and Human Factors Researcher Dr. John Deaton, Director of Research and Chair, Human Factors
Harris Corporation	Amar Patel, F22 Program Manager
IBM	Jelmer Letterie, External Relations, IBM Benelux Kees Verweij, CIO, IBM Netherlands
LaQuSo	Dr. Harold Weffers, Director
Limburg Media Group	Hans Thijssen, Human Resources Manager Huub Paulissen, Editor-in-Chief
Xerox	Bill McKee, Corporate Public Relations Caroline Privault, Project Leader, Xerox Research Centre Europe Chris Holmes, CEO, Meshin (a start-up funded by Xerox)

We appreciate Dr. Arjen Verhoeff's assistance in arranging the interviews with IBM and with the Limburg Media Group as well as his input on the text of the interviews themselves. Dr. Michael Gould was a skillful facilitator for the Xerox interview, as Dr. Svafa Grönfeldt was for the Alvogen interview, and we appreciate their assistance.

During the final stages of manuscript preparation, the following people contributed their editing skills, and we are grateful for their assistance: Rolanda Gallop, Bill Leach, Maya Oluyesi, Kay Rettich, Jenna Sheldon, Edwin Strother, Murtaza Syed, and Michelle Verkooy. Ivana Kostadinovska created the artwork for Chapter 4, and Dean Faithfull, Katheryn Broderick, and Jillian Knight provided assistance during the final manuscript preparation.

Such a monumental undertaking could only have been completed with support from our families and friends. Edwin Strother, Judy's husband; Pieta Ulijn, Jan's wife; and Murtaza Syed, Zohra's husband, have provided unflagging encouragement and enduring understanding. For that, and so much more, they have our love.

A NOTE FROM THE SERIES EDITOR

The IEEE Professional Communication Society (PCS), with Wiley-IEEE Press, is excited to launch its book series titled *Professional Engineering Communication* with this collection of insightful essays. Judith Strother, Jan Ulijn, and Zohra Fazal have gathered a chorus of knowledgeable voices that will contribute greatly to the ways that we think about the plethora of information channels crowding our working and leisure hours. While this may be the Age of Information, the information we get is now filtered, sponsored, edited, firewalled, blogged, tagged, hashtagged, shared, liked, deleted, filed, stored, archived, tweeted, and pinned—many of these permutations for information delivery could not have been predicted even 5 years ago. Understanding and controlling information input has become yet another task professionals need to fit into their daily workflow. In truth, the stability of this printed book may be a nice reprieve from the electronic influx sitting in your e-mail box right now, even if you are reading it on a Kindle®.

We begin the series with an examination of "information overload," a term which can have various interpretations, all of them seemingly negative. To engineers, an overload of anything is detrimental and to be avoided. Information overload, however, is an interesting problem to have. Is too much information bad? Is increasing transparency within and around organizations somehow a detriment? Most of us would argue that more information is better, regardless. Thus, this book examines from many angles the ways in which we have come to manage the now-constant flow of information presented in the twenty-first century industrialized workplace.

I want to back up a bit and talk about this new PCS-sponsored project. As a series, *Engineering Professional Communication* (ECP) has a mandate to explore areas of communication practices and application as applied to the engineering, technical, and scientific professions. Including the realms of business, governmental agencies, academia, and other areas, this series will develop perspectives about the state of communication issues and potential solutions when at all possible.

The books in the ECP series will keep a steady eye on the applicable while acknowledging the contributions that analysis, research, and theory can provide to these efforts. Active synthesis between on-site realities and research will come together in the pages of this book as well as other books to come. There is a strong commitment from PCS, IEEE, and Wiley to produce a set of information and resources that can be carried directly into engineering firms, technology organizations, and academia alike.

At the core of engineering, science, and technical work is problem solving and discovery. These tasks require, at all levels, talented and agile communication practices. We need to effectively gather, vet, analyze, synthesize, control, and produce communication pieces in order for any meaningful work to get done. It is unfortunate that many technical professionals have been led to believe that they are not effective communicators, for this only fosters a culture that relegates professional communication practices as somehow secondary to other work. Indeed, I have found that many engineers are fantastic communicators because they are passionate about their work and their ideas. This series, planted firmly in the technical fields, aims to demystify communication strategies so that engineering, science, and technical advancements can happen more smoothly and with more predictable and positive results.

TRACI NATHANS-KELLY, PH.D.
Series Editor

CONTRIBUTORS

Debashis "Deb" Aikat, Ph.D., a former journalist, is an Associate Professor and media futurist with the University of North Carolina at Chapel Hill School of Journalism and Mass Communication. An expert on people, products, and perspectives shaping the digital revolution, he researches digital media and the future of communication. He is an award-winning teacher and researcher on the impact of communication technologies and social aspects of interactive technologies. His research has been published in ACM SIGDOC, Microsoft Corporation publications, *Global Media and Communication, Electronic Journal of Communication*, and *Convergence: The Journal of Research into New Media Technologies* among other refereed publications. He completed a Certificate in American Political Culture from New York University and earned his Ph.D. degree from the E. W. Scripps School of Journalism at Ohio University.

Noël T. Alton has been a graduate student of linguistics at Brigham Young University and a teaching assistant for undergraduate level classes in syntax and translation applications. Her thesis topic deals with information overload in visual design and applications of basic semiotic theory to improve design. She currently works as an editor with Western Governors University, where she helps write articles on educational practices for online universities.

Prasanna Bidkar, M.A., is a graduate in English (technical writing) from Oklahoma State University. He has over 9 years of industry experience, having worked on several end-user documentation projects for leading organizations. He now assists various small and medium businesses communicate to their technical and non-technical audiences, as a partner at RightMix Technologies LLP.

Anne Caborn is a digital communication and digital governance consultant; Founder and Director of Make it and Mend it Ltd., a sustainable living website; a Director of Brighton Housing Trust Enterprises Ltd., an organization that supports social enterprises; and a steering group member of Mentoring Digital Minds, a mentoring organization for digital industries. A journalist and editor by training, she moved into digital communications over 10 years ago. She regularly speaks at conferences as well as develops and delivers training courses and workshops related to digital communication, digital governance, and information handling.

Toon Calders, Ph.D., is an Assistant Professor with the Information Systems Group at the Eindhoven University of Technology, The Netherlands. He graduated in 1999 from the University of Antwerp with a Diploma degree in mathematics. He received his Ph.D. degree in computer science from the same university in May 2003, in the database research group ADReM. From May 2003 to September 2006, he continued working with the ADReM Group as a Post-Doctoral Researcher. He has been with the Eindhoven University of Technology since October 2006. He has published over

50 papers on data mining in conference proceedings and journals, served on several program committees of the most important data mining conferences, including ACM SIGKDD, IEEE ICDM, ECMLPKDD, and SIAM DM, was the Conference Chair of the BNAIC 2009 and EDM 2011 conferences, and is an Editorial Board Member of the *Springer Data Mining* journal and the Area Editor of the *Information Systems* journal.

Cary L. Cooper, CBE, is a Distinguished Professor of organizational psychology and health at Lancaster University Management School, England. He is also the Chair of the Academy of Social Sciences (a body composed of 40 learned societies in the social sciences numbering over 86,000 members), the President of the British Association of Counselling and Psychotherapy, the President of RELATE, and the President of International Stress Management Association (U.K.). He is a fellow of the British Psychological Society, the Royal Society of Medicine, the Academy of Management (U.S.), the British Academy of Management, and many more. He has a number of Honorary Doctorates (e.g., Aston University, Sheffield University), was awarded The Times Higher Education Lord Dearing Life Time Achievement Award for contributions to U.K. higher education, and in 2001, was honoured by the Queen with a CBE for his contribution to organizational health.

Martin J. Eppler, Ph.D., is a Professor of communication management at the University of St. Gallen, where he directs the Institute for Media and Communication Management. Besides his research on information overload, he focuses on knowledge management, knowledge visualization, and knowledge communication. He has been a Guest Professor at various universities in Asia, Europe, and South America, and an advisor to organizations such as the United Nations, Philips, UBS, the Swiss Military, Ernst & Young, KPMG, Swiss Re, BMW, and Daimler. He studied communications, business administration, and social sciences at Boston University, the Paris Graduate School of Management, and the Universities of Geneva and St. Gallen. He has published more than 100 academic papers (in journals such as *Organization Studies*, *LRP*, *Harvard Business Manager*, *TIS*, *EMJ*, and *IV*) and 10 books, mostly on knowledge communication, management, and visualization.

George H. L. Fletcher, Ph.D., is an Assistant Professor with the Databases and Hypermedia Research Group. He received a Doctorate degree in computer science from Indiana University, Bloomington, in 2007. After completing his Ph.D. degree, he worked for 2 years as an Assistant Professor at Washington State University. He has been at the Eindhoven University of Technology since July 2009. His research focuses on the study of database query languages for data integration, XML, and graph data. His research has resulted in publications appearing in top-tier conferences (e.g., PODS, ICDT, EDBT, SIGMOD, and CIKM) and journals (e.g., *Information Systems*, *ACM Transactions on Database Systems*, and *IEEE Transactions on Knowledge and Data Engineering*).

Eduard Hoenkamp, Ph.D., is a Professor of information sciences, Queensland University of Technology. He received the Doctorate degree in psychology (psycholinguistics) from the University of Nijmegen, The Netherlands, and the Master's degree in mathematics (formal languages and automata theory) from Delft University, The Netherlands. Hoenkamp has worked extensively on the artificial intelligence (AI) approach to natural language processing at the AI Departments of Yale University, MIT, and UCLA. During the last decade, his research has focused on information retrieval (IR). He has made contributions to the theoretical foundations of IR as well as improvements to search algorithms on the basis of a cognitive, human-centered, approach. Hoenkamp is a member of the Association of Computing Machinery (ACM), IEEE, the Cognitive Science Society, Computer Professionals for Social Responsibility, and the International Association of Privacy Professionals.

Faisal Kamiran, MSCS, received his Master's degree in science and computer science from the University of the Central Punjab (UCP), Lahore, in 2006. He has been a doctoral candidate with the Databases and Hypermedia (DH) Group of the Eindhoven University of Technology under the supervision of Prof. Dr. Toon Calders and Prof. Dr. Paul De Bra since February 2008. His research interests include constraints-based classification, privacy preserving, and graph mining.

Alan Manning, Ph.D., is a Professor of linguistics and English language at Brigham Young University. He received his Doctorate degree from Louisiana State University in 1988 in linguistics with an emphasis in technical communication, subsequently teaching at Stephen F. Austin University in Texas and Idaho State University. He currently teaches graduate courses in writing and research design and undergraduate courses in linguistics, information design, and editing. He has served as an Associate Editor of the *IEEE Transactions on Professional Communication* from 1998 to 2004. He is a co-author of the book *Revising Professional Writing in Science and Technology, Business, and the Social Sciences, 3rd ed.* (Chicago, IL: Parlay Press, 2011).

Jeanne Mengis, Ph.D., is an Assistant Professor of communications management at the University of Lugano, Switzerland, and an Associate Fellow at Warwick Business School, U.K. Prior to this, she was a research fellow at Warwick Business School and a visiting fellow at the School of Management, Boston University, and at the Kennedy School of Government, Harvard University. Her main publications (e.g., in *Organization Science, Organization Studies, Social Science, Medicine,* and *Harvard Business Manager*) deal with integrating knowledge in cross-disciplinary collaborations and knowledge communication, phenomena that she has studied in healthcare settings, scientific laboratories, financial institutions, and public organizations. She has teaching experience at the universities of St. Gallen, Lausanne, and at the Swiss Federal Institute of Technology in Zurich.

Mykola Pechenizkiy, Ph.D., is an Assistant Professor with the Department of Computer Science, Eindhoven University of Technology, The Netherlands. He received his Ph.D. degree from the Department of Computer Science and Information Systems at the University of Jyvaskyla, Finland, in 2005. He has broad expertise and research interests in data mining and data-driven intelligence, and its application to various (adaptive) information systems serving industry, commerce, medicine, and education. He has co-authored over 60 publications and has organized several workshops (HaCDAIS at ECML/PKDD2010, LEMEDS at AIME2011), conferences (IEEE CBMS 2008, BNAIC 2009, and EDM 2011), and tutorials (at ECML/PKDD 2010, IEEE CBMS 2010, and PAKDD 2011) in these areas. Recently, he co-edited the *Handbook of Educational Data Mining* and served as a Guest Editor of two special issues in Elsevier *DKE* and *AIIM* journals.

David Remund, Ph.D., APR, is an Assistant Professor of public relations with the School of Journalism and Mass Communication at Drake University in Des Moines, Iowa. He also conducts research on communication leadership and the effectiveness of communication programs. Prior to entering academia, he designed and managed strategic communication programs for four Fortune 250 corporations over the span of nearly 20 years. He has also worked in the public relations agency setting as the Director of strategy and planning. He completed his Ph.D. degree at the University of North Carolina at Chapel Hill. He is nationally accredited by the Public Relations Society of America.

ArjenVerhoeff, Ph.D., is a Senior Advisor at AWVN on social innovation, labor relations, and the interface between firms and academics. The mainstream of his recent projects and academic research in a great variety of national and international companies focuses on improving productivity and labor participation. He is an interlocutor of management of companies and branches for the development, implementation, and evaluation of HR policy. He has developed many HR instruments related to labor productivity and labor participation, and he has executed projects in a great variety of market-oriented companies.

Paulus Hubert Vossen is with the Software Quality and Innovation Research Institute (SQUIRE). He started his career as an experimental–mathematical psychologist at the Radboud University of Nijmegen, The Netherlands. He continued as a Project Coordinator at the Erasmus University of Rotterdam, managing interdisciplinary projects on topics such as optimal decision making and conflict resolution. His interest in the personal and professional use of computers caused him to leave The Netherlands and join the Fraunhofer Institute for Work and Organization in Stuttgart, Germany, where he became one of the founding members of the local Research Group on Human-Computer-Interaction (HCI) and Interactive Systems. After 15 years, he swapped for a while to industry, applying HCI wisdom and standards to Web design as well as learning more about the craft of software testing and software quality management. Since 2003, he has been a Freelance Research and Development Manager as well as an engaged Lecturer of software and knowledge engineering at several vocational universities in the south of Germany.

ABOUT THE EDITORS

Judith B. Strother, Ph.D., is the Chair of the Graduate Program in Global Strategic Communication and a Professor of Communication at the Florida Institute of Technology (Florida Tech) in Melbourne, Florida. She is a founding member of the Center for Communication Excellence at Florida Tech. She is also a Visiting Professor at the Eindhoven University of Technology in The Netherlands, where she teaches Technical Report Writing and Editing in two doctoral programs. She earned her Ph.D. degree in Psycholinguistics at the Eindhoven University of Technology. She has a Master's degree in Applied Linguistics from Antioch International University and an M.B.A. degree from Florida Tech.

Dr. Strother's research has covered a broad spectrum of topics in professional communication. She has written four books and several book chapters, in addition to journal articles and conference proceedings. She is the Vice President for Content Development for Aviation English Training, Inc., where she both wrote content and led a team of writers to produce online training modules in Aviation English for international pilots and air traffic controllers. In addition, she co-edited (with Dr. Julia Williams) a special issue for *Transactions*. She serves as a reviewer of the *Journal of Business Communication,* the *International Journal of Applied Aviation Studies*, IASTED (International Association of Science and Technology for Educational Development), and several other journals.

Jan M. Ulijn, Ph.D., is an early emeritus on an endowed Jean Monnet Chair in Innovation, Entrepreneurship, and Culture at the Eindhoven University of Technology (Organization Science and Marketing); Professor of International Entrepreneurship, Innovation, and Culture at the Open University of The Netherlands (since 2006); and a part-time Visiting Professor at the Delft University of Technology in European Aviation Entrepreneurship and Culture. He holds a CLUSTER (Consortium Linking Universities of Science and Technology for Education and Research) Chair at the INP de Grenoble (Grenoble Institute of Technology), France. He is a fellow of the Eindhoven Center of Innovation Studies (ECIS) and the American Society of Technical Communication. He has published about 150 internationally refereed journal articles (including 10 special issues), contributed chapters to about 10 books, authored 2 books, and recently edited 2 with others. In 1998, he earned the Association of Business Communication Outstanding Researcher Award. In 2003, one of his journal articles (with Antoon van Luxemburg and Nicole Amare) won the IEEE PCS Best Article Award.

Since his post-doctoral year at Stanford University (1979–1980) and the inception of his Jean Monnet Chair (1995), psycholinguistics and technical communication have been his focus. Dr. Ulijn's most recent (since 2004) experience with edited books has been four volumes with Edward Elgar Publishers on topics other than communication. In addition, he has edited several special issues of journals related to intercultural communication and management, especially linked to innovation and entrepreneurship.

Zohra Fazal, M.S., is an Instructor of Humanities and Communication and a founding member of the Center for Communication Excellence at the Florida Institute of Technology (Florida Tech). She is currently pursuing her Ph.D. degree in Science/Math Education. Her research interests include the use of new media in communication, the role of linguistics in the science classroom, interdisciplinary collaboration in writing, and communication in multicultural settings.

INFORMATION OVERLOAD: AN INTERNATIONAL CHALLENGE TO PROFESSIONAL ENGINEERS AND TECHNICAL COMMUNICATORS

Judith B. Strother, Jan M. Ulijn, and Zohra Fazal

> *People have become information junkies, and the very tools that cause Information Overload are also the tools that feed their habit.*
>
> Jonathan B. Spira
> CEO and Chief Analyst of Basex

1.1 Definitions, Causes, and Consequences of Information Overload

In today's information-driven economy, the ability to efficiently find, critically analyze, and intelligently use reliable information is a major factor, if not the key, to profit making. Ironically, however, overexposure to that critically valuable resource, information, leads to *information overload* and its detrimental effects.

1.1.1 Definitions of Information Overload

We have all experienced information overload and we know it when we encounter it—but how do we define it? Businessdictionary.com defines information overload as the "stress induced by reception of more information than is necessary to make a decision (or that can

Information Overload: An International Challenge for Professional Engineers and Technical Communicators, First Edition. Edited by Judith B. Strother, Jan Ulijn, Zohra Fazal.

be understood and digested in the time available) and by attempts to deal with it with outdated time management practices" [1]. In addition to using the standard definition of "too much information for one person to absorb," *PC Magazine* says that information overload includes "the excessively intricate and often indecipherable manuals that must be read to operate everything from a handheld device to a software application" [2].

1.1.2 Causes of Information Overload

PC Magazine cites the oft-quoted observation that "the volume of information that crossed our brains in one week at the end of the Twentieth Century is more than a person received in a lifetime at the beginning of it" [2]. IEEE *Spectrum* notes that "Information, the very thing that makes it possible to be an engineer, a doctor, a lawyer, or any other kind of modern information worker, is threatening our ability to do our work. How's that for irony? The global economy may run on countless streams, waves, and pools of information, but unrestrained, that tidal wave of data is drowning us" [3].

The development of the Internet quickened the pace at which information could be transmitted and received, and the advent of Web 2.0 went even further. With social networking tools, such as Facebook™ and numerous blog sites, anyone with a thought, an idea, or an opinion and access to the Internet can now become not only an author but also a publisher of information—and Web 3.0 and 4.0 are on the horizon. As an example, a simple Google™ search for *information overload* results in over 22,800,000 hits, while an article search using the more selective Google Scholar engine yields about 446,000 hits. The searcher is then left to sift through all the results to find the information being sought. While some forms of information and communication technologies provide solutions for information overload, for example through the use of search engines and gatekeepers, the further development of all kinds of communication support devices and software, such as Skype™, Twitter™, LinkedIn™, Facebook, YouTube™, iPhones™, and iPads™, provides additional sources of communication leading to information overload as well.

The general openness of Web communication poses a problem for all information seekers, who face the task of looking for valuable facts in a vast sea of information. For example, social media provide new avenues for corporations to communicate with their supply chain and their customers, but these same media can also engender a flood of feedback that corporations must cope with. Sometimes, businesses have access to so much information that they fail to distinguish between the information they can or cannot access. A commonly reported complaint at Hewlett-Packard was "If only HP knew what HP knows," reflecting the "growing gap between the potential and actual value of HP's collective intellectual assets . . . " [4]. This clearly demonstrates that large organizations face the challenge of keeping up with and organizing the volume of information that is produced, sent, received, and archived daily. As Edmunds and Morris [5] pointed out, the main problem is how to deal with the "paradox of a surfeit of information and a paucity of useful information."

Zeldes *et al.* [6] believe that one of the issues is that technology and organizational culture have not evolved at the same rate and therefore are often out of sync. Table 1.1 shows what is possible technologically versus what is actually done within the paradigm of most corporate cultures.

To add to the problem, much of the information that constantly surrounds today's information seeker is not reliable or comes from limited sources. While the system of careful peer review is still in place in academia, no longer are academic or technical publications the only information source one can find, nor are they the primary source for

Table 1.1 The Chasm Between Technology and Corporate Culture—Between Possibility and Practice [9]

Technological Reality (What Is Possible)	Cultural Paradigm (What Is Done)
Unlimited *accessibility* of everyone to everyone by many communication channels	Everyone is expected (by managers, peers, and self) to be *available* to everyone 24×7
Sending messages is easy to do and perceived as practically free of cost (monetary or other)	We sanction the unlimited sending of unsolicited messages ("freedom of speech")
Free, asynchronous access to everyone's attention queues	Interruption-driven, unnegotiated task management replacing plan-driven methodology
Queued messaging is available for most communication modes (e-mail, voice mail)	Expectation that message queues be emptied (including unsolicited messages)
Work from home technology is "as good as being in the office"	No clear understanding, much less a policy, of where to place the work–life barrier
Computers allow multitasking and rapid switching from task to task	Implicit expectation that all people are good at multitasking and can switch rapidly

even highly educated managers. In fact, too often, solid research is not accessed and used inside firms, where employees often depend on vendors or industry information to the exclusion of other sources, resulting in a biased view of the landscape. A practicing engineer, for example, may turn to "The Weld Guru" site instead of doing deep research for his or her needs in technical manuals or reports. Similarly, although news media continue to thrive, news is often broken, not by major newspapers but by individuals on social media such as Twitter and blogs. Commentary about the news, be it political, technical, or economic, is no longer limited to designated "experts"; rather, almost anyone can express opinions via the Web. Thus, the role of the information gatekeeper has been reduced significantly, for better or for worse.

In the workplace, information overload occurs not only when knowledge workers search for information but also when information searches for them. E-mail was once considered to be the efficient and cost-effective replacement for other nonelectronic forms of communication. Now, however, solicited and unsolicited e-mails fill up and clog inboxes, often significantly reducing the knowledge worker's efficiency. Smart phones are like minicomputers that allow employees and executives to receive not only calls and voice and text messages but also digital media, including full-length documents. Computer monitors and smart devices stream real-time business data as well as international news events. In short, knowledge workers are simply bombarded by information that constantly demands their attention. Indeed, information overload creates a problem for all information users, including practicing engineers, technical communicators, experts in information technology and computer science, and managers and business communication practitioners.

1.1.3 Consequences of Information Overload

Information overload "places knowledge workers and managers worldwide in a chronic state of mental overload. It exacts a massive toll on employee productivity and causes significant personal harm, while organizations ultimately pay the price with extensive financial loss" [6]. Hundreds of thousands of hours are lost in a typical organization—up to 25% of a workday—just from workers' attempts to cope with the flood of information [7].

As Spira notes, "Information Overload decimates work-life balance, decreases knowledge workers' effectiveness and efficiency, and causes diminished comprehension levels, compromised concentration levels, and reduced innovation" [7]. In their study of computer-mediated communication, Jones *et al.* [8] analyzed over 2.65 million postings to 600 Usenet groups. They found that the more overloaded the users were, the more likely they were to respond only to simpler messages, to generate increasingly simpler responses and to eventually terminate their active participation. (For a recent update of computer-mediated communication studies, see [9].) Zeldes *et al.* [6] added that information overload causes decreased mental acuity, leading to a reduction in "thinking, generating creative ideas, and effectively solving problems."

To prevent information overload, people sometimes avoid certain avenues of information, thereby missing opportunities. Although it is not easy to quantify these indirect costs of information overload, they add to the already significant direct costs, making information overload an expensive problem that must be addressed. Zeldes *et al.* attempted to quantify the costs of what they call *infomania*, "the mental state of continuous stress and distraction caused by the combination of queued messaging overload and incessant interruptions." Through their research at Intel, they concluded that infomania costs about US$1 billion per year for a 50,000-employee knowledge-intensive company—and noted that this calculation was conservative [8]. According to a Basex estimate, the total cost of information overload to the U.S. economy amounted to almost US$ 1 trillion in 2011 [10].

Resources for managing information overload currently include search engine filters and spam blockers, but not much more. While the computer industry tries to deal with the technical aspects of the problem, knowledge workers cannot do much except try to deal with the overload as well as they can while they also attempt to prevent themselves from contributing to the problem. Considering that end users tend to employ avoidance strategies, it is imperative that communication professionals employ a variety of strategies to avoid contributing to information overload for their clients and other audiences. At the same time, they themselves face the challenges of dealing with the ever-increasing flood of information that constantly surrounds them, increasing their own level of frustration.

Although there has been little solid research on the causes and consequences of information overload, it affects all of us. It can be especially troublesome and costly for practicing engineers and professional communicators who are considered subject matter experts (SMEs) and whom others tend to rely on for information. This volume addresses the needs of engineering and technical communication professionals in both academia and industry by focusing on the causes and consequences of as well as presenting strategies to deal effectively with the challenges of information overload, one of the most pervasive problems of the electronic information age.

1.2 Perspectives on the Concept of Information Overload

Decision makers of all kinds are often overwhelmed with information and lack the time management techniques to cope with the problem. The definitions of information overload given earlier provide a good start for studying the subject, but how do we deal with the elements that make information overload such an international and intercultural challenge for professional communicators in the engineering, scientific, technical, and business fields? To conceptualize the challenge, we consider information overload in terms of information and time management. How can professional communicators assist? In this book, we argue that communicators need to know how their products are processed by their clients to help them minimize and/or avoid information overload. To gain this understanding, communicators must be intimately familiar with the supplier/producer/ writer and client/user/reader perspective. In addition, the intercultural perspective has to be taken into account. Finally, a focus on innovation needs to be maintained to try and manage information overload.

1.2.1 An Information and Time-Management Perspective

If we assume that most consumers of information, for instance, as engineers or business and technical managers, have to find and use information within the real constraints of time to make decisions of all kinds, information overload is reduced to a matter of the management of *data* (facts without any interpretation), *information* (data interpreted meaningfully in a communicative chain of writers and readers), and *knowledge* (information that refers to a learning cycle) [11]. Knowledge comes in two basic forms relevant to engineering and technical communication: declarative, which addresses *what*, and procedural, which addresses *how* [12]. When it comes to decision making within the constraints of time, frustration may arise not only from information overload but also from its opposite— information underload—which occurs when there is not enough information available to make the right decision.

Information overload is closely linked to high *cognitive load*. According to Kelsey and St. Amant [9], "Cognitive load is the burden on working memory during information processing. It can be augmented by the individual's characteristics, but also by the content of the form and structure of the message. For example, a complex integration of modalities can elicit a cognitive load on the working memory so that the information is less remembered." A decision maker faces a cognitive load more when processing information for declarative knowledge than for procedural knowledge. An engineer or technician, for instance, might convert what he or she reads into an immediate action, requiring relatively little cognitive load. If the same information has to be stored in the brain to use for later application, then the cognitive load will be heavier. Therefore, reliance on memory can lead to the perception of information overload.

It is important for producers of information to keep in mind the time constraints that users face and create communication products that provide the right balance of information to prevent information overload—as well as information underload. At the same time, producers of information also need to package communication products according to the needs of the users, so as to prevent the perception of information overload.

1.2.2 A Supplier/Producer/Writer and Client/User/Reader Perspective

To prevent information overload, the suppliers, producers, or writers of information need to know who their audience is and how that audience processes a particular communication product. (For the sake of simplicity, we do not distinguish between the roles of suppliers, producers, and writers, as all are creators of information, or between the roles of clients, users, and readers, as all are audiences or receivers of information.)

Most technical communicators are trained well to keep the needs of their audience in mind, but with respect to cognitive load in particular, communicators must remember that the audience may be composed of multitaskers and cater to their needs. Biddix discusses how individuals relate in contemporary "networks," thanks to the possibility of multi-tasking through computer-mediated communication (CMC) [9]. Such multitasking changes the wiring of our brain through the exposure to several simultaneous information flows throughout our lifetime [13]. The *homo zappiens*, natives of the electronic digital age [14], might think they can keep up with information overload in an easier way by multitasking; however, recent studies [15–18] strongly indicate that multitasking may not be the best road to efficient performance, at least for students. Nass and Brave [17] suggested that multitasking even affects our ability to be emphatic or expressive:

> New research is showing that such immersion can cause multitaskers to have more fractured thinking and trouble shutting out irrelevant information, and that even when they are offline, those problems persist.

Producers of information can implement certain strategies to help their multitasking audience. Two strategies were field tested to reduce information overload in student Web searches in biology and history. The first used work sheets and modeling examples; the second used rich representation such as mind mapping techniques [15]. Both appear to work equally well for students. We can reasonably expect these strategies to transfer to engineering and professional communication practice as well.

There is also an increasing need for information producers to write for audiences that do not have a clear human face, such as for search engines, as if they were information robots allowing easy data mining. A study by Killoran [16] explores the extent to which technical communication businesses with websites are attempting to reach an audience of prospective clients through an intermediary audience of search engines. It draws on a survey of 240 principals of these businesses, brief interviews with half of them, and an analysis of their sites. Results show that search engines are among the most helpful methods leading people to those business sites. In addition, higher levels of search engine helpfulness are strongly associated with higher percentages of clients who originate through these sites. Most businesses take search engines into account, and search engine optimization (SEO) is a booming business. However, meaningfully pursuing a specific audience requires ongoing investments that some professional communication businesses are reluctant to make.

Fragoso [18] made a strong plea based upon a discussion of the available search engines to return to the mass distribution mode in a Web many-to-many information distribution model, so that professional communicators can avoid creating information overload for those people who do not belong to an interested group. This implies that writers should clearly distinguish between information presented to the masses versus that presented to a target audience or community of professionals who share the same interests, professions, and culture. In this way, professional communicators could act as organizational gatekeepers [19] who monitor and regulate the flow of information in and out of an institution, through all media, including mass media [18, 20], to help prevent information overload for their audience.

The current business and technical communicator faces the challenge of preventing information overload not only in writing but also in the visual aspects of written text (think of color and image use in websites or in technical illustrations) and in oral modes, such as in presentations. Written, visual, and oral modes of communication need concise information that is clear and that has the right density so as to be useful to the audience.

Last, but not least, information producers should not forget readability measures that can help predict potential overload caused by complex writing. Presently, engineers and technical writers not only write but also manage information in the broadest sense. In this role, they have a special responsibility toward the information loads of their audiences, both individuals and institutions.

1.2.3 An International/Intercultural Perspective

All those working in today's global economy must be prepared to face intercultural differences. They might already face such differences in their own countries if they have to address the needs of diverse populations with cultural minorities, which is the case in many countries today. In addition to the national culture, communicators must adapt to the corporate or professional culture in which they work. Belonging to the same corporate or professional culture can actually help to reduce overload as those who are exchanging information share common vocabulary and frames of reference. Because the intercultural perspective is so important, this volume includes a full chapter on this topic. In addition, our authors bring diverse and international perspectives, as do our *Practical Insights* contributions from multinational corporations.

1.2.4 An Innovation Perspective

The information economy is based on the collection and exchange of data and ideas. We all either contribute to or use materials from the information economy in almost all aspects of our everyday lives. Few of us, however, understand all of the nuances of the information economy or the communication factors that affect its operations.

Innovation of products, services, or processes is an important source of change, be it radical or incremental, both in its exploration or R&D phase, where the engineer and/or technical communicator operates, and in the exploitation phase, where both technical and business communicators operate. There is a clear overlap between professional communication activities and information-management practices, and there are multiple opportunities to regulate the information flow through careful innovation [21].

Ulijn and his colleagues [22–25] have argued that professional communicators often play a crucial role in formulating strategic innovation targets of companies and helping them implement and assess this formulation process with guidelines suitable for the company's culture. They then use the various functions of the strategic targets with different professional cultures, such as engineering and marketing, and act as cross-pollinators in exchanging ideas and information between top, middle, and lower management. They might act then not only as external gatekeepers but also as internal regulators of information load (avoiding both underload and overload). Since information overload seems to be a matter of information and time management, all communicators should also act as information and knowledge managers for themselves and their companies, dealing with time constraints by multitasking as efficiently as possible.

1.3 Readers of this Book

This book addresses the ever-increasing problem of information overload from both the corporate and the academic perspectives. We hope to serve the following groups of readers.

- Practicing engineers and other technical professionals, who, as SMEs, would use the strategies provided in the book to reduce information overload, both for themselves and for their audiences
- Engineering communication and other technical communication practitioners, who would use the book to keep up with current strategies for reducing information overload

- Managers, who would use the book to reduce the negative effects of information overload in their organizations by making informed decisions regarding information management
- Educators who teach courses in engineering, management, technical communication, or business communication, who would use the book as a textbook or supplementary reading material in courses that address information overload
- Academics interested in information overload, who would use the book as a resource for research
- Students, who would use the book as a textbook or supplementary reading material and continue to use it as a professional reference or resource

1.4   Structure of this Book

In addition to this introductory chapter by the editors, this collection has 11 chapters from an international group of authors. Contributing authors represent various nationalities (including Germany, India, The Netherlands, Pakistan, Switzerland, Tanzania, the United Kingdom, and the United States) and have a variety of backgrounds and affiliations including academia, research organizations, and private corporations. Strategies and techniques that can be used to reduce information overload and minimize its effects are presented. For the academic audience, the concept and theoretical framework of information overload are presented along with suggestions for future research in this critical underresearched area.

In addition to the chapters, *Practical Insights* appear throughout this book to illustrate the best practices of various technical corporations for dealing with the challenges of information overload. These companies include IBM, Xerox, and Harris Corporation.

This volume is divided into two sections centered on key issues in the field of information overload.

- *Section I*: Causes and Costs of Information Overload
- *Section II*: Control and Reduction of Information Overload

1.4.1   Section I: Causes and Costs of Information Overload

Before a productive discussion can ensue on the topic of information overload, it is important that we define the term itself. In Chapter 2, on the evolution of information overload theories and concepts, Aikat and Remund situate information overload research within a larger intellectual context of communication and document the theoretical and conceptual development of information overload as it relates to communication in its many forms. Based on archival studies of communication history, meta-analyses of information concepts, and theoretical reviews of information overload, the chapter explicates the theory and concept of information overload and explores its historical evolution as well as identifies theoretical trends in information overload research, evaluates strategies for managing it, and delineates communication strategies to reduce and control it. In addition, the chapter assesses significant trends, important research findings, critical issues, and some knowledge management techniques. Finally, it identifies research directions to investigate information overload problems at a global level.

In Chapter 3, Vossen argues that human beings need information, continuously and increasingly, in order to master their environment. To satisfy this innate need, they are not only equipped with perceptual and cognitive functions but have also developed auxiliary facilities, from languages via books to modern digital media. However, the success of these artifacts has led to information explosion and pollution, making people feel overwhelmed and paralyzed by the very systems that should help them master their world. Vossen explores some of the symptoms, causes, and consequences of information overload from a multidisciplinary perspective and argues that current approaches to alleviate the problem— either educating human agents to become better information processors or developing more auxiliary systems to cope with the previous ones—are doomed to fail because they accept and thus aggravate the original problems. According to Vossen, the fallacy is to assume that information overload can be cured by still more information. Vossen suggests that basic research on all-inclusive information resource management beyond the simple solutions proposed up to now is urgently needed.

In Chapter 4, Caborn and Cooper argue that information overload is a mental mindset and a practical challenge for technical communicators. The chapter gives technical communicators insight into how their attitudes to information and information overload in the medical field might be influenced by developments in information technology and management and how these attitudes might influence the way they personally handle and disseminate information. The authors also consider how the human brain engages with information on screen as opposed to in print.

The influence of culture on information overload is the theme explored by Ulijn and Strother in Chapter 5. Too often, all communication genres are analyzed without recognizing the unavoidable impact of a communicator's cultural framework on either producing or receiving and attempting to comprehend the document. This chapter analyzes the impact of culture, primarily using Hall's high-context versus low-context model. In addition, Hinds' theory of reader versus writer responsibility for comprehension, the impact of professional culture, and the relative benefits of globalization and localization are also explored. Suggestions for dealing with cultural issues in technical communication are presented.

In Chapter 6, on the effect of color, visual form, and textual information on information overload, Alton and Manning observe that there is a distinct trend in current menu design, both in print and online, to overload the menu with too many separate entries, colors, flashing images, and so on. They argue that viewers have difficulty navigating such designs since information they may wish to locate is too often buried. Alton and Manning describe a coherent theory of information design and offer empirical evidence in support. They argue that there is no clear-cut boundary between "mere" document design issues and actual information-content issues. Color and form convey a spectrum of emotional information that, if deployed carelessly, interferes with propositional information conveyed by text. Furthermore, color, form, and text also provide *indicative* information; these elements tell viewers to perform specific actions of looking, focusing, dividing, and contrasting one visual element from another. The chapter offers specific advice on how to keep emotional, indicative, and propositional information in balance in order to avoid information overload.

Bidkar, in Chapter 7, puts forward a framework for calculating the cost of information overload in end-user documentation based on the concept of quality costs. Drawing upon Wilson's information behavior model and data from a user survey, Bidkar establishes end-user behavior. He then puts forward scenarios showing how information overload can increase the cost of failure for both the organization producing the documentation as well as the end user. The chapter suggests a new approach to evaluate the factors that increase cognitive load and hinder information use. Engineers, technical communication practitioners, and managers can use the cost framework to analyze and calculate the costs of failure because of errors and deficiencies in the manual that result in increased cognitive load, and therefore in information overload.

1.4.2   Section II: Control and Reduction of Information Overload

In Section II, authors provide strategies for dealing with information overload.

Chapter 8 by Hoenkamp, on a human-centered approach to surviving the information deluge, starts with the assertion that the fear of imminent information overload predates the World Wide Web by decades. The chapter then focuses on the overload issues that have emerged with the advent of the Web. It further focuses on three ways to mitigate the issues, especially as they relate to search engines: being specific in what we ask, amending our requests when we do not find what we need, and making our retrieval techniques more human centered. Hoenkamp presents research that has been conducted in these three areas and shows that a human-centered approach substantially improves the retrieval results of algorithms underlying the search engine. All examples discussed have actually been implemented. Having built the applications, Hoenkamp presents an operational, not just an aspirational approach.

Calders, Fletcher, Kamiran, and Pechenizkiy focus on technologies for dealing with information overload in Chapter 9. The authors survey storage and querying techniques for semistructured data, data mining, and information retrieval for analyzing large data collections as well as stream processing techniques for online handling of continuously flowing data. They purport that since storage of the massive amounts of data is no longer an option, we need to rely on immediate processing to be able to distill information from the data. The chapter surveys the main challenges and gives some insight into recent developments in these areas.

Mengis and Eppler focus on reducing information overload through visualization in Chapter 10. The authors address the qualitative aspects of information, systemize existing findings from the literature on how these aspects cause information overload, and show how they can be addressed to reduce overload. The authors propose a solution to information overload that can be adopted by information producers and show, in particular, how visualization (the graphic rendering of information) can be used to improve the quality of information so that information overload can be reduced. They discuss why visualization is still not frequently used by communication professionals, even though its potential for reducing information overload is explicitly acknowledged. Mengis and Eppler also present the results of a survey of 636 International Association of Business Communicators (IABC) members regarding their views on visualization as a counterstrategy against information overload.

In Chapter 11, Remund and Aikat examine the effects of information overload within organizations and present strategies for improving knowledge transfer and decision making. The authors acknowledge that research about information overload largely focuses on individuals and that little attention has been paid to the aggregate effect of information overload within organizations. Drawing insight from scholarly and professional journals, this chapter emphasizes the principles of effective communication as a strategy for mitigating or reducing information overload within an organization. The chapter also stresses the need for organizations—and the individuals within organizations—to focus on information sharing that is situation specific and decision making that is actionable.

Finally, in Chapter 12, Verhoeff deals with the information paradox as an important issue that needs to be addressed and uses the backbone model to deconstruct the information paradox. He stresses that companies need managers who are capable of negotiating increased information symmetry between a company's internal and external stakeholders. In addition, he shows that the orientations of managers in daily life are not quite consistent with the new information reality. He recommends that managers and other readers of this book reflect on integrated relational and technical solutions for the information paradox.

REFERENCES

[1] *BusinessDictionary.com* [Online]. Available: http://www.businessdictionary.com/definition/information-overload.html Sep. 1, 2011 (last date accessed).

[2] PC. (2010, Oct. 19). *Definition of Information Overload* [Online]. Available: http://www.pcmag.com/encyclopedia_term/0,2542,t=information+overload&i=44950,00.asp Sep. 1, 2011 (last date accessed).

[3] N. Zeldes. (2009). "How to beat information overload," *IEEE Spectrum* [Online]. Available: http://spectrum.ieee.org/computing/it/how-to-beat-information-overload Sep. 1, 2011 (last date accessed).

[4] C. G. Sieloff, "'If only HP knew what HP knows': The roots of knowledge management at Hewlett-Packard," *J. Knowledge Manage.*, vol. 3, no. 1, pp. 47–53, 1999.

[5] A. Edmunds and A. Morris, "The problem of information overload in business organizations: A review of the literature," *Int. J. Inform. Manage.*, vol. 20, no. 1, pp. 17–28, 2000.

[6] N. Zeldes *et al.*, "Infomania: Why we can't afford to ignore it any longer," *First Monday*, vol. 12, no. 8, Aug. 2007.

[7] J. Spira, *Overload! How Too Much Information Is Hazardous to Your Organization*. Hoboken, NJ: Wiley, 2011.

[8] Q. Jones, *et al.*, "Information overload and the message dynamics of online interaction spaces: A theoretical model and empirical exploration," *Inform. Syst. Res.*, vol. 15, no. 2, pp. 194–210, 2004.

[9] S. Kelsey and K. St. Amant, Eds., *Handbook of Research on Computer Mediated Communication*, vol. 2. New York: Information Science Reference, 2008.

[10] J. Spira, *Information Overload: Now $900 Billion—What Is Your Organization's Exposure?* [Online]. Available: http://www.basexblog.com/2008/12/19/information-overload-now-900-billion-what-is-your-organizations-exposure/ Jan. 10, 2012 (last date accessed).

[11] I. Nonaka and H. Takeuchi, *The Knowledge-Creating Company: How Japanese Companies Create the Dynamics of Innovation*. New York: Oxford Univ. Press, 1995.

[12] F. Machlup, *Knowledge: Its Creation, Distribution, and Economic Significance*, vol. 1. Princeton, NJ: Princeton Univ. Press, 1980.

[13] A. Gazzaley and M. Desposito, "Top-down modulation and normal aging," *Ann. NY Acad. Sci.*, vol. 1097, pp. 67–83, Feb. 2007.

[14] W. Veen and B. Vrakking, *Homo Zappiens: Growing up in a Digital Age*. London, U.K.: Network Continuum Education, 2006.

[15] A. Walraven, "Becoming a critical websearcher: Effects of instruction to foster transfer," Ph.D. dissertation, Open Univ. The Netherlands, Heerlen, The Netherlands, 2008.

[16] J. B. Killoran, "Targeting an audience of robots: Search engines and the marketing of technical communication business websites," *IEEE J. Prof. Commun.*, vol. 52, no. 3, pp. 254–271, Sep. 2009.

[17] C. Nass and S. Brave, *Wired for Speech: How Voice Activates and Advances the Human-Computer Relationship*. Cambridge, MA: MIT Press, 2005.

[18] S. Fragoso, "Seek and ye shall find," in *Handbook of Research on Computer Mediated Communication, Part 2*, S. Kelsey and K. St. Amant, Eds., 2008, pp. 740–754.

[19] D. M. White, "The 'gatekeeper': A case study in the selection of the news," *J. Quart.*, vol. 27, no. 4, pp. 383–390, 1950.

[20] M. Boardman, *The Language of Websites*. Abingdon, U.K.: Routledge, 2005.

[21] K. St. Amant and J. M. Ulijn, "Examining the information economy: Exploring the overlap between professional communication activities and information management practices," *IEEE Trans. Prof. Commun.*, vol. 52, no. 3, pp. 225–228, Sep. 2009.

[22] J. Ulijn, "Introduction," *J. Business Commun.*, vol. 37, no. 3, pp. 202–208, 2000.

[23] J. Ulijn and R. Kumar, "Technical communication in a multicultural world: How to make it an asset in managing international businesses, lessons from Europe and Asia for the 21st century," in *Managing Global Discourse: Essays on International Scientific and Technical Communication*, P. J. Hager and H. J. Scheiber, Eds., New York: Wiley, pp. 319–348, 2000.

[24] J. Ulijn, *et al.*, "Innovation, corporate strategy, and cultural context: What is the mission for international business communication?" *J. Business Commun.*, vol. 37, no. 3, pp. 293–316, 2000.

[25] J. Ulijn and K. St. Amant, "Mutual intercultural perception: How does it affect technical communication, some data from China, The Netherlands, Germany, France and Italy?" *Tech. Commun.*, vol. 47, no. 2, pp. 220–237, 2000.

Section I

CAUSES AND COSTS OF INFORMATION OVERLOAD

2

OF *TIME MAGAZINE*, 24/7 MEDIA, AND DATA DELUGE: THE EVOLUTION OF INFORMATION OVERLOAD THEORIES AND CONCEPTS

Debashis "Deb" Aikat and David Remund

> *What information consumes is rather obvious: It consumes the attention of its recipients. Hence a wealth of information creates a poverty of attention and a need to allocate that attention efficiently among the overabundance of information sources that might consume it.*
>
> Herbert Simon
> Social and Political Scientist

ABSTRACT

Originating as a debilitating trend in the 1990s, information overload has exacerbated in the digitally disseminated information age of the twenty-first century. In sharp contrast to the limited sources of information in the mid-1990s before the exponential growth of the Internet, the twenty-first century has witnessed a deluge of data, 24/7 media, and digital information that inundate professionals as well as common citizens, leading to the phenomenon of information overload.

Based on meta-analyses and state-of-the-discipline review of information overload research paradigms, this chapter traces the theoretical and conceptual evolution of information overload in four parts. The first part explicates the theory and concept of information overload. The second part delineates the evolution of information overload as a twentieth century concept and credits the founders of *Time Magazine*, American media pioneers Henry Luce and Briton Hadden, for precociously recognizing information overload in the early decades of the twentieth century. It also traces the works of scientist Vannevar Bush, futurist Alvin Toffler, and designer Richard Saul

Information Overload: An International Challenge for Professional Engineers and Technical Communicators, First Edition. Edited by Judith B. Strother, Jan Ulijn, Zohra Fazal.
© 2012 Institute of Electrical and Electronics Engineers. Published 2012 by John Wiley & Sons, Inc.

Wurman, who posited critical perspectives that advanced information overload theory. The third part features a historical timeline of important innovations that have contributed to the evolution and proliferation of information in our society. The fourth and final part presents definitions, concepts, and theories related to the context and causes of information overload.

In tracing the conceptual evolution of information overload, this chapter identifies a significant opportunity for strategies to combat information overload even as we are drowned by it. Such opportunities highlight the importance of the information overload research paradigm, which has inspired academic researchers and professionals to explore the surging significance of information overload research and practice in the twenty-first century.

2.1 Introduction

Information overload occurs when the information available exceeds the user's ability to process it [1, 2]. Allen and Wilson [1] emphasized that information overload is not just a matter of there being more information available or of technology's ability to provide us with more information than we actually need. Human factors also enter into the situation in terms of the innate propensity of people to demand information and to distribute it to others [1].

By situating information overload research within a larger intellectual context of communications, this chapter documents the theoretical and conceptual development of information overload as it relates to professional engineers and technical communicators. Based on meta-analyses of information concepts and theoretical reviews of information overload, this chapter covers four parts. First, it explicates the theory and concept of information overload. Second, it delineates the evolution of information overload as a twentieth century concept and the surging significance of information overload in research and practice. Third, it features a historical timeline that traces the evolution of information and its proliferation in six epochs: Early Quest for Information and Knowledge (320 BCE Thirteenth Century); Age of Renaissance (Fourteenth–Seventeenth Century) and the Printing Press; Industrial Revolution (Eighteenth–Nineteenth Century) and Its Information Innovations; Era of the Mind and the Machine (Twentieth Century); Internet Boom and Information Explosion of the 1990s; and Data Deluge and Information Overload in the Twenty-first Century Digital Age. In the fourth and final part, this chapter presents concepts related to the context and causes of information overload. In tracing the conceptual evolution of information overload, we highlight the importance of information overload as a research paradigm and a professional challenge as information, news, and entertainment messages proliferate in our society.

2.2 Theory and Concept of Information Overload

In the current age of digitally disseminated information, the runaway growth of technology has complicated and catapulted information overload in our society [3, 4]. The world of 24/7 media is composed of information content that mass media diffuse 24 hours a day, 7 days a week, in a myriad of aspects consisting of social media, news entities, news feeds, television networks, e-mail exchanges, and other information modes that comprise an endless array. In multitudinous ways, 24/7 media inundate the consumer with a plethora of choices. For instance, the number of available media channels exceeds the 1440 minutes in a day. Social media and other information sources collectively redistribute millions of partially overlapping information content items each day. This superabundance of information at home, work, and beyond has caused information overload and a general feeling of being too busy to notice you are too busy [2–4].

As digital media proliferate new ways of disseminating information, news, and entertainment, information overload has emerged as a research paradigm and a professional problem. The problem has been so acute that the leading peer-reviewed journal, *Science*, published a special issue in February 2011 on "Dealing with Data," which explores issues surrounding the increasingly huge influx of research data and highlighted both the challenges posed by the data deluge and the opportunities to better organize and access the data [5]. The special issue reported two dominant themes: one, most scientific disciplines "are finding the data deluge to be extremely challenging," and two, "tremendous opportunities can be realized if we can better organize and access the data" [5]. However, scientists are not the only group dealing with the data flood. People in other segments of society say they are drowned by information overload [6–8].

Information overload refers to exposure to too much information and related consequences such as cognitive dissonance or mental stress from seeking to assimilate excessive amounts of information from the media, the Internet, or at work. Some researchers have defined information overload simply as "the notion of receiving too much information" [8].

2.3 Information Overload as a Twentieth Century Phenomenon

The founders of *Time Magazine*, American media pioneers Henry Luce and Briton Hadden, precociously recognized information overload in the early decades of the twentieth century. In those days, most newspapers were complex compendiums of information, including shipping schedules, stock prices, transcripts of political speeches, and reports of accidents and fires [9, 10].

Luce and Hadden founded *Time Magazine* in 1923 on the promise that they would end information overload for the majority of Americans. They stated in their 1922 prospectus that no one had the time (or the patience) to sift through all the available information and figure out what was important, so their new magazine would do it for them. Thus, in 1923, *Time Magazine* was launched with the proclamation, "People are uninformed because no publication has adapted itself to the time which busy men are able to spend on simply keeping informed" [9,10].

In its attempt to adapt itself to the "busy" people of the 1920s, *Time Magazine* proposed three innovations: one, news could be "completely organized," in a fixed pattern of departments, such as National Affairs, Sports, and Foreign News; two, *Time Magazine* would report the news with balance and focus ("*Time* gives both sides, but clearly indicates which side it believes to have the stronger position"); and three, stories told in flesh and blood terms would get into the readers' minds when stories told in journalistic banalities would not [9, 10]. The "world's movers and shakers," stated the prospectus, are "something more than stage figures with a name. It is important to know what they drink. It is more important to know to what gods they pray and what kind of fights they love" [9, 10]. The success and legacy of *Time Magazine's* mandate "to keep busy people informed" and thereby alleviate information overload has provided new lessons for information overload researchers in the information age of the twenty-first century.

Implicit in the prospectus was another idea, which was not directly stated: "The busy man," for *Time Magazine's* purposes, was regarded as an expert on nothing. "The National

Affairs department was not written for politicians, nor Foreign News for cosmopolites, nor Books for bookworms, nor Sport [sic] for sport fans" [9, 10]. The whole magazine was supposed to be comprehensible to one "busy man"—a vastly different notion from its contemporary daily newspaper departments (women's, sports, finance, etc.), each appealing to special groups [9, 10]. Luce and Hadden founded *Time Magazine* on the notion that a "surplus of news" had to be packaged into usable shape and, therefore, *Time Magazine* felt no need to gather its own news until the 1930s, when it began building up its own reporting system [9, 10]. To that end, this new publication emerged as a significant source of news and information during the global military conflict of World War II from 1939 to 1945, which claimed nearly 68 million lives.

In terms of lives lost and material destruction, World War II was among the most devastating incidents in human history. However, progress made by scientists during and just after the war ultimately contributed to the development of information overload. Throughout World War II, scientist and administrator Vannevar Bush (1890–1974) worked closely with the 32nd President of the U.S., Franklin D. Roosevelt, as his Scientific Adviser. Roosevelt's long tenure in office (1936–1945) gave Bush ample opportunities to pursue his scientific work with various government organizations. Bush convinced Roosevelt of the need to harness technology for war and shepherded to fruition the United States government's complex Manhattan Project (1942–1945). Science played a significant role in the Manhattan Project (1942–1945), which produced the atomic bombs that were used during World War II and caused tremendous loss of lives and material destruction.

Toward the end of World War II in 1945, Bush voiced deep concerns about the scientific efforts toward destruction in the "devastating" war in a prescient essay "As We May Think," published in *The Atlantic Monthly* in July 1945 [11]. In this article, Bush analyzed answers to the question, "What are the scientists to do next?" Bush's answer was that scientists should turn their attention to devices that could help store, manipulate, and retrieve information. He argued that "new and powerful instrumentalities" could help humans push to a new level of intellectual development. In his essay, Bush urged scientists to turn their energies from war to the task of making the vast store of human knowledge accessible and useful. Bush stated, "Thus far we seem to be worse off than ever before for we can enormously extend the record, yet even in its present bulk we can hardly consult it" [11].

Widely regarded as one of the most portentous essays about computing, Bush's essay, which was later reprinted with pictures in *Life* [12], delineated the remarkable clarity of Bush's vision. It marked a paradigm shift in scientific theories and the American psyche. Foreboding the limitations of the human mind, which "operates by association," Bush proposed the concept of "memex," a hypothetical information retrieval and annotation system, which later emerged as a precursor to the World Wide Web [11]. He conceptualized memex as a collective memory machine that would make knowledge more accessible and help transform an information explosion into a knowledge explosion. Bush even conceptualized memex to tame the then-unexplored problem of information overload by enhancing human memory (hence its name). He stated, "A memex is a device in which an individual stores all his books, records, and communications, and which is mechanized so that it may be consulted with exceeding speed and flexibility. It is an enlarged intimate supplement to his memory" [11].

Bush thus referred to information overload in the 1940s and proposed ways to combat it. Although never built, the memex was a conceptual machine that could store vast amounts of information and enabled users to create information trails consisting of links of related texts and illustrations stored and used for future reference [11]. This concept resembles the modern-day World Wide Web. Similar to the present day Internet's capability of customized content, Bush's memex promised the added benefit of letting its user link together disparate pieces of information, thus automating a process of retrieving associated ideas and data. Bush's description of memex eerily resembles today's Internet-linked personal computer. Bush visualized memex with ". . . slanting translucent screens, on which material can be projected for convenient reading" [11]. These personal associations or "trails" could be shared among people, Bush thought, even passed down from parent to child, giving their creators a measure of immortality. For Bush, the memex was a universal library that used microfilm to store vast amounts of text, crammed onto a desktop.

The legacy and impact of Bush's ideas about information overload influenced several scholars. In his 1964 book, American social scientist Bertram Gross credited Bush with these words, "As far back as 1945, Vannevar Bush described the problem of information overload" [13]. Writing on the "administrative struggle" while managing organizations, Gross repeatedly referred to the term "information overload" in his book [13]. The rise of the Internet cemented Bush's reputation as a prophet of computing innovations. For instance, in his 1992 article, "World Wide Web: The Information Universe," Tim Berners-Lee, inventor of the World Wide Web, credited Bush's "As We May Think" for outlining the intellectual seeds for the Web [14].

With striking similarity to Bush's vision, American writer and futurist Alvin Toffler predicted in his 1970 book *Future Shock* a rising flood of technological, social, and economic change, largely emanating from the increasing influence of science and technology into every area of contemporary life. In examining the effects of rapid industrial and technological changes upon the individual, the family, and society, Toffler described the term *information overload* "as the difficulty a person can have understanding an issue and making decisions that can be caused by the presence of too much information" [15]. Toffler conceptualized information overload more than two decades before the Internet emerged as a household name in the 1990s. Tofler's allusion to "too much information" has a portentous relevance to the formidable array of information sources that inundates common people in the twenty-first century. People are well aware of information overload and know what it feels like: overwhelming, frustrating, and even self-defeating.

Designer Richard Saul Wurman advanced Toffler's concept in his 1989 book, *Information Anxiety* [16]. According to Wurman, "Information anxiety is produced by the ever-widening gap between 'what we understand' and 'what we think we should understand.' Information anxiety is the black hole between data and knowledge. It happens when information 'doesn't tell us what we want or need to know'" [16]. Wurman estimated in 1989 that the weekday edition of *The New York Times* contained more information than the average person in the seventeenth-century England was likely to come across in a lifetime [16]. This personalized the oft-cited estimate that more information has been published and disseminated in the past 30 years than in the previous 5000 years. Comparative figures like these highlight the phenomenon of an information explosion and its consequences such as information overload or information anxiety. Wurman explained that information anxiety limits people to being only seekers of knowledge because no time is left over for them to be *reflectors of knowledge* [16].

In 1984, Jacoby concluded from his research of marketing studies related to information overload that although consumers can be overloaded with information, they are usually highly selective in how much and just which information they will access, and thus tend to stop well short of overloading themselves [17].

In 1984, Malhotra classified research studies on information overload in relation to psychology and marketing [18]. Malhotra divided the psychological research into two categories: one, information overload that pertained to individual decision-making situations (in the context of human information processing) and two, information overload related to nondecision-making, uncontrollable situations (urban life, crowding, noise, sensory overload, etc.).

In the marketing context, Malhotra authored a consumer branding position paper that demonstrated how consumers can, indeed, be overloaded. Malhotra gave subjects information about particular consumer brands and found that when they received too much information about the products, they were susceptible to choosing a product other than the "best-fit brand" to meet their needs (as previously determined on an individual basis) [19].

Focusing on information and sensory overload as it pertains to psychology and marketing/consumer behavior, Malhotra conducted studies where participants were overloaded with information about consumer products [18]. His research suggested that the phenomenon of information overload was mediated by several individual differences, was task related, and had situational variables. Individual differences include age, sex, intelligence, experience, authoritarianism, dogmatism, cautiousness and tolerance for ambiguity, and cognitive complexity/simplicity. Pertinent situational variables included factors such as intensity and duration.

Nearly two decades after Malhotra's 1984 studies explored information and sensory overload, Allen and Wilson's 2003 study researched the organizational effects of information overload, which they believed had been overlooked in the topic's research history [1]. Allen and Wilson emphasized that information overload is not just a matter of there being more information available or of technology's ability to provide us with more information than we actually need. Human factors also enter into the situation in terms of the innate propensity of people to demand information and to distribute it to others [1]. Their findings suggested that the organizational climate (e.g., of a business) may have a greater responsibility for pathological consequences of information overload than either the sheer volume of information or the power of technology.

In 2004, Eppler and Mengis (see Chapter 10) compiled a comprehensive scholarly literature review of information overload [8]. They pooled 30 years of management-related academic publications (in areas such as organization science, marketing, accounting, and management information systems) to consolidate the research paradigm related to information overload. Eppler and Mengis attributed the societal causes of information overload to the accelerated production of information through institutions and the more efficient distribution of media content through improved technologies for information and communication [8]. Their management-related studies of information overload focus on how performance (in terms of adequate decision making) of individuals varies with the amount of information they are exposed to. Eppler and Mengis concluded that performance (i.e., the quality of decisions or reasoning in general) of individuals correlated positively with the amount of information they received—up to a certain point. If further information were provided beyond this point, it ceased to be a factor in the decision-making process, and the performance of the individual would rapidly decline due to information overload. Eppler and Mengis also observed that the burden of heavy information overload confused the individuals and affected their ability to set priorities and made prior information harder to recall.

Information overload has affected individuals and organizations alike [17–20], and its effects include debilitating difficulty in making decisions due to the time invested and wasted in searching for and processing information [21], seemingly insurmountable inability to parse and pick among multiple information sources on the same topic [22], and persistent psychological issues caused by excessive interruptions by too many information sources [23]. The surfeit of unattended information also causes significant stress [24]. Some countermeasures that Eppler and Mengis [8] provided for avoiding information overload include the following: intelligent information management (prioritization), voting structures to make users evaluate the information, facilitator support through e-tools, decision support systems to reduce a large set of alternatives to a manageable size, information quality filters, and systems that offer various information organization options (e.g., filing systems).

2.4 Evolution of Information and Its Proliferation in Society

Although information overload seems to be a modern trend, the human quest for collecting, collating, and curating information dates back to ancient times. It is important, if not critical, to understand and analyze the evolution of information and its proliferation in society [25–27]. With brief facts about key events, this section features a historical timeline that provides a chronological vignette of the epochal events and perspectives that contributed to the evolution of information that began long before the days of movable type, memex, and microchips. The timeline is divided into six epochs classified by their distinctive character. Both in content and design, this timeline is not intended to be exhaustive because it encapsulates a complex chronicle of events that trace decades of progressive proliferation of information overload challenges [28, 29].

2.4.1 The Early Quest for Information and Knowledge (320 BCE–Thirteenth Century)

Information overload did not exist in the early civilizations of the world because information was, at best, sparse. The invention of paper and writing tools facilitated easier documentation of information and led to an arduous quest for knowledge.

> *320 BCE.* At the Royal Library of Alexandria in Egypt, the largest and most significant ancient library in the Western world, bibliographers charged with collecting and organizing the world's knowledge worked on a comprehensive collection of books from beyond the country's borders [30].
>
> *8 BCE.* The invention of paper and paper technologies in China greatly contributed to the spread and development of civilization. The technologies spread east, to Korea and Japan, and west, along the Silk Road, to Central Asia, eventually reaching Europe in the thirteenth century [31]. As the technology propagated and diffused, paper effected profound changes in each society it touched, becoming one of the most important of all cultural media, a status that it retains to the present [32]. Before the invention of paper, bones, tortoise shells, and bamboo slips were used as writing surfaces [31].
>
> *79 CE.* Codex was initially developed as a part of the Eurasian culture. A codex (Latin *codex*, *codic*, "tree trunk, wooden tablet, book") was a book in the format

used for modern books, with separate pages normally bound together and given a cover [33].

220 CE. Woodblock printing, which originated in China in antiquity, was developed as a technique for printing text, images, or patterns on textiles and later on paper. It was widely used throughout East Asia [34].

700 CE. The quill pen was introduced around 700 CE in Europe. As a writing instrument, the quill pen dominated for over a millennium as a tool for exquisite presentation of the written word [35].

1040 CE. The first known movable-type system for printing was created in China around 1040 CE by Bi Sheng (990–1051) [36].

1088 CE. Realizing that information about the daily positions of the sun, moon, and planets was important, astronomers organized this information in almanacs [37]. An almanac is an annual publication containing tabular information, such as astronomical tables, often arranged according to the calendar. The earliest known almanac in the modern sense was the *Almanac of Azarqueil* in 1088 in Toledo, al-Andalus, a region comprising areas of modern Spain and Portugal. This almanac published the true daily positions of the sun, moon, and planets from 1088 to 1092, as well as other related tables [37].

1154 CE. Several inventors developed ways to perpetuate information records by using rag paper because of its archival quality [38]. Rag paper was prevalent in medieval times as a durable tool for archiving information. The demand for rag paper increased with a surge in literacy in the eighteenth century and the resultant explosion of print publications [38]. In 1927, *The New York Times* began printing an edition on enduring rag paper but discontinued it due to its high cost [39].

2.4.2   The Age of Renaissance (Fourteenth–Seventeenth Century) and the Printing Press

The Age of Renaissance (fourteenth–seventeenth century) defined the power of information in society. The invention of the printing press played a key role in disseminating knowledge during the Renaissance, Reformation, and the Scientific Revolution. It heralded widespread appreciation for knowledge and the spread of learning for the masses. The invention of the printing press in 1436 and the founding of the first truly public library in England in 1598 facilitated diffusion of information and knowledge for a larger audience.

1436. Johannes Gutenberg, a goldsmith and businessman in Mainz, Germany, invented the printing press with replaceable and moveable wooden or metal letters in 1436 (and completed it by 1440) [40]. This method of printing led to a revolution in the production of books and fostered rapid development in the sciences, arts, and religion through the transmission of texts. Gutenberg's epochal invention was a clever combination of pragmatic printing elements such as a process for mass production of metal movable type, the use of oil-based ink, and the design of a wooden printing press similar to the screw olive and wine presses of the period [41].

1550. Presentation of information made another leap when pagination was developed as a system of presenting information on a newspaper, book page, manuscript, or other handwritten, printed, or displayed documents [42]. In the strict sense of the word, pagination connotes consecutive numbering to indicate the proper order of the pages, which was rarely found in documents predating 1500. It became common practice only in 1550, when it replaced foliation, which numbered only the front sides of folios [42].

Mid-1500. Bookbinding emerged as a popular process of physically assembling a book from a number of folded or unfolded sheets of paper or other material. It also involved attaching covers to the resulting text block [43]. Early and medieval codices were bound with flat spines, and it was not until the fifteenth century that books had rounded spines, which are associated with hardcovers today [43].

1566. The art of oratory defined punctuation. Back in ancient Greece and Rome, when a speech was prepared in writing, marks were used to indicate where, and for how long, a speaker should pause. Although used since 9000 BCE, punctuation was not standardized until after the invention of printing [44]. Credit for introducing a standard system is generally given to Aldus Manutius, who popularized the practice of ending sentences with the colon or a period, invented the semicolon, made occasional use of parentheses, and created the modern comma by manipulating the virgule [44].

1598. Established in 1598, the Francis Trigge Chained Library in Grantham, Lincolnshire, was the earliest example in England of a library to be endowed for the benefit of users who were not members of an institution such as a cathedral or college [45]. The library still exists and can justifiably claim to be the forerunner of later public library systems [45].

1604. The *Elementarie* created by Richard Mulcaster in 1582 was an early nonalphabetical list of 8000 English words [46]. The first purely English alphabetical dictionary was *A Table Alphabeticall of Hard Usual English Words*, written by English schoolteacher Robert Cawdry in 1604 [47].

1605. German publisher Johann Carolus's established *Relation aller Fürnemmen und gedenckwürdigen Historien* (Account of All Distinguished and Commemorable News), which The World Association of Newspapers recognizes as the world's first newspaper [48,49].

1626. Jesuit scientist Christopher Scheiner published the *Rosa Ursina sive Sol*, which used graphics to reveal his astronomical research on the sun. He used images to explain the rotation of the sun over time [50].

1665. The Royal Society of London published the first issues of *Philosophical Transactions*, generally considered to be the first scientific journal [51].

1696. The modern footnote was standardized by French philosopher Pierre Bayle in his *The Dictionary Historical and Critical* [52].

2.4.3 The Industrial Revolution (Eighteenth–Nineteenth Century) and Its Information Innovations

Information innovations and knowledge tools facilitated the spread of technologies that transformed life at home and work in the eighteenth and nineteenth century. The industrial revolution began in the United Kingdom and rapidly spread through Europe, the U.S., and other global regions. Spurred by growth of learning in the previous era, this period witnessed many firsts such as copyright laws, encyclopedias, magazines, and mass media.

1710. Copyright originated in Britain as a reaction to printers' monopolies. The unregulated copying of books led to the passing of the Licensing Act of 1662, which predated the Statute of Anne, widely regarded as the first copyright law [53].

1728. English writer Ephraim Chambers published his *Cyclopaedia* in 1728 [54]. It included a broad scope of subjects, used an alphabetical arrangement, relied on many different contributors, and included the innovation of cross-referencing other sections within articles [54].

1731. English printer Edward Cave first published *The Gentleman's Magazine*, which is considered as the first general-interest magazine, in January 1731 [55]. Under the pen name Sylvanus Urban (an anagram of the Latin words *Urbanus* for city and *Sylva* for forest), Cave edited *The Gentleman's Magazine* as a renowned and respected publication in the English language. Cave was the first to use the term "magazine" (meaning "storehouse") as a periodical and defined its role as "A Collection of all Matters of Information and Amusement" [56].

1735. Swedish scientist Carl Linnaeus was the first to establish taxonomy conventions for the naming of living organisms [57]. These became universally accepted in the scientific world and were the starting point of binomial nomenclature. The great eighteenth century expansion of natural history knowledge was based on Linnaean taxonomy, the system of scientific classification now widely used in the biological sciences. The first edition of *Systema Naturae* was printed in the Netherlands in 1735 [57].

1822. A convenient writing tool, the mechanical pencil was first invented in Britain in 1822 by Sampson Mordan and Gabriel Riddle [58]. The earliest Mordan pencils were thus hallmarked SMGR and were manufactured until World War II. Between 1820 and 1873, more than 160 patents were registered pertaining to a variety of improvements to mechanical pencils [58].

1848. Associated Press (AP) was founded in May 1848 in New York City to address the costly collection of news by telegraphy. Six newspapers decided to work together to get news from the west and from abroad to serve the public with increasingly wider coverage of the United States and the world. Associated Press serves more than 1500 newspapers and 5000 broadcast outlets in the United States alone. Worldwide, the AP serves more than 15,000 news organizations in 112 countries [59].

1860–1865. The U.S. Civil War sparked a demand for news. The value of telegraphic wire services with speedy war information spurred news readership and circulation [60].

1860s. The U.S. Civil War also marked a growing demand for pencils as a portable tool to record information [61]. During the 1860s, Jonathan Dixon introduced the mass production of pencils with erasers. This replaced the cumbersome construction of pencils by hand. As an innovative tool for convenient recording of information and subsequent communication, the pencil replaced pre-existing writing tools such as the stylus, which was developed in Roman times [58].

1898. The vertical filing cabinet, similar to the modern one, was invented by Edwin G. Seibels. Previously, businesses kept papers in envelopes stored in arrays of pigeonholes, often lining a wall. Seibels reasoned that folding was not necessary; papers could be kept in large envelopes standing on end vertically in a drawer, thereby revolutionizing recordkeeping [61].

2.4.4   The Era of the Mind and the Machine (Twentieth Century)

Information overload was identified as issue in the interplay of the mind and the machine during the twentieth century. The early years of the twentieth century ushered in an era of information and media for the masses even as information overload was recognized as a possible problem. The broadcast industry transformed mass media into a dominant source of news, information, and entertainment.

1915. The 3-hour silent film, *The Birth of a Nation*, marked the start of modern movie industry [62].

1923. Time Magazine was first published March 3, 1923, as a newsmagazine that summarized and organized the news so that "busy men" could stay informed [9]. *Time Magazine's* founders Henry Luce and Briton Hadden stated in their 1922 prospectus, "Although daily journalism has been more highly developed in the U.S. than in any other country of the world–people in America are, for the most part, poorly informed . . . . The reason people are uninformed is that no publication has adapted itself to the time which busy men are able to spend on simply keeping informed" [9, 10].

1920s. Promotion and persuasion assumed creative forms in the first advertisements on the radio [63].

1925. On March 25, 1925, Scottish inventor John Logie Baird conducted the first public demonstration of televised silhouette images in motion at Selfridge's Department Store in London [64].

1926. The National Broadcasting Company (NBC) was established in 1926 as the first U.S. commercial broadcasting network [65].

1936. Henry Luce, cofounder of *Time Magazine*, launched *Life*, a weekly news magazine, with a strong emphasis on photojournalism [66].

1938. During Halloween, Orson Welles's "War of the Worlds" broadcast shocked millions of U.S. radio listeners when radio news alerts announced the arrival of Martians [67]. The radio listeners panicked when they learned of the Martians' ferocious and seemingly unstoppable attack on Earth. Many left their homes screaming while others packed up their cars and fled [67].

1939. Television was demonstrated at New York's World Fair and marked an era of moving pictures in the living room [68].

1941. Orson Welles's 1941 American drama film *Citizen Kane* was released and eventually gained renown as the best movie of all time. It was praised for innovative cinematography, music, and narrative structure [69].

1945. In a 1945 article, scientist Vannevar Bush proposed that science be put to use in organizing the vast record of human knowledge [11,12]. Inspired by his previous work in microfilm mass storage, Bush envisioned an information workstation, the *memex*, capable of storing, navigating, and annotating an entire library's worth of information [11,12].

1949. Xerography (or electrophotography) was developed as a photocopying technique by Chester Carlson in 1938 and patented in 1942 [70]. Although dry electrostatic printing processes were invented in 1778, Carlson's unusual and nonintuitive innovation combined electrostatic printing with photography. The first commercial units were superseded by fully automatic units in the 1960s [70].

1946–1961. "The Golden Age of Television" in the United States marked a 16-year period when many hour-long anthology drama series received critical acclaim [68]. As a new medium, television introduced many innovative programming concepts and prime time television drama showcased both original and classic productions [68].

1960. The Kennedy-Nixon debates marked television's grand entrance into presidential politics [68,71]. They afforded the first real opportunity for voters to see their candidates in competition, and the visual contrast from radio reports was dramatic [68,71].

1964. In his 1964 book, *The Managing of Organizations: The Administrative Struggle*, American social scientist Bertram Gross cited Vannevar Bush's reference to the "problem of information overload" [13].

1968. In 1965, Ted Nelson defined the concept of "hypertext" as "a body of written or pictorial material interconnected in [such] a complex way that it could not be conveniently represented on paper" [72] as part of Project Xanadu, inspired by "As We May Think," an essay that Vannevar Bush published in 1945. Bush described a microfilm-based machine (the memex) in which one could link any two pages of information into a "trail" of related information [12]. Independently, Douglas Engelbart (with Jeff Rulifson) was the first to implement the concept for scrolling within a single document (1966) and for connecting separate documents (1968) [5]. The principle of hypertext was based on associating information through "links" into a coherent organization [5,11,72].

1960s–1994. The Internet was initiated with U.S. government-funded computer networking efforts. In 1969, the precursor to the Internet began with the U.S. Defense Department's ARPAnet. In 1985, the U.S. National Science Foundation considered how it could provide greater access to the high-end computing resources at its supercomputer centers by linking many research universities to the centers [73].

1970. Futurist Alvin Toffler popularized the term "information overload" in his 1970 book *Future Shock* [15]. He described information overload as the difficulty a person can have understanding an issue and making decisions that can be caused by the presence of too much information [15].

1971. In December 1971, the American Telephone and Telegraph (AT&T) Corporation proposed a cellular service to the Federal Communications Commission, which, after years of hearings, approved the proposal for advanced mobile phone system (AMPS) in 1982 [74]. One of the first successful public commercial mobile phone networks was the ARP network in Finland, launched in 1971 [74,75]. Wireless technologies pumped up collaboration among various information workers, making it mobile and personal [75].

1977. Fiber-optic technology was introduced to expand point-to-point audio and video transmissions [76]. During the 1990s, the success of the Internet sparked massive investment in transnational fiber-optic cables [76].

1979. Compact discs (CDs) were originally developed to store and play back sound recordings exclusively. Audio CD players became commercially available in 1982. CDs were later expanded to encompass data and information storage in its many forms [77].

1980. Upon its launch in 1980, CNN was the first all-news television channel in the United States and the first channel to provide 24-hour television news coverage. CNN changed the notion that news could be reported only at fixed times throughout the day [78]. At the time of CNN's launch, TV news was dominated by three major networks, ABC, CBS and NBC, and their nightly 30-minute broadcasts [78].

1981. Music Television (MTV) was launched with "Video Killed the Radio Star" [79]. MTV had a profound impact on the music industry and popular culture. It successfully launched the idea of a dedicated video-based outlet for music [79].

1989. The fall of the Berlin Wall on November 9, 1989, signified two important trends. One, the fall of the Berlin Wall was among the first historical events that unfolded on television worldwide with a visual impact more powerful than traditional news reports [80]. Second, the fall of the Berlin Wall ended the Cold War conflict and led to revolutionary changes in the diffusion of information in Eastern Europe. The fall of the Berlin Wall was a culminating point that marked the fall of Communism in Europe and the Soviet Union. It ended restrictions in travel and information policies and liberated the people in Communist regimes with a spirit of freedom and unfettered information. The fall of the Berlin Wall thus tipped the worldwide balance of power toward democracies and free markets, leading to a free flow of information through scientific and cultural exchanges [81].

2.4.5   Internet Boom and Information Explosion of the 1990s

The agents of information overload emerged in the 1990s when the Internet diffused into society and became a household name. During the 1990s, the advent of the World Wide Web spearheaded the popularity and commercial growth of the Internet. Besides attaining notoriety as a persistent purveyor of information overload, the Internet emerged as a tool for bringing together the small contributions of millions of people and making them matter.

1991. On August 6, 1991, British physicist and computer scientist Tim Berners-Lee, the inventor of the World Wide Web, posted a summary of the Web project on the alt.hypertext newsgroup. This date also marked the debut of the Web as a publicly available service on the Internet [82].

1990s. With the ease of Web access for the general public, the Internet ushered in widespread changes in work and play. The Internet enabled people at home, work, and beyond to forge robust services that increased business efficiency and developed social exchanges [83]. Until the end of the 1990s, networked computers were connected through expensive leased lines and/or dial-up phone lines. High-speed Internet-linked computer networking in the 1990s led to closer coordination and collaboration among people in far-flung areas. Search engines enabled everyone to use the Internet as a personal ready reference of learning and lore [83]. Open-source applications, such as Linux, created self-organizing communities, launched a collaborative revolution, and redefined globalization [84].

1993. Internet media emerged with newspapers and magazines offering free online content and perpetuated the "free" mentality among consumers of online content [85].

1995. Billing itself as the "world's largest bookseller," Amazon.com launched online in 1995 [86]. Amazon stated that "a physical bookstore as big as Amazon.com was economically impossible because no single metropolitan area was large enough to support such a mammoth store. Were Amazon.com to print a catalog of all of its titles, it would be the size of seven New York City phone books" (each of which lists more than eight million phone numbers) [86].

1996. Beginning in 1996, Stanford University graduate students Larry Page and Sergey Brin built a search engine called "BackRub" that used links to determine the importance of individual Web pages. By 1998, they had formalized their work and launched the search engine, Google, named after a play on the word "googol," the mathematical term for a 1 followed by 100 zeros [87]. Even though it was not the first Internet search engine, Google quickly became the most popular.

2.4.6 Data Deluge and Information Overload in the Twenty-First Century Digital Age

The information overload phenomenon has exacerbated in the digitally disseminated information age of the twenty-first century. Empowered by digital tools, people with a myriad of motivations have launched innovative products such as YouTube, new forms of television, and electronic gadgets. Social media, mobile communication, and other forms of new communication have swamped individuals and small groups with a global glut of information. Even as scientists in a wide range of disciplines are overwhelmed with data, people download and distribute megabytes of data and have become principal players in the twenty-first century marketplace.

2004. A 19-year-old Harvard sophomore, Mark Zuckerberg, launched Facebook (then called "thefacebook.com") as a social networking service on February 4, 2004. In fewer than 5 years, Facebook transformed itself from a dorm-room novelty to a company with more than a billion users in 2011 [88]. Many Facebook users spend several hours a day on this site and admit that Facebook causes information overload with increased challenges of maintaining social relationships.

2005. In February 2005, three former PayPal employees launched a video-sharing website, YouTube, where users can upload, share, and view videos. Before the launch of YouTube, there were few easy methods available for ordinary computer users who wanted to post videos online. With its simple interface, YouTube made it possible for anyone with an Internet connection to post a video that a worldwide audience could watch within minutes [89].

2006. Twitter, the real-time information network and microblogging service, was developed in 2006 by three programmers, Jack Dorsey (@Jack), Evan Williams (@Ev), and Biz Stone (@Biz), in San Francisco. At the heart of Twitter are small bursts of information called tweets, which are limited to 140 characters, with a details pane that provides additional information, deeper context, and embedded media. Since its launch in July 2006, Twitter has rapidly gained worldwide popularity, with more than 400 million users in nearly every country in the world [90]. Although Twitter has helped open exchange of information, most users complain that it contributes to rising expectations to keep up with information [9].

2009. Since June 12, 2009, full-power television stations in the United States have been broadcasting exclusively in a digital format [92]. U.S. television stations started airing 3-D television serials based on the same technology as 3-D movies, possibly opening a new era in TV programming [9].

2010. With more than two billion views per day, YouTube's viewership in May 2010 surpassed all three major broadcast networks (ABC, CBS, and NBC) combined during their "primetime" evening time slot [89].

2011. The online retailer, Amazon.com, announced in January 2011 that "Kindle books have now overtaken paperback books as the most popular format on

Amazon.com" [93]. Since 2011, for every 100 paperback books sold, Amazon has sold 115 Kindle books [93].

2011. Mobile phones emerged as a near-ubiquitous tool for information seeking and communicating. A U.S. survey conducted by the Pew Internet and American Life Project found that 83% of Americans reported owning mobile phones and of these, more than half said they have used their phones at least once to get information they needed right away. The report stated: "Mobile devices help people solve problems and stave off boredom, but create some new challenges and annoyances" [94].

2011. In a special issue, "Dealing with Data," the renowned research publication, *Science*, explored the issues related to the increasingly huge influx of research data [5]. Two themes dominate: most scientific disciplines are finding the data deluge to be extremely challenging, and tremendous opportunities can be realized if we can better organize and access the data [5].

In tracing the evolution of information and its proliferation in society, this timeline chronicles the growth of information and highlights four important issues. One, the information overload phenomena originated as a debilitating trend in the 1990s before the exponential growth of the Internet. Two, as media, entertainment, and other sources of information proliferate in our society, information overload has exacerbated in this digital age. Three, the twenty-first century has witnessed a deluge of data, 24/7 media, and other information that inundate professionals as well as common citizens, leading to the phenomenon of information overload. Four, information overload has emerged as an important research paradigm and a challenge to professional engineers and technical communicators.

2.5 Information Overload Concepts

Information overload concepts play a pivotal role in our understanding of its causes and effects. Based on a comprehensive state-of-the-discipline review of scholarly research and meta-analysis of concepts, this section presents concepts related to the context and causes of information overload. Each conceptual construct is explicated with a short definition.

2.5.1 Definitions of Information Overload and Related Concepts

Every day, media consumers and professionals tussle to adequately process and comprehend the unending flow of information content and media messages they encounter. Such overabundance of media and information burdens the brain's ability to process information and adversely affects users. This section explicates concepts related to information overload and related theories.

Interruption Overload is defined as a plethora of interruptions that afflicts people at work, home, and beyond [23, 95–98]. In contrast to the traditional modes of interruption such as unsolicited phone calls and unscheduled visits, interruption overload today may be caused by modern modes such as social media messages, mobile phones, and personal digital assistants [13, 15, 23, 99–101]. Some individuals have adopted the time-tested strategy to combat interruption overload by simply ignoring messages and data that inundate and disrupt productivity.

Cognitive Overload refers to a surfeit of messages that strains and weakens the message receiver's working memory [18, 95–98, 101, 102]. For instance, students face *cognitive overload* in a classroom when teachers profess too much information or assign too many tasks simultaneously. Cognitive overload occurs when the processing demands of an activity are disproportional to the processing limits of the learner. Similar to information overload, cognitive overload leads to anxiety and stress. Cognitive overload may hamper and hinder learning and communication [99–102].

Sensory Overload. Often confused with cognitive overload, the concept of *sensory overload* explains why individuals find it difficult to focus on a task when one or more of their senses are strained. Sensory overload also refers to an internal feeling of an overload of the senses that prevents us from focusing on a task at hand [7, 102].

Filter Failure. As a counter argument to information overload, *filter failure* posits that an abundance of information is nothing new because it has affected us since the advent of the printing press. So, the problem of lost productivity is caused not by information abundance, but by the failure to filter information as it is either published or consumed. A proponent of this concept, technologist Clay Shirky delivered a 2008 Web 2.0 Expo talk titled "It's not information overload. It's filter failure" [103]. However, it can be argued that *filter failure* is simply one aspect of information overload, as we are battling an abundance of unfiltered information anyway.

Infomania. Coined in the 1980s by Elizabeth M. Ferrarini in her books, *Confessions of an Infomaniac* [104] and *Infomania: The Guide to Essential Electronic Services* [105], the concept of *infomania* relates to the psychologically debilitating condition of information overload, caused by an obsessive fixation with unattended and pending information (such as e-mail) and a compulsive interplay of interruptions from instantaneous communication technologies such as social media, phones, and e-mail. Infomania also refers to the distraction and dissonance caused by a fixation to access information at the expense and neglect of work obligations and family commitments [104, 105].

2.5.2 The Context of Information Overload

An ever-expanding world of technological excesses creates diverse contexts of information overload ranging from essential information to spam among other unsolicited information. Information overload can also be conceptualized in a *spatial context* (such as having a quantity of data that is too voluminous to read) [95], a *temporal context* (such as having too much information about the same topic in a short time period) [96–98], and a *personal context* (such as being too busy to notice you are too busy) [99–106], just to name a few. This section explicates conceptual constructs related to the context of information overload.

Post Scarcity (also termed Post-Scarcity Economics, Post-Scarcity, or Postscarcity). Information overload has been aided and abetted by information content that is free. When information is free, more people use it [107]. The term *post scarcity* refers to a hypothetical form of economy or society in which commodities such as goods, services, and information are free or practically free [107, 108]. This would be due to an abundance of fundamental resources (matter, energy, and intelligence) in conjunction with sophisticated automated systems capable of converting raw materials into finished goods, which allows manufacturing to be as easy as duplicating information. Information content may be free because the traditional economics of scarcity do not apply to information any more. Some information products are promoted with "free" models such as *cross-subsidies* (a free television set or a free phone to sell a paid service) and *freemiums* (offering free information services such as Flickr for free while selling premium services such as FlickrPro to users for a fee). Thus, free information fosters widespread use and in some instances an overabundance of information [108, 109].

Learning Society. Information overload is both a boon and a bane of the rising expectations for education. International agencies such as the Organisation for Economic Co-Operation and Development (OECD) and the United Nations Educational, Scientific, and Cultural Organization (UNESCO) advocate the educational philosophy of a *learning society* [110, 111]. By both intent and design, a *learning society* positions education as the key to economic development and holds that education should extend beyond formal learning (based in traditional educational institutions—schools, universities, etc.) into informal learning centers to support a knowledge economy, also known as a "world education culture." Such aspirations have contributed to a widespread level of information overload caused by an unending quest of knowledge and expertise worldwide [111–114].

2.5.3 Causes and Consequences of Information Overload

A surging growth of information sources such as the Internet has inevitably amplified preexisting information overload issues [115–117]. This section provides conceptual constructs related to causes and consequences of information overload. The general causes of information overload include the following:

- *Voracious Appetite for Information.* The economic and social impact of information overload will continue to rise as society's voracious appetite for information increases in its many forms [4, 6, 117–120].
- *Free and Unfettered Information.* Easy digital duplication and instantaneous transmission of data across the Internet has led to a proliferation of new information exchange that is free and unfettered in nations that foster free flow of news and information [6–8, 27, 117–120]. In nations such as China, Iran, Iraq, Egypt, and other regions that censor information, digital media have helped bypass information controls [5, 29, 83–85, 88, 89].
- *The Role of the Internet as a Medium Without Gatekeepers.* The Internet, being a medium without gatekeepers, offers a low signal-to-noise ratio that provides legitimacy to redundant communication [28, 29, 119]. Even in nations that seek to control free exchange of information, Internet communication has empowered people and protesters to share their voice worldwide [5, 8, 29, 83–85, 88, 89].
- *Unreliable Information.* Dubious information sources perpetuate contradictory data, inaccuracies, and other forms of unreliable information that is disseminated with no guarantee of importance, trust, or accuracy [26, 28, 120].
- *Manifold Message Modes.* A plethora of channels of information (e.g., social media messages, telephone, e-mail, instant messaging, and various forms of digital feeds) create manifold message modes [3–6, 8, 23].

- *Customized Content.* Large amounts of searchable archival information with customized content have empowered users. However, content that is self-centered and indiscriminate or pieces of information that are unrelated or that do not have any overall structure to reveal their relationships cause problems [6–8, 27–29, 117, 118, 121].

Exposure to too much information from 24/7 media and the deluge of data leads to information overload effects such as apathy [7, 27, 117], indifference [18, 119, 120], mental exhaustion [103, 104, 117, 118], and other complex communicative challenges [21–24, 26, 98–101, 118–120].

2.6 Conclusion and Four Lessons Learned

Information overload inundates common people as well as professionals [115–120]. In our digital age, scientists, engineers, technical communicators, information technology workers, scholars, managers, and communication practitioners are seeking ways to combat the data deluge [6, 8, 25–27, 121]. In tracing the conceptual evolution of information overload, this chapter highlights its importance as a research paradigm and a professional challenge as information, news, and entertainment messages proliferate. Four important themes emerge from the meta-analyses of information concepts and theoretical reviews of information overload.

The first lesson identifies striking contrasts in the twenty-first century age of digitally mediated information. The exponential growth of technology, while empowering people and professionals with vast and easy access to knowledge, has also complicated and catapulted information overload in a myriad of ways. Even as we are overwhelmed by information overload and find it to be extremely challenging, there is a stupendous opportunity for people and professionals who can devise failsafe strategies to master the data deluge.

The second important lesson from this chapter relates to the evolution of information overload as a twentieth century concept based on primary research resources. This chapter credits the founders of *Time Magazine*, American media pioneers Henry Luce and Briton Hadden, for precociously recognizing information overload in the early decades of the twentieth century. In later years of that century, scientist Vannevar Bush, futurist Alvin Toffler, and designer Richard Saul Wurman posited critical perspectives that advanced the theory and concept of information overload. Such pioneering efforts have inspired academic researchers to explore the surging significance of information overload research and practice in the twenty-first century digital age.

The third lesson relates to a complex range of important innovations that have contributed to the evolution of information and its proliferation in society. The evolution occurred in six epochs: Early Quest for Information and Knowledge (320 BCE–thirteenth century); Age of Renaissance (fourteenth–seventeenth century) and the Printing Press; Industrial Revolution (eighteenth–nineteenth century) Ushers Information Innovations; Era of the Mind and the Machine (twentieth century); Internet Boom and Information Explosion of the 1990s; and Data Deluge and Information Overload in the Twenty-First Century Digital Age. Besides attaining notoriety as a persistent purveyor of information overload, the Internet has forged communities that foster contributions of people in this global glut of information.

The fourth and final lesson from this chapter derives from locating the information overload research paradigm within a larger intellectual context of communication. To that end, this chapter presents definitions, concepts, and theories related to the context and causes of information overload.

Information overload theories and concepts explicate how the average person today is swamped by information from various Internet applications, 24/7 media, digital smog, and junk mail [4–6, 8, 25–28, 115–121]. Research studies delineate how excessive information and the intrusion of data sources into our lives sap productivity and performance [21–24, 98–101, 117–120]. Future research should compare and contrast information overload problems from local and global perspectives. In tracing the conceptual evolution of information overload, this chapter highlights the importance of information overload as a research paradigm and a professional challenge as information, news, and entertainment messages proliferate in our society.

Acknowledgment

The authors gratefully acknowledge their debt to Dr. J. B. Strother, Dr. J. M. Ulijn, Z. Fazal, and Dr. T. Nathans-Kelly for their critiques to earlier versions of this chapter; B. McAuley and J. Grainger for research help; and Dr. J. Aikat for research feedback and ideas to enhance this chapter.

REFERENCES

[1] D. Allen and T. D. Wilson, "Information overload: Context and causes," *New Rev. Inform. Behaviour Res.*, vol. 4, no. 1, pp. 31–44, 2003.

[2] S. Bergamaschi, *et al.*, "Guest editors' introduction: Information overload," *IEEE Internet Comput.*, vol. 14, no. 6, pp. 10–13, Nov.–Dec. 2010.

[3] M. Baez, *et al.*, "Addressing information overload in the scientific community," *IEEE Internet Comput.*, vol. 14, no. 6, pp. 31–38, Nov.–Dec. 2010.

[4] J. Turow and L.S Tsui, Eds., *The Hyperlinked Society: Questioning Connections in the Digital Age.* Ann Arbor, MI: Univ. Michigan Press, 2008.

[5] Science Staff, "Challenges and opportunities," *Science*, vol. 331, no. 6018, pp. 692–693, Feb. 2011.

[6] E. Simperl, *et al.*, "Overcoming information overload in the enterprise: The active approach," *IEEE Internet Comput.*, vol. 14, no. 6, pp. 39–46, Nov.–Dec. 2010.

[7] J. D. Falk and M. S. Kucherawy, "Battling spam: The evolution of mail feedback loops," *IEEE Internet Comput.*, vol. 14, no. 6, pp. 68–71, Nov.–Dec. 2010.

[8] M. Eppler and J. Mengis, "The concept of information overload: A review of literature from organization science, accounting, marketing, MIS, and related disciplines," *Inform. Soc.*, vol. 20, no. 5, pp. 325–344, Nov.–Dec., 2004.

[9] Time, Inc. History. (2009, May 30). *History of Time Archive Collection* [Online]. Available: http://www.timeinc.com/aboutus/history.php Jan. 20, 2012 (last date accessed).

[10] Time Collection. (2009, May 30). *History of Time Archive Collection* [Online]. Available: http://www.time.com/time/archive/collections/0,21428,c_time_history,00.shtml Jan. 20, 2012 (last date accessed).

[11] V. Bush, "As we may think," *Atlantic Monthly*, vol. 176, no. 1, pp. 641–649, Jul. 1945.

[12] V. Bush, "As we may think," *Life Mag.*, vol. 19, no. 1, pp. 112–114, 116, 121, 123, 124, Sep. 1945.

[13] B. M. Gross, *The Managing of Organizations: The Administrative Struggle*, vol. 2. New York: Free Press of Glencoe, 1964.

[14] T. Berners-Lee, *et al.*, "World-Wide Web: The information universe," *Internet Res.*, vol. 2, no. 1, pp. 52–58, 1992.

[15] A. Toffler, *Future Shock*, New York: Random House, 1970.

[16] R. S. Wurman, *Information Anxiety*, New York: Doubleday, 1989.

[17] J. Jacoby, "Perspectives on information overload," *J. Consumer Res.*, vol. 10, no. 4, pp. 432–435, Mar. 1984.

[18] N. K. Malhotra, "Information and sensory overload," *Psychol. Market.*, vol. 1, nos. 3–4 pp. 9–21, 1984.

[19] N. K. Malhotra, "Reflections on the information overload paradigm in consumer decision making," *J. Consumer Res.*, vol. 10, no. 4, pp. 436–40, Mar. 1984.

[20] N. Kock, *et al.*, "Can Hofstede's model explain national differences in perceived information overload? A look at data from the US and New Zealand," *IEEE Trans. Prof. Commun.*, vol. 51, no. 1, pp. 33–49, Mar. 2008.

[21] A. F. Farhoomand and D. H. Drury, "Managerial information overload," *Commun. ACM*, vol. 45, no. 10, pp. 127–131, 2002.

[22] X. Yang, *et al.*, "Summarizing relational databases," in *Proc. Very Large Data Base Endowment*, vol. 2, no. 1. 2009, pp. 634–645.

[23] E. Russell, *et al.*, "Describing the strategies used for dealing with email interruptions according to different situational parameters," *Comput. Human Behavior*, vol. 23, no. 4, pp. 1820–1837, 2007.

[24] J. Wajcman, "Life in the fast lane? Towards a sociology of technology and time," *British J. Sociol.*, vol. 59, no. 1, pp. 59–77, 2008.

[25] D. M. Levy, "To grow in wisdom: Vannevar Bush, information overload, and the life of leisure," in *Proc. 5th ACM/IEEE-CS Joint Conf. Digital Libraries*, Jun. 2005, pp. 281–286.

[26] W. H. Shaw, "Information overload," *IEEE Eng. Manag. Rev.*, vol. 38, no. 1, p. 2, Mar. 2010.

[27] L. K. Langford, "Surf's up: Harnessing information overload," *IEEE Eng. Manag. Rev.*, vol. 38, no. 1, pp. 164–165, Mar. 2010.

[28] E. L. Berman, "Relationships and technology: Can they coexist?" *IEEE Eng. Manag. Rev.*, vol. 38, no. 1, pp. 162–163, Mar. 2010.

[29] R. Levine, "Geography of busyness," *Social Res., Int. Quar.*, vol. 72, no. 2, pp. 355–370, 2005.

[30] E. D. Johnson, *A History of Libraries in the Western World.* New York: Scarecrow Press, 1965.

[31] T. Weber, *The Language of Paper: A History of 2000 Years.* Bangkok, Thailand: Orchid Press, 2008.

[32] D. Mudd, *All the Money in the World: The Art and History of Paper Money and Coins from Antiquity to the 21st Century.* New York: Collins, 2006.

[33] L. Munkhammar, *The Silver Bible: Origins and History of the Codex Argenteus.* Uppsala, Sweden: Selenas, 2011.

[34] K. Kawase, *An Introduction to the History of Pre-Meiji Publishing: History of Wood-Block Printing in Japan.* Tokyo, Japan: Yushodo Booksellers, 1973.

[35] S. Eldon, *From Quill Pen to Satellite: Foreign Ministries in the Information Age.* London, U.K.: Royal Institute of International Affairs European Programme, 1994.

[36] B. Gu, *From Oracle Bones to Computers: The Emergence of Writing Technologies in China.* West Lafayette, IN: Parlor Press, 2009.

[37] L. C. Taub, *Ancient Meteorology.* London, U.K.: Routledge, 2003.

[38] J. E. Smith, *A History of Paper: Its Genesis and Its Revelations, Origin and Manufacture, Utility and Commercial Value of an Indispensable Staple of the Commercial World.* Holyoke, MA: Clark W. Bryan, 1881.

[39] T. Hughes. (2009, Jul. 9). *History's Newsstand Blog* [Online]. Available: http://blog.rarenewspapers.com/?tag=rag-paper Jan. 20, 2012 (last date accessed).

[40] H. G. Fletcher, *Gutenberg and the Genesis of Printing*. New York: Pierpont Morgan Library, 1994.

[41] S. Füssel, *Gutenberg and the Impact of Printing*. Aldershot, U.K.: Ashgate Pub., 2005.

[42] J. Drucker and E. McVarish, *Graphic Design History: A Critical Guide*. Upper Saddle River, NJ: Prentice-Hall, 2009.

[43] M. Foot, *The History of Bookbinding as a Mirror of Society*. London, U.K.: British Library, 1998.

[44] P. Burnhill, *Type Spaces: In-House Norms in the Typography of Aldus Manutius*. London, U.K.: Hyphen, 2003.

[45] J. Glenn, *et al.*, *Catalogue of the Francis Trigge Chained Library, St. Wulfram's Church, Grantham*. Cambridge, U.K.: D. S. Brewer, 1988.

[46] R. Mulcaster and E. T. Campagnac, *Mulcaster's Elementarie*. Oxford, U.K.: Clarendon Press, 1925.

[47] R. Cawdry, *A Table Alphabeticall of Hard Usual English Words (1604); the First English Dictionary*. Gainesville, FL: Scholars' Facsimiles & Reprints, 1966.

[48] P. J. Dunnett, *The World Newspaper Industry*. London, U.K.: Croom Helm, 1988.

[49] W. Heide, *Die alteste Gedruckte Zeitung*. Mainz, Germany: Verlag der Gutenberg-Gesellschaft, 1931.

[50] C. Scheiner, *et al.*, *Rosa Vrsina; Sive, Sol Ex Admirando Facvlarvm & Macularum Suarum Phoenomeno Varivs*, Bracciani Apud Andraeam Phaeum typographum ducalem, 1630.

[51] D. Atkinson, *Scientific Discourse in Sociohistorical Context: The Philosophical Transactions of the Royal Society of London, 1675–1975*. Mahwah, NJ: L. Erlbaum Associates, 1999.

[52] P. Bayle, *et al.*, *The Dictionary Historical and Critical*, London, U.K.: Hunt & Clarke, 1826. Printed for J. J. and P. Knapton, D. Midwinter, J. Brotherton, and A. Bettesworth, 1734–1738.

[53] L. Bently, *et al.*, *Global Copyright: Three Hundred Years Since the Statute of Anne, from 1709 to Cyberspace*. Cheltenham, U.K.: Edward Elgar, 2010.

[54] P. Shorr, *Science and Superstition in the Eighteenth Century; a Study of the Treatment of Science in Two Encyclopedias of 1725–1750: Chambers' Cyclopedia, London (1728); Zedler's Universal Lexicon, Leipzig (1732–1750)*. New York: Columbia Univ. Press, 1932.

[55] E. A. Reitan, *The Best of the Gentleman's Magazine, 1731–1754*. Lewiston, NY: E. Mellen Press, 1987.

[56] C.L. Carlson, *The First Magazine; a History of the Gentleman's Magazine, with an Account of Dr. Johnson's Editorial Activity and of the Notice Given America in the Magazine*. Providence, RI: Brown Univ., 1938.

[57] A. Polaszek, *Systema Naturae 250: The Linnaean Ark*. Boca Raton, FL: CRC Press, 2010.

[58] J. I. Whalley and J. Irene, *Writing Implements and Accessories: From the Roman Stylus to the Typewriter*. Detroit, MI: Gale Research Co., 1975.

[59] Associated Press. (2011, Mar. 31). *Associated Press Facts and Figures* [Online]. Available: http://www.ap.org/pages/about/about.html Jan. 20, 2012 (last date accessed).

[60] T. Wheeler, *Mr. Lincoln's T-mails: The Untold Story of How Abraham Lincoln Used the Telegraph to Win the Civil War*. New York: Collins, 2006.

[61] S. Khoshafian and M. Buckiewicz, *Introduction to Groupware, Workflow, and Workgroup Computing*. New York: Wiley, 1995.

[62] D. Platt, *Celluloid Power: Social Film Criticism from the Birth of a Nation to Judgment at Nuremberg*. Metuchen, NJ: Scarecrow Press, 1992.

[63] S. Smulyan, *Selling Radio: The Commercialization of American Broadcasting, 1920–1934*, Washington, DC: Smithsonian Institution Press, 1994.

[64] J. L. Baird and M. H. Baird, *Television and Me: The Memoirs of John Logie Baird*. Edinburgh, U.K.: Mercat Press, 2004.

[65] R. Campbell, *The Golden Years of Broadcasting: A Celebration of the First 50 Years of Radio and TV on NBC*. New York: Scribner, 1976.

[66] E. L. Doss, *Looking at Life Magazine*. Washington, DC: Smithsonian Institution Press, 2001.

[67] J. Gosling and H. Koch, *Waging the War of the Worlds: A History of the 1938 Radio Broadcast and Resulting Panic, Including the Original Script*. Jefferson, NC: McFarland & Co., 2009.

[68] D. A. Heller, *The Great American Makeover: Television, History, Nation*. New York: Palgrave Macmillan, 2006.

[69] J. Naremore, *Orson Welles's Citizen Kane: A Casebook*. New York: Oxford Univ. Press, 2004.

[70] D. Owen, *Copies in Seconds: How a Lone Inventor and an Unknown Company Created the Biggest Communication Breakthrough Since Gutenberg: Chester Carlson and the Birth of the Xerox Machine*. New York: Simon & Schuster, 2004.

[71] E. A. Hinck, *Enacting the Presidency: Political Argument, Presidential Debates, and Presidential Character*. Westport, CT: Praeger, 1993.

[72] T. H. Nelson, "Complex information processing: A file structure for the complex, the changing and the indeterminate," in *Proc. 20th Nat. Conf. ACM*, 1965, pp. 84–100.

[73] J. Abbate, *Inventing the Internet*. Cambridge, MA: MIT Press, 1999.

[74] L. Cauley, *End of the Line: The Rise and Fall of AT&T*. New York: Free Press, 2005.

[75] J. Agar, *Constant Touch: A Brief History of the Mobile Phone*. Cambridge, U.K.: Icon, 2003.

[76] H. B. Killen, *Digital Communications with Fiber Optics and Satellite Applications*. Englewood Cliffs, NJ: Prentice-Hall, 1988.

[77] S. Knopper, *Appetite for Self-destruction: The Spectacular Crash of the Record Industry in the Digital Age*. New York: Free Press, 2009.

[78] P. M. Smith, *How CNN Fought the War: A View from the Inside*. New York: Carol Pub. Group, 1991.

[79] P. Temporal, *The Branding of MTV: Will Internet Kill the Video Star?* Singapore: Wiley (Asia), 2008.

[80] S. Manghani, *Image Critique and the Fall of the Berlin Wall*. Bristol, U.K.: Intellect, 2008.

[81] J. A. Engel, *The Fall of the Berlin Wall: The Revolutionary Legacy of 1989*. Oxford, U.K.: Oxford Univ. Press, 2009.

[82] T. Berners-Lee and M. Fischetti, *Weaving the Web: The Original Design and Ultimate Destiny of the World Wide Web by Its Inventor*. San Francisco, CA: HarperSanFrancisco, 1999.

[83] J. Ryan, *A History of the Internet and the Digital Future*. London, U.K.: Reaktion Books, 2010.

[84] L. Torvalds and D. Diamond, *Just for Fun: The Story of an Accidental Revolutionary*. New York: HarperBusiness, 2001.

[85] M. A. Winget and W. Aspray, *Digital Media: Technological and Social Challenges of the Interactive World*. Lanham, MD: Scarecrow Press, 2011.

[86] Amazon Media Room. (1995, Oct. 4). *World's Largest Bookseller Opens on the Web: Amazon. com Offers Million+ Titles, Orders Pour in from 45 Countries in First 4 Weeks* [Online]. Available: http://phx.corporate-ir.net/phoenix.zhtml?c=176060&p=irol-newsArticle&ID=1250170 Jan. 20, 2012 (last date accessed).

[87] D. A. Vise and M. Malseed, *The Google Story: For Google's 10th Birthday*. New York: Delta Trade Paperbacks, 2008.

[88] D. Kirkpatrick, *The Facebook Effect: The Inside Story of the Company That Is Connecting the World*. New York: Simon & Schuster, 2010.

[89] D. Gauntlett, *Making Is Connecting: The Social Meaning of Creativity from DIY and Knitting to YouTube and Web 2.0*. Cambridge, U.K.: Polity Press, 2011.

[90] T. O'Reilly and S. Milstein, *The Twitter Book*. Sebastopol, CA: O'Reilly Media, Inc., 2009.

[91] A. Beard and A. McNayr, *Historical Tweets: The Completely Unabridged and Ridiculously Brief History of the World*. New York: Villard Trade Paperbacks, 2010.

[92] J. Bennett and N. Strange, *Television as Digital Media*. Durham, NC: Duke Univ. Press, 2011.

[93] Amazon Media Room: News Release (2011, Jan. 27). *Amazon.com Announces Fourth Quarter Sales Up 36% to $12.95 Billion* [Online]. Available: http://phx.corporate-ir.net/phoenix.zhtml?c=176060&p=irol-newsArticle&ID=1521090 Jan. 20, 2012 (last date accessed).

[94] A. Smith. (2011, Aug. 15). *Americans and Their Cell Phones*. Pew Internet & American Life Project, Washington, DC, Rep. [Online]. Available: http://pewinternet.org/Reports/2011/Cell-Phones.aspx Jan. 20, 2012 (last date accessed).

[95] N. Belkin and W. B. Croft, "Filtering and information retrieval: Two sides of the same coin?" *Commun. ACM*, vol. 35, no. 12, pp. 29–39, Dec. 1992.

[96] P. Breivik, "Putting libraries back in the information society," *Am. Libraries*, vol. 16, no. 11, p. 703, Nov. 1985.

[97] P. S. Breivik, *Student Learning in the Information Age*. Phoenix, AZ: American Council on Education/Oryx Press Series on Higher Education, 1997.

[98] G. Marchionini, *Psychological Dimensions of User-Computer Interfaces. ERIC Digest, ERIC no. ED 337 203*. Syracuse, NY: ERIC Clearinghouse on Information Resources. Oct. 1991.

[99] J. Naisbitt and P. Aburden, *Megatrends 2000: Ten New Directions for the 1990's*. New York: Morrow, 1990.

[100] K. Obraczka, *et al.*, "Internet resource discovery services," *IEEE Comput.*, vol. 26, no. 9, pp. 8–22, Sep. 1993.

[101] K. A. Miller, *Surviving Information Overload: The Clear, Practical Guide to Help You Stay on Top of What You Need to Know*. Grand Rapids, MI: Zondervan, 2004.

[102] J. Sweller, *et al.*, "Cognitive architecture and instructional design," *Educational Psychol. Rev.*, vol. 10, no. 3, pp. 251–296, 1998.

[103] C. Shirky, (2008, Sep. 19). *It's Not Information Overload. It's Filter Failure, Web 2.0 Expo NY, Sebastopol, CA, O'Reilly Media* [Online]. Available: http://www.youtube.com/watch?v=LabqeJEOQyI Jan. 20, 2012 (last date accessed).

[104] E. M. Ferrarini, *Confessions of an Infomaniac*. Berkeley, CA: SYBEX, 1984.

[105] E. M. Ferrarini, *Infomania: The Guide to Essential Electronic Services*. Boston, MA: Houghton Mifflin, 1985.

[106] J. B. Appleberry,"Changes in our future: How will we dope?" *Faculty speech presented at California State University*, Long Beach, CA, Aug. 1992.

[107] P. Sadler, *Sustainable Growth in a Post-scarcity World: Consumption, Demand, and the Poverty Penalty*. Farnham, Surrey, U.K.: Gower, 2010.

[108] M. Bookchin, *Post-Scarcity Anarchism*. Montreal, QC, Canada: Black Rose Books, 1977.

[109] C. Anderson, *Free: The Future of a Radical Price*. New York: Hyperion, 2009.

[110] D. Buckingham and R. Willett, *Digital Generations: Children, Young People, and New Media*. Mahwah, NJ: Lawrence Erlbaum Associates, 2006.

[111] P. Jarvis, *Globalisation, Lifelong Learning and the Learning Society: Sociological Perspectives*. London, U.K.: Routledge, 2007.

[112] O. Strietska-Ilina, *A Clash of Transitions: Towards a Learning Society*. New York: Peter Lang, 2007.

[113] J. Gershuny,"Busyness as the badge of honour for the new superordinate working class," Inst. Social Economic Res., Univ. Essex, Colchester, Essex, U.K., Working Paper 2005-9, 2005.

[114] U. A. Kelly, *Migration and Education in a Multicultural World: Culture, Loss, and Identity.* New York: Palgrave Macmillan, 2009.

[115] J. Gershuny, *Changing Times: Work and Leisure in Post-Industrial Society.* Oxford, U.K.: Oxford Univ. Press, 2000.

[116] T. Veblen, *The Theory of the Leisure Class: An Economic Study in the Evolution of Institutions.* New York: Macmillan, 1899.

[117] J. B. Spira, *Overload! How Too Much Information Is Hazardous to Your Organization.* Hoboken, NJ: Wiley, 2011.

[118] F. Douglis, "Thanks for the fish . . . but I'm drowning!" *IEEE Internet Comput.*, vol. 14, no. 6, pp. 4–6, Nov.–Dec. 2010.

[119] C. A. Ahern, *Beyond Individual Differences: Organizing Processes, Information Overload, and Classroom Learning.* New York: Springer, 2011.

[120] B. Kovach and T. Rosenstiel, *Blur: How to Know What's True in the Age of Information Overload.* New York: Bloomsbury, 2010.

[121] H. A. Simon, "Designing organizations for an information-rich world," in *Computers, Communication, and the Public Interest*, M. Greenberger, Ed. Baltimore, MD: Johns Hopkins Press, 1971.

PRACTICAL INSIGHTS FROM IBM

IBM has recently celebrated its 100th anniversary, and it has plans to continue making an enduring impact over the long term. Part of this plan includes streamlining the way IBM produces and handles information. With over 426,000 employees worldwide and vast numbers of customers around the world, the company has had to develop technologies and strategies to deal with both its back-office operations as well as its customer-centered communications. The sheer amount of data within IBM has provided a challenge that the company has had to face constantly.

The company has dealt with internal information overload with three main strategies: standardizing as many of the business processes as possible, developing tools to facilitate information handling, and empowering employees to work independent of time and place.

The process of transforming *data* into *information* and information into *knowledge* is at the core of any enterprise's business operations. Simplifying that process can improve operations significantly. So, the first strategy IBM is using to decrease information overload is to standardize data. IBM's *Roadmap to 2015* incorporates a number of programs to help all parts of the company reach the goal of simplifying all processes through standardizing the way data is presented and making sure that data is understood and analyzed correctly. Perhaps, one of the biggest efforts toward standardizing business reports is a program called Blue Harmony. Blue Harmony is IBM's major transformation program to radically simplify the design and operation of three globally integrated support processes—finance, opportunity-to-order, and order-to-cash. Previously, 80%–90% of all reports within these processes were generated and compiled on an ad hoc basis, which has

Information Overload: An International Challenge for Professional Engineers and Technical Communicators, First Edition. Edited by Judith B. Strother, Jan Ulijn, Zohra Fazal.
© 2012 Institute of Electrical and Electronics Engineers. Published 2012 by John Wiley & Sons, Inc.

made the process very time consuming and resulted in numbers that could be unclear and difficult to validate. The goal, through *Blue Harmony*, is to reduce the number of ad hoc reports to fewer than 20% and to automate the generation of many routine reports. This will enable IBMers to more quickly and accurately understand each report, no matter which corner of the globe it comes from. By providing a standard report framework, IBM is reducing the cognitive overload that comes from struggling with a variety of data reporting styles, making it easier to reach better decisions.

Second, IBM continues to improve its tools for searching, finding, and sharing information. The current deployment of a new search technology developed by IBM Research customizes the search to the user, filtering search results by geography, language, and other attributes. A new Expertise Locator tool allows employees to register as experts in specific business areas and topics. These experts are made available both within IBM and on the Internet. They publish their blogs, microblogs, and wikis so that people can follow them and they are available to consult, using e-mail or chat.

Most importantly, IBM's deployment of a social computing platform (based on the IBM Connections product) has changed how information is handled within the company. Connections integrate communities, wikis, blogs, microblogging, shared files, employee profiles, and light project management in a single platform. The amount of information on Connections is indeed overwhelming—58,000 communities, 17,000 blogs, 370,000 shared files, and so on—but the platform has multiple features to help users focus on what they need. In addition to the search feature, the use of social tags, ratings, and e-mail notifications help users find the content and the people they need. Users can also create watchlists for the people they want to follow and join communities they want to participate in.

The third major strategy focuses on individual responsibility of all employees within a strong corporate culture. In the time and place independent culture that IBM has, IBMers enjoy a considerable amount of trust and personal responsibility from their management to successfully fulfill their daily job. IBMers are measured by their performance instead of by their presence. This stimulates entrepreneurial thinking and a culture in which employees themselves are responsible for searching for the corporate information they need, instead of being overloaded by partly irrelevant information. This has resulted in an environment where corporate information is very rarely pushed through mass communication channels, such as e-mail or newsletters. Information remains limited to an absolute minimum of "need to know," leaving it up to individual IBMers to search for the information they personally need.

IBM continues to focus on dealing with the pervasive problem of information overload. It is being proactive in implementing strategies to standardize its processes and reports, creating tools to share and find data more effectively, and empowering its employees to work independently yet efficiently. While IBM is generating as much data as ever, its employees are able to control and use the data more effectively.

Contributors to Practical Insights from IBM

- Jelmer Letterie, External Relations, IBM Benelux
- Kees Verweij, CIO, IBM Netherlands

3

THE CHALLENGE OF INFORMATION BALANCE IN THE AGE OF AFFLUENT COMMUNICATION

Paulus Hubert Vossen

> *The human brain takes in 40 million bits of information per second. . . . Second by second, we are each creating our own reality. Second by second, we are unconsciously filtering out information we have somehow determined irrelevant.*
>
> Mary Madsen

ABSTRACT

Information processing is a basic characteristic of human beings and other social organisms. In fact, both need information, continuously and increasingly, in order to master their natural and artificial environments. To satisfy this innate need, people are equipped with specialized emotional, perceptual, and cognitive functions. In addition to these basic functions, they have developed auxiliary facilities for information processing, from languages via books up to modern digital media.

However, the undeniable success of these artifacts has its downside. It has led to an information explosion and pollution that reminds us of the famous fable of Goethe, *The Sorcerer's Apprentice,* with the significant difference that there is no sorcerer to stop the inundation. People feel overwhelmed, confused, frightened, or paralyzed by the very information systems that should help them master their world and solve practical, technical, or theoretical problems.

I set out to explore some of the symptoms, causes, and consequences of information overload from an interdisciplinary system perspective. I show that current approaches to alleviate the problems—either educating human agents to become better information managers or developing

Information Overload: An International Challenge for Professional Engineers and Technical Communicators,
First Edition. Edited by Judith B. Strother, Jan Ulijn, Zohra Fazal.
© 2012 Institute of Electrical and Electronics Engineers. Published 2012 by John Wiley & Sons, Inc.

still more auxiliary systems and tools to cope with the information flood—are doomed to fail, because they leave the original problem—a fundamental imbalance between information production and consumption—untouched. I conclude that basic practice-oriented research on professional information management systems beyond the simple solutions proposed up to now is urgently needed.

3.1 Introduction

Information overload? An urgent problem, some say, having become especially acute since the introduction of Internet technology, but we are working on solutions, using the latest Web technologies in support of collaborative communities. Information overload is not a new problem, others tell us. People have complained about it at least since the widespread availability of books and magazines. Current technologies have finally given us immediate access to almost any public information, free or paid. People should just learn to cope with the new media and use the information made available through them. No problem at all, not in principle at least, evolutionary biologists point out. It is the natural state of being in the world for living creatures—such as human beings—equipped with multiple sense organs but without fixed response repertoires. We steadily filter information coming to us through our senses. As William James put it, "the world is a buzzing, pulsating, formless mass of signals, out of which people try to make sense, into which they attempt to introduce order, and from which they construct against a background that remains undifferentiated" [1]. It may take some time, but in the end, *Homo sapiens* will adapt to the new information environment.

Information overload—is it real or are we dealing with a myth [2] put into the world by irresponsible philosophers, researchers, or consultants looking for another topic that is worth telling about and selling? Or do the terms "information overload" and its many synonyms [3] indeed refer to a state of the world that has quantitatively and/or qualitatively changed in a way that deserves our attention and action in order to solve current problems and prevent worse?

The least we can say is that the term has found its way into Wikipedia, which, for many of those growing up now, is proof enough that it denotes something substantial and objective (i.e., proof by concept naming). In addition, the number of articles, books, and conferences devoted to the phenomenon of information overload, some more serious than others, is rapidly growing, a clear sign that we have convinced each other that there is something out there properly called information overload (proof by belief sharing). And, of course, there is the first Internet forum (www.iorgforum.org) exclusively devoted to information overload research (i.e., proof by community building). On the other hand, we find voices that downplay or deny the existence or urgency of such a problem, and it will not be long before someone will publish a paper about the nonsense of the information overload debate, as has already occurred for the closely related field of knowledge management [4].

This chapter takes as a *working hypothesis* that there is such a phenomenon as information overload—or whatever we like to call it: cognitive overload, sensory overload, communication overload, knowledge overload, information explosion, or information anxiety [3]. It also assumes that information overload has both increased (quantitatively) and changed (qualitatively) substantially over the past decades, if not centuries. This would explain the recent increased interest in it as evidenced by numerous publications, conferences, and research on the theme, including some innovative proposals as to how to best deal with it (e.g., using the computational knowledge engine Wolfram|Alpha).

This chapter presents evidence for both assumptions from an interdisciplinary perspective. This is necessary because information overload manifests itself quite differently on different levels of reality and thus may require different theoretical, empirical, and practical approaches (including countermeasures, if required). For example, from a philosophical perspective, we might be interested in the question of how the growth of science and technology has contributed to a steady proliferation of disciplines and cross-disciplines and how this has contributed to the information explosion and overload in science, research, and technology. Or, from the political and governmental perspectives, how does the development of states and, most recently, a global system of states lead to an overabundance of information, which can overwhelm and confuse citizens and paralyze the bureaucracies that are supposed to run the government? The question also arises as to how individuals are able to cope with the steadily increasing flow of information in their private as well as professional lives through multiple forms of media such as ads, newspapers, brochures, TV, phone, e-mail, books, magazines, journals, and other professional media sources.

First, however, it is necessary to clarify the distinction between quantitative and qualitative aspects of the phenomenon of information overload. Part of this discussion concerns the apparent increase of low-quality information, so-called information smog or glut, due to the lack of quality standards and control, especially in the new media. Then, after presenting a multidisciplinary view on the concept of information overload, closely related issues are discussed, especially those having to do with the reverse of information overload, i.e., lack of information due to unequal distribution or accessibility of information sources.

The chapter closes with the proposal to work out a systemic approach to professional information management based on the apparently paradoxical premise that valuable and trustworthy information is not free, although it is the only commodity that is not scarce. Information supply should closely follow information demand, not the other way around. In other words, we have to restore some sort of information balance, or *homeostasis*, in our systems so as to facilitate instead of inhibiting professional communication and collaboration.

3.2 Quantitative Aspects of Information Overload

Before answering any question pertaining to the presence (or absence) of *information overload* in our own lives or in science and society at large, it is important first and foremost to clarify (as far as possible) what is meant by *information load* on human beings or professional communities, and how it is going to be measured, if that is at all possible. Furthermore, once these preliminary questions have been settled, we should specify which level or bandwidth of information load may be considered normal, not in the statistical sense of the word, but from the perspective of a goal-seeking system—which human beings undoubtedly are—with more or less strict bounds on proper functioning leading to useful outcomes. Finally, on the basis of such a normative view on effective and efficient information processing, we are able to describe, and hopefully understand, the symptoms of information overload or underload, as the case may be, and think about ways to arrive at a genuine information balance in the systemic sense.

Quite generally, information load may be defined as the amount of information (data) that is being processed, in one way or another, by an information processing system (IPS). This is a fairly general and uncontroversial definition, which can be found in any scholarly source dealing with information processing in physical (e.g., computers) or biological (e.g., human) systems. In order to escape an infinite regression loop, we adopt the notion of an information processing system as an open system—natural or artificial—that exchanges data (i.e., symbols) with its environment for the purpose of knowing, evaluating, and acting upon its environment. Of course, this exchange is usually enabled by some sort of energy exchange, but that does not bother us here. In the above sense, human beings are just a special kind of IPS. For an in-depth, although not completely uncontroversial, logical, and philosophical treatment of the notion of IPS, the interested reader can refer to the well-known monograph, "Sciences of the Artificial," by Nobel laureate Herbert Simon [5].

Recently, this notion of IPS has been taken up again under the more inclusive name of intelligent agent in the branch of artificial intelligence aptly called agent technology [6]. Therefore, in this chapter, I occasionally refer to agents, especially when I want to emphasize that information is being processed in order to enable, initiate, or select some action. Typically, within the field of agent technology, an agent is understood to be a software agent (perhaps, a member of a collection of agents, in the case of multiagent technology). Of course, an agent may also be a single human being or a collaborative community. Unfortunately, to the best of my knowledge, there is no equally general and neutral word denoting an IPS that only deals with information for its own sake, e.g., to create a higher order of knowledge out of it, although "philosopher" and "scientist" come very close to it.

Information load is a particular aspect of agents living in an environment with which they exchange data (in addition to energy and matter) in order to satisfy basic survival needs and the less basic needs of self-actualization (see Maslow's famous hierarchy of needs [7]). In particular, it denotes the total amount of data that impinges on such an agent during a certain time interval causing that agent to feel obliged to further process the data.

However, given that agents may possess different channels of information, such as one for visual information and another for auditory information, we may as well differentiate among several specific information loads, one for each channel of information processing. Also note that, once we accept such a definition, it becomes increasingly unclear where in the chain of information processes we should actually measure this information load. For example, should this take place at the very beginning (sensory input), or somewhere during its intermediate inner processing (cognitive throughput), or at the very end of the information processing chain, where decisions are mapped to actions (motor output)?

Furthermore, information load is a dynamic concept in the sense that the amount of processed information may vary from time to time, so that it critically depends upon the time frame chosen as to how much information our human agent is actually processing. However, we assume that agents have a fixed and well-defined architecture that sets strict physical upper bounds on the maximally possible information load per unit time. This upper bound is usually denoted by channel capacity of the respective information processes. It is with respect to those channel capacities that information overload is most easily defined, but we see that this is only one way—a rather simplistic one—of approaching the phenomenon of information overload.

It may well be the case that, for any particular agent type (e.g., a human agent) and any particular information process (e.g., interpreting several succeeding visual signals), there is an optimal level of information load that is definitely lower than the theoretically possible channel capacity. Thus, information overload should be defined with respect to this optimal level instead of the maximal level. If there happens to be such a thing as optimal information load, then we should not be surprised to find that an agent may also suffer from too little information [8, 9]. Indeed, that is what psychologists and neurophysiologists have found for human agents during conditions of sensory deprivation [10] and what organizational and industrial psychologists, as well as human factors researchers, have found for certain monotonous office or production work.

The lack or even loss of ready-to-use information leading to *information poverty* or *information illiteracy* and the inability to act and live adequately and effectively is as much an issue as the abundance of information leading to information overload and anxiety [11]. Even in an age of general information abundance, there are still *islands of disinformation and misinformation*, intended or not. This occurs, for example, when individuals or communities are excluded from receiving relevant information, as is typically the case in nondemocratic societies or when information is available but not accessible because of economic or technical reasons.

3.3 Qualitative Aspects of Information Overload

Although interesting in its own right, the quantitative approach to optimal versus suboptimal information load obscures a whole range of fundamentally qualitative issues with respect to the phenomenon of temporary or chronic communication and information processing disorders between or within human agents. Moreover, it is not clear yet whether we can reliably measure one or the other or whether it will remain forever an elusive, subjective experience. Apart from some basic psychophysical facts about channel capacity or some very concrete application fields such as e-mail load [12], it will not be easy to find a generally applicable objective measuring framework. Therefore, let us now focus on some qualitative observations and analyses regarding facilitating and inhibiting conditions of information overload from a multidisciplinary perspective. In particular, we take up the following six perspectives on the problem of information overload, more or less corresponding to ever-lower levels in the natural hierarchy of human affairs.

- Philosophical perspective: information in science and technology
- Political perspective: information in modern society and a global world
- Economic perspective: information as a commodity on the market
- Societal perspective: information as the glue between communities
- Psychological perspective: information as a basis for knowing and acting
- Ecological perspective: information as a prerequisite for living creatures

Figure 3.1 gives a hierarchical representation of those perspectives.

3.3.1 Philosophical Perspective: Information in Science and Technology

Information processing is a basic characteristic of human beings and other intelligent agents. In general, individuals who live and operate in a free and open system equipped with the desire to adapt and survive and, when possible, to control and dominate their environment have to collect a wide variety of data about their natural and self-made *umwelt* (environment) and must act on it or react to it adequately in a timely manner. In short, we need information, continuously and increasingly, on the personal, organizational, societal, and political levels, as well as on the cultural level.

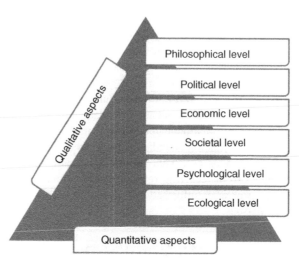

Figure 3.1 A multidisciplinary hierarchical view on human information: six perspectives on the problem of information overload more or less correspond to ever-lower levels in the natural hierarchy of human affairs.

In order to satisfy this need, humankind has developed, over the past few millennia, all kinds of communication devices, information appliances, and associated practices. However, as none of those developments have been planned in any strict sense, the processing power of the resulting techniques and corresponding ways of life have already far surpassed the natural limits of what normal human beings and communities can reasonably accommodate without experiencing the effects of what is aptly called information overload.

The consequences are both diverse and challenging. They range from a once-in-a-while suboptimal buying decision when overlooking a better alternative to a permanent struggle to cope with a never-ending stream of e-mails and other requests for our attention. At the organizational level, challenging consequences can arise following the failure of an organization to effectively communicate pertinent information during a natural disaster or other catastrophic incident (i.e., appropriate crisis communication—see [13, 14]). There has been a tremendous increase in and demand for information about how the physical world is structured, how it behaves, and how we may control it—through science and technology. This has led to a proliferation of disciplines and cross-disciplines as the amount of information—in the form of empirical and experimental data, theories, methods, and techniques—has increased so much that not a single person is able to oversee and understand it all.

Furthermore, because nature does not always easily disclose its regularities to us, we sometimes have to confront a range of partially conflicting theories about one and the same class of phenomena. For example, although there is one standard model—or paradigm—of physics, there are a number of competing theories and models, each with its own bag of evidence and adherents.

3.3.2 Political Perspective: Information in Modern Society and a Global World

The information era has created high expectations in humankind about the age-old wish to be "master of the universe" through mastery of all knowledge. This hope is closely coupled with the equally age-old wish to be mighty and powerful, as knowledge implies power. Of course, this will always remain a dream, as the sheer amount of potential knowledge about the world (including ourselves!) and the sheer amount of time it would take for a person to just gather, process, store, and relate all that information ("know what and how") far outstrips an individual's limited resources [15]. The same holds true for society at large—apart from the fact that at that level, additional complicated issues about potential conflicts between privacy and publicity arise. Information comes in various types. That is, some information may be made publicly available for everyone without disturbing a natural power balance while other information may be as dangerous as nuclear power in the hands of evil people.

There is no universal right to all information for every citizen. The belief in such a universal right—as implied or expressed by some Internet communities—is grossly misleading. Generally, it would not be a good idea to make all information available for whoever wants to peek into the repository of that information. In fact, whole islands of potentially dangerous information are already hidden from the public at large [16]—in sharp contrast with the dreams of most pioneers [17, 18] of the information society.

3.3.3 Economic Perspective: Information as a Commodity on the Market

Needless to say, information has become an important commodity on the global market. Due to advances in computer and network technology, it has become relatively easy and cheap to reproduce, distribute, and store information on a global scale directly to any potential consumer, whether private or professional. However, it is important for the present discussion to make a sharp distinction between the original information as produced by its author(s) and the packaged digital version of it, which is transmitted through the world's networks and eventually stored on private, public, commercial, or governmental computers. It is the latter sort of information, in the form of what is usually called digital data, which is the proper substance of economic transfer.

These items of digital data—largely stripped of their meanings—can be handled as if they were an old-fashioned material commodity. It is here that the classical economic laws of production and consumption apply. Digital data can be reproduced almost at will and at little or no cost in the hope of finding consumers for it, which is a much less risky business than, for instance, producing cars no one is interested in. By clever marketing and merchandising, enough consumers may be created to make them profitable. Information has become a gigantic mass market of products that are usually consumed only once. Therefore, the rate of new information products should be kept very high, as long as the demand can also be kept at a high level. Because of the innate curiosity of human beings and the ease with which they can be convinced of the utility of new information, achieving and maintaining a high demand level is often easy.

What goes largely unnoticed is that the economic process of information production and distribution is itself embedded in a larger intellectual and cultural process in which the original information is first created (by its "author") and finally re-created (by its "reader"). The latter process, however, is extremely slow compared to the former process, on which the new information economy is built. That is, the amount of information that people can or may consume is not increasing at the pace at which the bare data is produced (see Chapter 8). No wonder that we observe information overload on the interface between the inner "economic affairs" and the outer "human affairs."

Thus, the undeniable success of all technologies for information production and dissemination has its downside. We feel more and more overwhelmed, confused, frightened, and paralyzed by these very information systems that should help us master our world. The situation reminds us of Goethe's famous fable, *The Sorcerer's Apprentice*, with the difference that there is no sorcerer to stop the inundation.

Furthermore, it is not always easy for us to decide no attend to offered information (i.e., still in the form of digital data), which might give us some added value. The problem is that we cannot know for sure if there is any added value in digital data until we have gone through the process of re-creating the encoded information. By then, it is already too late because we have probably already paid for the data, and, in any case, we have spent time and mental effort on decoding the message. If we are lucky, we will have received new and valid information that we can use for our problem solving or decision making. If not, the information may turn out to be a paraphrase of what we already knew before, it may be of very low quality, or it may even be invalid or misleading (which may not be obvious immediately).

3.3.4   Societal Perspective: Information as the Glue Between Communities

At the societal level, information is mainly used to support community life in its manifold forms, on different levels, from small collaborative communities within modern enterprises via local sports clubs or civic organizations to large international organizations with thousands of members and employees.

Traditionally, much of the information exchange on the social level occurs in the context of what is called group dynamics (e.g., rituals, habits, rules, and taboos) and group structures (e.g., tribes, clans, clubs, and states). In a certain way, these group dynamics and structures support and represent "external" memories used to provide the right background for a range of common social goals such as survival of the fittest or a response to external threats and opportunities. Most information at this level is transmitted and absorbed in a natural way, by imitation, instruction, repetition, and learning-by-doing, and is more or less adapted to the mental capacities of human beings.

Nevertheless, recent technological developments have changed some of the habits that we have traditionally thought to be "natural" and "stable." For example, "networking" has become an entirely new way of community life for the younger generations. (Of course, many in the older generations are catching up, eager not to lose contact with the new world of networkers.) Barabási [19] has studied in depth how this networking functions and may be radically changing the way we perceive each other, find out about each other, and eventually work together.

Perhaps, in hindsight, we realize that these societal developments had already started with the invention and introduction of e-mail in the early days of the Internet, somewhat later supplemented by other forms of "publication" on the Internet, such as discussion forums, personal websites, and weblogs. Next, we got the completely unexpected but explosive rise of the mobile phone, the invention *par excellence* to make the dream of instant reachability—at home and at work—finally come true. What contribution and effects do these technologies have on perceived information overload? It appears extremely difficult for most people not to conform to the above-mentioned expectation of permanent reachability. Thus, a technological capability has become a social norm and sometimes even a case of information addiction [20].

3.3.5 Psychological Perspective: Information as a Basis for Knowing and Acting

In previous ages, people took a lifetime to acquire personal information and to store it in their own private memories, where, once stored, it could be found, accessed, or recalled by seemingly effortless "acts of will." Of course, none of this is really effortless or simple. Cognitive psychologists and neurophysiologists are still seeking to explain all the separate mechanisms that are involved. However, it looks as if the capacity of human memory is almost limitless, at least up to a certain age.

It follows that information load may also be viewed as a special form of mental (cognitive) workload. This has long been an area of study in cognitive psychology and its application-oriented subfields such as human factors research (ergonomics) or human-computer interaction (HCI). Furthermore, from a holistic and top–down perspective, both are again special quantitative features of human information processing agents.

The general concept of human information processing subsumes several intertwined and codependent specialized information processes—attention, search, filtering, reception, storage, indexing, abstraction, verification, validation, integration, transmission, publication, deletion, and many more. Information overload may or may not have to do with or be particularly affected by this or that process. For instance, information overload may be said to negatively affect the human capability for attention for or awareness of new information, which then itself may become a cause for still more information overload. The situation is similar for storing-indexing-abstracting. If we spend too much time on these processes, which are clearly required and legitimate the problem of overload will not vanish. On the contrary, it may become more critical because information providers (senders) are not waiting for us to be ready with basic information management before they send their next batch of information. In other words, the idea of information buffers in between the many senders and receivers raises its own problems because these buffers and corresponding processes have to be managed themselves, which costs capacity (workload) and time (resources).

The basic underlying fallacy is to assume that information overload can be cured by still more information and faster information processing. However, there is no evidence that the capacity of human agents to attend to and process so much information has increased or will increase at the same pace as the information explosion surrounding us. Under such circumstances, we have no choice other than to follow Herbert Simon's advice of satisficing [5], i.e., knowingly reducing our ill-conceived information needs and quality standards and learning to live and cope with the inevitable consequence of our bounded rationality. The alternative is not viable—enduring stress, ending up with burnout, and suffering from other symptoms of exhaustion.

But what does it mean to have all the information we might ever need at our fingertips? Does it mean that we just have to enter a question, and *voilà*, the correct answer will appear on our computer screen, mobile phone, or television? But how do we arrive at the right question in the first place? How is it possible that sometimes we recognize important information, although we were not looking for it at all (i.e., serendipity)? Many of these questions pertain to the subjective side of information load and overload, which are difficult to measure, but are still very real.

On the next cognitive level is private long-term memory, which is important in order to be prepared for future cases that resemble current cases, e.g., repetition of similar actions or recognition of someone we know. Our long-term memory also deals with our own life—who are we, where were we born, etc. —as well as knowledge about facts and procedures important for higher-level activities, mainly to do with validating and learning. For instance, in order to learn new facts, such as relationships between previously learned facts, we first have to validate that the assumed relationships are indeed real, which in turn requires us to compare the stored facts with each other and any other pertinent information, new or old. Although we are usually not aware of it, these higher-level activities form the essence of human intelligence.

The basic trick of remembering and recalling seems to be repetition. For example, the more often we think of and actively work with pieces of information, whether declarative or procedural, the easier it becomes to recognize and recall it later in new but similar contexts. The information will be there in a fraction of a second, as soon as we need it. We just have to think of it. The explanation—which is not really a scientific explanation—is that the recall process works through association within a large network of pieces of information, and this has also become the dominant metaphor for information storage, dissemination, and retrieval (going back to ideas already formulated in Schema Theory and taken up by Roger Schank, e.g., in his scripting approach to artificial intelligence [21]). If this psychological model of information processing is correct, it explains why we can in fact recognize, recall, and access a huge amount of information, once we have dealt with it and stored it "correctly." Of course, the price we have to pay for that is our active engagement with this steadily growing network of information, expanding and restructuring it as needed. Thus, in times of information overload, we may rapidly approach the limits of our processing capacity.

3.3.6 Ecological Perspective: Information as a Prerequisite for Living Creatures

As human beings, we are always exposed to data, stimuli, and information from all directions, but in order to live and survive, we usually need only a fraction of it. Nature has equipped us with subconscious mechanisms to filter out most of that data before it reaches conscious deliberation and decision making. Otherwise, we would be paralyzed were we to become aware of all this information and were we to be required to respond to it. So, in a way, information overload is a natural state of the world for agents, including human beings, and other biological organisms equipped with versatile, basic senses, which "inform" them about their inherently dangerous environment. On a conscious level, it still may be the case that what we receive, unintentionally, by natural processes, is far too much to sift out unless we have powerful mechanisms to quickly find any piece of information that is either a potential threat or an opportunity for us as living beings.

This kind of information is fluid and ephemeral, steadily arriving at our sensory organs. We cannot escape it without becoming "senseless" and in a way "helpless." This is because we are open systems, continuously responding to outer stimuli as well as inner stimuli that enable us to be better prepared for the moment and for the immediate future.

This processing of information has much to do with basic, primary biological mechanisms including short-term memory, which are given and can only partly be trained to do more than to store basic kinds of data (training has to do with our ability to see and recognize certain assemblies of data as so-called chunks of data, which become primary data themselves). On the contrary, we wish to be able to filter out as much as possible, because we know and expect that many of the signals have little or no significance for us.

On the biological level, the most important goal is to survive in a potentially hostile environment, and information is mainly made up of signals of the following three kinds: (1) insignificant and irrelevant; (2) significant and positively valued, indicating a source that is useful for the purpose of surviving or opportunities such as sheltering, eating and drinking, or having sex; and (3) significant but negatively valued, indicating a source that may threaten us—threats by nature (e.g., fire, floods, storms), threats by our own kind (e.g., dominance, war, competition), or threats by other animals (e.g., being attacked or treated as food). As all of this basic biological information has a potential intrinsic survival value, it is difficult to see how we would not wish to pay attention to it. Luckily, nature has equipped us with basic information processing structures and procedures that usually take care of the bulk of our information needs on this level. However, in uncommon or extreme situations, we may very well experience a kind of information overload on this ecological level, too.

3.4 Conclusion

The overwhelming evidence for lasting and growing problems with information processing in human beings and their communities of all sizes and at all levels leads us to the conclusion that we are not dealing with a myth or just another hype of the information age. There are real issues and tough challenges, which we are increasingly becoming aware of, although most of them existed long before the electronic information age, albeit in other forms and on different scales. Almost all have, in some way or another, to do with qualitative or quantitative information overload. However, it is also fair to point out that information *overload* is just one of several interwoven downsides of this precious commodity we call *information* and the many divergent processes we use to manage it.

The expectation is that there will remain gross inequalities between those who have access to reliable sources of information as well as to effective means of dealing with massive amounts of it and those who do not. On the contrary, it is not likely that these inequalities will disappear or diminish [16]. So the inherent *power of information* is not and will not be equally distributed among all members of the global community, which is another form of *information imbalance*, or even *information injustice*.

A further issue that should be taken up has to do with synchronous versus asynchronous information sharing. Asynchronous information (in written or digital form) is a way of coping with the evident problem that the sender cannot directly (in person) communicate simultaneously with hundreds of recipients, and that many recipients may not have time, occasion, and channel capacity to listen to hundreds of senders at the same time. So senders and recipients have to be separated in space and time by the introduction of a huge central reservoir in between. This reservoir serves for "dropping" and "getting" information at a place and time that suits both senders and recipients. In this sense, it acts as a kind of external memory in much the same way brick and mortar libraries do.

The problem is that we cannot scan this huge reservoir regularly in order to find out if there is something of interest to us on which we should react immediately or in due course. Instead, we step back and consult this resource only if and when we have an urgent need for a particular piece of information—an answer for a clearly defined and specific question. Now this mechanism—similar to a huge blackboard between all human beings and communities—has itself become problematic, not being an optimal solution for the original communication problem. On the contrary, we have created an information problem on top of the original information impasse. In addition to that, an increasing number of senders have begun to aggravate the problem even more, as they have invented additional methods to "flood" the Internet with their own (often self-serving) agendas and "shove" certain kinds of information onto the intended recipients, in this way counteracting the prime purpose of the external reservoir.

In the end, we have to question whether we have any viable, scalable information management approaches, methods, or techniques that have the potential to solve our self-made problems. What we see today is an unstructured collection of technical (e.g., hundreds of search and meta-search engines), political (e.g., security and privacy laws), organizational (e.g., communities of practice and learning, gatekeeping), social (e.g., social Web, Facebook), and psychological (e.g., how-to books and seminars) *coping mechanisms*. Although each of these in itself is nice, taken together, they are not strong or convincing enough to lift information resource management to a new level.

On the contrary, most current approaches to handling information overload problems—whether by educating human agents (individuals or communities) to become better information processors or by developing still more information systems to cope with the previous ones—are doomed to fail, because they are based on a fallacy. The fallacy is to assume that information overload can be cured by still more information, i.e., information about information (meta-information). For example, there are extracting services and huge repositories, or information intended to improve and increase the information processing capacity and capability of "information workers" in the form of self-help courses or procedures to follow. By virtue of this fallacy, the original problems are not solved, but taken for granted and, worse, aggravated.

We collectively need to work hard to find intelligent ways to deal with this permanent communication and information surplus. Some solutions may be rather technical, e.g., smart adaptive filtering methods on top of enhanced Web search engines (Chapters 8 and 9). Equally, if not more effectively, there will be radical, lasting changes in our communication and information processing habits and rituals, solving the problem at the root. We will have to critically reconsider the tacit assumption, "the more information the better." We should become masters instead of slaves of our own communication devices and practices, and we should develop explicit strategies and policies for denial of superfluous and unwanted communication. To accomplish this will require that we not accept all information, not store all information, and not generate and send unnecessary information. This will not be easy, as it implies *proactive* conscious judgment in each and every single case instead of a *re-active* automatic reaction taken care of unconsciously. However, this kind of information denial may be necessary for the sake of a healthy and sustainable information age and society, both privately and professionally.

3.5 A Call for Fundamental Research

What is needed, and urgently so, is fundamental research based on sound theoretical principles of *professional information resource management* beyond the rather simple, mostly scattered solutions proposed up to now. By information resource management, we mean a principled and thus systematic way of selecting and applying information resources for the purpose of information processing. Such information resources may be anything from our own common sense to whatever external devices, software applications, peers, or organizations are available to support us as human agents and the communities we belong to in fulfilling our information needs while minimizing information overload.

The most important new aspect of the sort of resource management that we envision will be to treat neither information (the stuff itself) nor information resources (the means, so to speak) as free and always available. If there is a myth of an information society, it is this: we can get all the information we want, and it is free [8]. This is denying that valid reliable information has a value, and that it costs a lot of time and money to produce it, transmit it, gather it, store it, distribute it, make it available, and search it.

As long as we do not have a theoretically grounded and empirically evaluated integral cost model of information processing, broken down into the very information resources available to us, we will remain victims of all the negative aspects of an unbridled information society, which results in information explosion, information inundation, and information anxiety.

REFERENCES

[1] G. Patriotta, "Sensemaking on the shop floor: Narratives of knowledge in organizations," *J. Manage. Studies*, vol. 40, no. 2, pp. 349–376, 2003.

[2] T. J. Tidline, "The mythology of information overload," *Library Trends*, vol. 47, no. 3, pp. 485–506, 1999.

[3] M. J. Eppler and J. Mengis, "The concept of information overload: A review of the literature from organization science, marketing, accounting, MIS and related disciplines," *Inform. Soc.*, vol. 20, no. 5, pp. 325–344, 2004.

[4] T. D. Wilson "The nonsense of 'knowledge management,'" in *Inform. Research*, vol. 8, no. 1, D Oct. 2002.

[5] H. A. Simon, *The Sciences of the Artificial*, 3rd ed. Cambridge, MA: MIT Press, 1996.

[6] S. Russell and P. Norvig, *Artificial Intelligence. A Modern Approach*, 3rd ed., Upper Saddle River, NJ: Prentice-Hall, 2009.

[7] A. H. Maslow, *Towards a Psychology of Being*, Hudson County, NJ: Wiley, 1968.

[8] C. Griffin. (2007). *Content is Not – and Should Not Be – Free* [Online]. Available: http://www.pardonthedisruption.com/?s=Content+is+Not+and+Should+Not+Be+Free Aug. 25, 2011 (last date accessed).

[9] C. Griffin. (2007). *The Real Problem with Information Isn't Overload or Underload* [Online]. Available: http://www.pardonthedisruption.com/2007/11/07/the-real-problem-with-information-isn%e2%80%99t-overload-or-underload/ Aug. 25, 2011 (last date accessed).

[10] J. C. Lilly, *et al.*, *The Quiet Center: Isolation and Spirit*, Berkeley, CA: Ronin Publishing, 2003.

[11] A.van Deursen and J.van Dijk, "Using the Internet: Skill related problems in users' online behavior," *Inter. Comput.*, vol. 21, nos. 5–6, pp. 393–402, Dec. 2009.

[12] S. Whittaker and C. Sidner, "Email overload: Exploring personal information management of email," in *Culture of the Internet*, S. Kiesler, Ed. Hillsdale, NJ: Lawrence Erlbaum Associates, 1997, pp. 277–295.

[13] W. T. Coombs, "Designing post-crisis messages: Lessons for crisis response strategies," *Rev. Business*, vol. 21, nos. 3–4, pp. 37–41, 2000.

[14] J. Strother, "Crisis communication put to the test: The case of two airlines on 9/11," *IEEE Trans. Prof. Commun.*, vol. 47, no. 4, pp. 290–300, Dec. 2004.

[15] B. Schwarz, *The Paradox of Choice: Why More is Less. How the Culture of Abundance Robs Us of Satisfaction*, New York: HarperCollins, 2005.

[16] R. B. Laughlin, *The Crime of Reason: And the Closing of the Scientific Mind*, New York: Basic Books, 2008.

[17] V. Bush, "As we may think," *Atlantic Monthly*, vol. 176, pp. 101–108, Jul. 1945.

[18] T. Nelson, *Computer Lib/Dream Machines*. Redmond, WA: Microsoft Press, 1987.

[19] A.-L. Barabási, *Linked: How Everything Is Connected to Everything Else and What It Means for Business, Science, and Everyday Life*. New York: Plume, 2003.

[20] M. Meckel, *Das Glück der Unerreichbarkeit. Wege aus der Kommunikationsfalle [The Luck of Unattainability. How to Avoid the Communication Trap]*. Munich, Germany: Goldman, 2009.

[21] R. Schank and R. Abelson, *Scripts, Plans, Goals, and Understanding: An Inquiry into Human Knowledge Structures*, Hillsdale, NJ: Lawrence Erlbaum Associates, 1977.

PRACTICAL INSIGHTS FROM XEROX

Xerox Takes on Information Overload

You could almost say that Xerox Corporation created the problem of information overload 50 years ago by making it possible to easily copy and distribute documents. The volume of information with which we are confronted on a daily basis has grown hugely in the past five decades.

Today, the average employee faces an annual barrage of thousands of e-mails and spends hundreds of hours in meetings. To help manage the inbox avalanche, Xerox Corporation has taken a number of initiatives to help its own employees improve their productivity. In addition, the company is offering many of its capabilities to customers to help them manage their information overload and thus concentrate on running their businesses.

Identifying the Problem

In May 2011, Xerox launched a free desktop application that represents a first step toward managing our inbox, while shedding light on the mental processes. The Business of Your Brain™ desktop application analyzes e-mails and calendar notices to identify those distractions that take us away from our real business.

Information Overload: An International Challenge for Professional Engineers and Technical Communicators, First Edition. Edited by Judith B. Strother, Jan Ulijn, Zohra Fazal.
© 2012 Institute of Electrical and Electronics Engineers. Published 2012 by John Wiley & Sons, Inc.

"I find that managing the barrage of 'inbound'—e-mail and meetings—is one of the most challenging aspects of running a business in our 24/7 information age," says Jay Lauf, Vice President and Publisher of *The Atlantic*. He went on to say,

Business of Your Brain is not only a fun and enlightening way to understand what underlies that challenge, but it's revealed things that have changed my habits and approach to communication—I've freed my mind to focus on more proactive elements of my work. It's actually the best time-management exercise I've done since I read and deployed some of David Allen's *Getting Things Done*.

Among other things, the application provides the following tools:

- *Meeting Buddy.* Did you show up to the office today dressed just like a colleague? Maybe it is because he is the person with whom you have the most meetings.
- *Quickest to Panic.* Find out who sends the most e-mails flagged with the high alert exclamation point.
- *Focus Drainer.* What are the latest buzzwords in your e-mail subject lines? Your Focus Drainer results will clue you in.
- *Motor Mouth and "Many Thanks."* We all have a few contacts who are notorious for their lengthy e-mails; find out who takes the prize in your inbox. On the flip side, you will also see who most frequently adds to your e-mail count with Reply All "thanks" notes.
- *Most Ignored.* Do you delete many e-mails without reading them? This tool will tell you which ones you are most likely to ignore.

Although this application is designed to focus attention on information overload, a great deal of work at Xerox is dedicated to delivering real solutions to this problem. Researchers at the Palo Alto Research Center (PARC) have worked with a company called Meshin—a start-up funded by Xerox—to develop a new search tool, which is designed to act like a human personal assistant. The new tool (also called Meshin) uses semantic technologies to reach beyond keywords and metadata and gather information based on meaning and context. This is the first time that semantic technology has been applied to a search tool that scans not only e-mail messages but also RSS feeds and communication in social networks.

As a result, when Meshin is downloaded onto a personal computer, it can help, e.g., a vice president of sales prepare for a last-minute customer meeting by pulling together relevant e-mails and document exchanges, tweets, blog postings, and the customer's company news. "Our challenge in the workplace is not only too much information. It's also finding, linking, reconciling and organizing multiple pieces of information logically and rapidly so we can use this information to do our jobs effectively," explains Chris Holmes, CEO of Meshin. "That's what Meshin does."

Sharing Information

Xerox has also embraced a number of collaboration tools to simplify the way works gets done. Company employees use an enterprise microblogging site called Yammer as a quick and easy way to connect with coworkers around the world. Yammer employs similar principles to conversations in Facebook, enabling more than 20,000 of Xerox's 134,000 global employees to capture quick, informal feedback. They can find out what is going on in other areas of Xerox or ask a question and get quick feedback, making this kind of retrieval technique extremely interactive and personal.

Xerox also uses two document repositories to facilitate sharing and storage of working documents. Thousands of SharePoint sites allow project workers around the globe to access and update shared documents. Xerox's own system, DocuShare, is a Web-based enterprise content management platform that enables users to capture, manage, share, and protect a wide range of paper and digital content in a single secure, central, and highly scalable repository.

In addition, all Xerox employees have access to employee information repositories, including all Intranet sites and all the DocuShare (ECM) servers. A global "Intellicentre" tool makes it simple and fast for them to find reports and publications purchased from many external sources, as well as intelligence reports and publications produced internally.

Xerox's internal approach is also being used to help customers reduce information overload. Scientists at the Xerox Research Centre Europe, Grenoble, France, and at PARC are working to develop technology for managing their document-intensive workflows.

Sorting Information

CategoriX is one example of a tool that helps people manage their documents. It uses the textual information in a document, and by employing machine learning, it learns vocabulary and can thus identify "responsive" documents with higher-than-human accuracy. The typical review process is automated by tagging the documents where it is sufficiently confident—leaving the question marks for human confirmation.

Xerox uses this technology internally to automatically sort and make sense of unstructured employee comments from global employee engagement and satisfaction surveys, as well as in litigation services, where it is combined with the knowledge and judgment of the attorneys most familiar with the cases.

Now the technology is moving to the next level: almost "going Hollywood," you could say, in the form of a prototype "smart document" review system, which allows users to gather around a large touch-sensitive tabletop display and sort documents merely by moving their fingers across the screen. It is very much like using your Smartphone touch screen—but more than a 150 times larger, making it easy for a number of people to work together.

Documents can be placed side-by-side for comparison, scaled up or down, or piled in a corner of the table, just like real sheets of papers, but with all the capabilities of digital document management. And, thanks to the large intuitive visual display, by simply pointing to a few relevant documents, the user can "teach" the system what information is important. The system then uses this knowledge to automatically categorize and sort millions of documents in a fraction of the time it would take working manually.

Cutting Through the Clutter

This technology has the potential to significantly improve document review—one of the biggest challenges for data-intensive industries such as insurance, pharmaceutical, medical, financial, and legal. The legal field is, in fact, probably one of the most intensive—it was definitely drowning in information overload. To give you an idea of the scale, Xerox offers a litigation service that can handle more than 2 billion pages of hosted data. "When you consider the vast number of documents involved in litigation, patent searching, government security, and intelligence analyst reports, it's easy to understand how some review efforts involve millions of documents," says Caroline Privault, Project Leader at Xerox Research Centre Europe.

But that is not all. Research in business analytics at Grenoble and PARC combines expertise in machine learning, statistics, natural language, visualization, and ethnography to create technology that provides business insight garnered from masses of structured and unstructured data to allow professionals to cut through the clutter and thus deliver real business advantage.

Life-Saving Software

In France, a 3-year research project, titled ALADIN, is being sponsored by the French Government in an effort to help detect hospital-acquired infections (HAI) more quickly and thus reduce infections. According to the Centers for Disease Control, hospital-acquired infections in the United States result in an estimated 1.7 million infections and as many as 99,000 deaths each year. The cost is pegged at $45 billion annually. In France alone, it is estimated that 4000 HAI-related deaths occur each year and that a third of these could have been prevented.

To help combat this serious problem, the ALADIN team is using text mining technology FactSpotter, developed at Xerox's European Research Center, to analyze medical records and identify specific terms and sequences of facts that indicate whether a patient may have contracted an HAI. The software pinpoints not only meaningful pieces of information, such as patient symptoms, drugs, and names of bacteria, but also how they are linked. When these links point to a potential risk of HAI, the system automatically alerts the staff, so that preventative action can be taken.

Urban Central Nervous System

In another example, PARC and Xerox global research are applying some of their initiatives to assist in projects developed by their business process outsourcing arm ACS. ACS has a significant transportation business and processes enormous amounts of data. Researchers are now working to create an "urban central nervous system" that will constantly check traffic flow, sensors, and data in order to understand and respond to a whole new class of business challenges.

So how do you manage the information overload a city receives? Based on data analysis, municipalities could determine where to put parking meters and meter attendants, how to ease congestion, where to consider dynamic pricing and demand management, as well as model public response behavior and predict when people will be more likely to use mass transit systems.

And in reverse mode, Xerox is working with customers to cut through the information overload involved in getting their messages out. Marketing and communications messages can make information visually stand out through personalized and customized capabilities in video, print, and on the Web.

One of the most visited pages on the Xerox Website is a video offering a humorous look at the dangers of information overload and its effect on the human brain. The message is that to deal with all the data we face every day, we first need to wipe our cerebral hard disks. Only then can we see the wood for the trees.

Contributors to Practical Insights from Xerox

- Bill McKee, Corporate Public Relations
- Caroline Privault, Project Leader, Xerox Research Centre Europe
- Chris Holmes, CEO, Meshin (a start-up funded by Xerox)

FROM CAVE WALL TO TWITTER: ENGINEERS AND TECHNICAL COMMUNICATORS AS INFORMATION SHAMAN FOR DIGITAL TRIBES

Anne Caborn and Cary L. Cooper

The fog of information can drive out knowledge.
Daniel J. Boorstin
U.S. Historian and Professor

ABSTRACT

Technical communication may pose specific challenges when professionals disseminate information. Technical communicators should therefore pay particular attention not only to what they deliver but also to how they deliver it, to ensure clarity and avoid information overload (for themselves and their recipients). Information handlers should also think about how information may be further dispersed by its initial recipients as a consequence of electronic delivery enabling easier/more rapid/wider dissemination. We consider whether positive and negative attitudes to, and engagement with, information are changing as information availability and delivery mechanisms change, specifically with the use of online channels. As a consequence of these availability and delivery changes, are those involved in information handling, as recipients and/or deliverers, more likely to feel overloaded, i.e., suffer from information overload?

This chapter is designed to give technical communicators insight into how their attitudes toward information and information overload may be influenced by developments in information technology and management, which is set in a broader historical context, and influence the way they handle and disseminate information. This chapter addresses the following questions. Is information in danger of

Information Overload: An International Challenge for Professional Engineers and Technical Communicators,
First Edition. Edited by Judith B. Strother, Jan Ulijn, Zohra Fazal.

being perceived as less valuable as it becomes more readily available and dispersible? Have developments in online accessibility, underpinned by technological advances such as personal computing and the Internet, altered the way we value information as well as how we engage with it? Are audiences and authors adapting in a digital age?

4.1 Introduction: The Dawn of the Information Shaman

It is hard to evaluate or determine the value of any given piece of information without first paying attention to the source and delivery mechanism of the information. If we know, value, or *trust* the person or organization (i.e., the "source") issuing the information, then we ascribe to the information a value based on that relationship. When removed from any *relationship*, the value of information is supported by its delivery mechanism. For example, an invitation on embossed paper and in copperplate handwriting may be deemed more important than a casually left message on a telephone answering machine or a cell phone text message asking "can U come?" Human beings are hardwired to evaluate information in the context of its delivery.

It can be argued that the *tools* for information communication *and* delivery were universally owned and valued in a prehistoric context, for example, the cave wall and pigments were generally available even if the ability to employ them was held by a specific few having artistic ability, social status, or religious standing. The information disseminated by cave paintings was perceived as divinely inspired as well as secularly motivated. Rock and pigment were used to both capture and augur, for example, a successful hunt for food, and in this respect, were highly valued, we suspect, in both a needs and a spiritual context.

To share these practical and spiritual values, both the creator and the viewer needed to develop a shared understanding of context and meaning—this is how the animal *was killed*; this is how the animal *will be killed*. In their history of graphic design, Meggs and Purvis [1] observed that early human markings and cave paintings were not the beginnings of art but the dawn of visual communication, created for survival and for utilitarian and ritualistic purposes. "The presence of what appear to be spear marks in the sides of some of these animal images indicates that they were used in magical rights designed to gain power over animals and success in the hunt." This desire to share understanding is, the authors would argue, the bedrock of *all* good communication endeavors.

Writing online about the history of visual communication, Whitley [2] referenced studies that contextualize these paintings in hunter-gatherer societies as the work of Cro-Magnon shaman. According to this view, "the shaman would retreat into the darkness of the caves, enter into a trance state, and then paint images of visions, perhaps with some notion of drawing power out of the cave walls themselves."

As modern disseminators of information, it is important that we do not lose sight of this responsibility and its power. Words deserve to be chosen as wisely as a particular shade of ochre.

As the codification of information (oral and written language) became more complex, it was increasingly extended to practicalities as well as religious texts. At the same time, the ability to decode and distribute became tightly controlled. We can all read some meaning into cave paintings, even if we have lost our ancestors' abilities to decode their deeper significance. We can all apply charcoal to a surface to create a picture, even if we choose not to because we feel we lack sufficient skill or inclination. Yet we do not have to go back too far into our recent past to find a time when reading and writing were restricted to elites (religious, political, hereditary). Information was knowledge, knowledge was power, and power was control. Control of both information content and delivery mechanisms (books, paper, and literacy itself) ensured that this power remained within dominant hierarchies, classes, and structures.

Graff [3] pointed out that, from the classical period onward, both lay leaders and religious leaders often viewed "untempered literacy" as a "threat to social order, political integration, and authority." In contrast, carefully controlled and formalized literacy through ordained channels could help maintain order, enforce moral values, and enable progress.

Human periods of evolution and revolution have often been accompanied not only by new information perspectives, for example, *The Republic* (Plato), *The 95 Theses* (Luther), and *Das Kapital* (Marx), but also by new methods of information emancipation, such as pictograms, cuneiform, paper production, and, more recently, the pre-digital mass production of information enabled by printing, as well as the mass decodifying of information enabled by increased literacy and translation into other languages. This language translation was initially more about emancipation than geography—for instance, translating the *Bible* from Latin to English. Interestingly, the only large-scale printing exercise undertaken by the inventor of the modern printing press, Johannes Gutenberg (1400–1468), the *Gutenberg Bible*, was still printed in Latin, in effect continuing the Church's control over both its study and its interpretation.

This information evolution and revolution has continued, expanded, and gathered pace; and with it, we have seen a seismic shift in terms of volume and reach. For example, the precursors of our modern newspapers were pamphlets, often handwritten, whose publication was restricted to matters of note (for the noteworthy). No news meant no publication. Effort was expended only when there was information of value to disseminate. Outside of elite groups of disseminators and recipients, largely illiterate (unnoteworthy) populations still relied on the transfer of information by word of mouth.

The advent of newspapers, coupled with enhanced printing and papermaking techniques, made the acquisition of information by general populations affordable. In the United States, cheap, one cent newspapers appeared in the early 1800s, giving rise to the name *The Penny Press*. Newspaper production, driven by production schedules rather than the availability of news, shaped a demand for information that had to be found rather than waited for, creating an entire industry in its wake.

In *The Commercialization of News in the Nineteenth Century*, Baldasty [4] observed that as editors diversified content to attract readership and assure advertising, their coverage of crime, courts, accidents, women, and leisure increased and their coverage of politics decreased, at least relatively. Audience size (reach) became significant, as could be seen by newspaper "circulation wars."

It can be argued that this content shift, coupled with decreased price and increased availability/frequency (the daily newspaper), also made newspapers more dispensable/disposable. With the advent of each fresh daily newspaper, the previous day's news and its delivery context became disposable. Newspapers became out of date—so-called "old news." Out-of-date newspapers were demoted to secondary uses of the most mundane type (cleaning, sanitation) or recycled into *new* news. The New York Benedetto family was credited with starting one of the earliest commercial newspaper recycling operations (along with rags and trash) in the United States in 1897 [5].

Over time, this appetite for news (by the news industry and the market it supplied) led to what can be interpreted as a recalibration of the equation between quality and quantity. The counterargument is that the sheer volume of information required to *feed the beast* (the newspaper industry and its drive for advertising and circulation) ensured the egalitarian nature of the content and made demands on early journalists that produced the benchmark for later investigative excellence. From Harris (1673–1716), it is but a small step to Woodward and Bernstein. Benjamin Harris' colonial newspaper, *Publick Occurrences Both Forreign and Domestick* [sic] published in Boston in 1690, was closed down by the authorities just after one issue, its content being "reflections of a very high nature: As also sundry doubtful and uncertain reports." The paper had been published without approval from the authorities, plus the first edition's content included stories alleging immorality within the French royal family and misconduct by English military forces.

Between 1972 and 1976, Bob Woodward and Carl Bernstein, reporters of *The Washington Post*, exposed the political "dirty tricks" being employed under the administration of the U.S. President Richard Nixon, which ultimately led to the President's resignation. The events began with the arrest of five men who had broken into the Democratic National Committee's office in the Watergate complex in Washington, DC, and were attempting to place listening devices [6]. Although investigative excellence does not speak directly to information overload, it does underpin the development of a press powerful enough to publish, *without approval*, information of value (that could topple a presidency), thus creating and sustaining sources of alternative/unauthorized information requiring assessment.

Gutenberg, Harris, Woodward, and Bernstein could not have predicted the information explosion that would be the by-product of the digital age. It was not simply the development of the Internet, but the plethora of platforms that built on the early technological emancipation brought about by the arrival of cheap desktop publishing (desktop computers and word processors, word processing and design software, plus more affordable, better quality, and smaller printers) in the 1970s and 1980s. As a result, we have all become potential mass consumers and communicators of information.

At the same time, information media have become more ephemeral and intangible. It was 1998 when Steve Jobs first produced the iMac, a Macintosh laptop with no floppy disk drive. Today, rewritable CD ROMs look like old technology. At present, we still may resort to a small, solid-state pen drive or memory stick to transport information between people and locations. More often, however, we are likely to rely on high-speed broadband connections and use remote, shared server space. Today, the content of some entire libraries might even fit comfortably on a single personal computer's multi-terabyte hard drive.

4.2 The Magic of Metaphor

In terms of our attitude to information, metaphor choices are a powerful indicator of how we feel about the subject the metaphor applies to. Research into how metaphors are used to describe and define information and its application and/or value includes work carried out by research physicists. These scientists were concerned that traditional metaphors were inadequate to describe advanced levels of high-energy elementary particle theory and behavior and thus were inhibiting the way students grasped some newer, higher level, and abstract scientific concepts. Brookes and Etkina [7] stated the following: "We found that physicists' language encodes different varieties of analogical models through the use of grammar and conceptual metaphor."

The central argument is that metaphors have a powerful influence on how we think about things. They can set us free to explore new concepts or they can inhibit new thinking. Brookes and Etkina also cited research showing how physicists use analogical models to construct new ideas and showing that these analogical models become, in time, encoded linguistically as conceptual metaphors. "The way physicists talk about already established knowledge is different from the way they talk about new and abstract ideas they are trying to comprehend themselves." Teacher and student (information disseminator and information recipient) need to "share the code" so that, for example, when a physicist states, "the electron is in the ground state," the meaning of "ground state" is understood by the student in energy terms rather than spatial [7].

Andriessen [8] noted that many Western metaphors for knowledge equate it to "stuff." The unconscious choice of metaphor has enormous impact on how we reason about knowledge, what is highlighted, what is hidden, what is seen in organizations as problems, and what is understood as solutions. Andriessen's central argument is that "knowledge as stuff" metaphors dehumanize organizations. But his supporting arguments open up interesting areas for further study, not the least of which is how our visualization and conceptualization of knowledge influences how we react to and deal with knowledge.

Our experience indicates that when organizations approach large digital projects, such as a new website or e-mail program, they often approach information as "stuff" that has to be moved around and loaded or uploaded, predisposing the task to appear to be daunting. This descriptive approach can also be seen, for example, in progress management methodology and tracking software, Agile, with its reference to "tasks," "progress," "sprints," and "scrum," which are grounded in physical states [9]. Would information appear so daunting if the metaphors we harnessed to describe it were more dynamic and fluid? Andriessen hypothesized that "knowledge as stuff" metaphors, which he found more predominant in Western organizations, strengthened the idea of organizations as machines with input and output, also influencing the culture and manager/worker power balance. Andriessen and Van den Boom [10] highlighted the difference in metaphors for knowledge in Arab countries and Asia, and knowledge as conveyed in Islam, Buddhism, Hinduism, and Confucianism, where the emphasis was on illumination or enlightenment. Alternative metaphors might be those that frame knowledge acquisition as a journey (or quest). "Knowledge is not a given and a prearranged body of data and information, but a discovery-like process with hindrances, obstacles, reluctances as well as positive stimuli: denial that the current and familiar position no longer holds, stepping out of the box, leaving one's position and the current knowledge environment, persuasion by others and the need of help from others, breakthrough experience of something completely new." Within an engineering context, it is important that authors are sensitive to language and metaphors that are suitable for describing engineering objects—a valve or motor—but less appropriate when describing the knowledge itself, such as of how an object is made.

4.3 The Audience: The Emergence of Digital Tribes

Given our experience and the research cited above, it is important that all practicing engineers and technical communicators consider the objectives of different audiences and how these may be influenced by professional and organizational cultures [11] and access and use of digital platforms, such as websites, e-mail, and social media, such as Twitter. For example, Vincenti [12] stated that for engineers, in contrast to scientists, knowledge is not an end in itself or the central objective of their profession; rather, it is a means to utilitarian ends. The emphasis is on the application of knowledge in problem solving. Elsewhere, Cullen [13] observed the emancipating relationship between healthcare professionals and patients predicated on shared decision making and informed consent. This has placed an increasing onus on healthcare information professionals to act as information interme-diaries and educators, exploiting digital channels in particular to achieve this. For example, "NHS Choices" is the online health resource provided by the U.K. National Health Service and, at the time of writing, the United Kingdom's biggest health website.

Therefore, the role of the doctor in the twenty-first century is not only to know what is best for the patient but also to communicate this to the patient in such a way that the patient concurs with the doctor's decision. At the same time, medical professionals are far from the only information source available to (would-be) informed and empowered patients, who increasingly use the Internet to access additional information about medicines and treatments. According to Cullen, "In this new role, health information professionals must develop an understanding of several key elements. . . the need for consumer health information, the principle of informed consent information literacy and critical thinking, the principles of effective teaching, and the criteria for evaluating consumer health information resources" [13].

These empowered patients may also create additional information overload for themselves and for the health professionals with whom they engage by random/untrained and unrestrained researching of diagnoses and treatments, which may or may not have been subject to academic rigor or clinical trial. Medical professionals may be using knowledge not only to inform their own decision making but also to inform or counter the knowledge seeking and decision making of "expert patients" [14–17].

All engineers and technical communicators may benefit from considering not only how they structure and deliver information to a known primary audience, peers, such as work colleagues, but also how that audience may employ that information or reorder and redeliver that information to secondary audiences of their own. An example here might be written information configured for a peer audience passed on in a learning context to an apprentice or student on work experience.

Simple questions to consider may include the following: How will this information be used? To whom may this information be passed on? If information is of value, then there is a strong argument that communicators have an obligation, moral if not actual, to ensure that the knowledge they communicate retains its integrity (value) when applied by the recipient or passed on by the recipient through subsequent iterations. This could be seen as a semantic application of information entropy/information theory [18].

Primary audiences may share vocabularies, skill sets, and levels of achievement. Secondary audiences may have lower or different knowledge or literacy parameters. The inexperienced apprentice may be unfamiliar with a chosen skill's vocabulary and technical language (and slang). Could considering these secondary audiences add value to the primary delivery? For example, Nielsen [19] demonstrated that writing for lower literacy audiences also decreased task time and increased satisfaction for higher literacy users. The research collected usability metrics before and after the website content for a major pharmaceutical product was rewritten and simplified. The before and after Web content was tested with higher and lower literacy users. Both sets of users scored higher in the rewritten and simplified content.

4.4   Quill to Keyboard: The Writer and New Media

Ease of production contributes to information and potential overload. The advent of smaller, non-mainframe, and personal computers, the Internet, and the World Wide Web has increased the ability to self publish to both traditional print media and online. Websites have become increasingly accessible to an increasingly broad author population. The range of online publications, options, and formats has also grown, from the brevity of Twitter and the informality of Facebook to blogs (opinion, fact, and fiction) and highly structured online networking groups with business or professional imperatives.

More than 133,000,000 blogs have been indexed by the blog collator and media agency Technorati since 2002. According to Technorati's December 2009 statistics, 72% of bloggers classify themselves as hobbyists, meaning that they report no income related to blogging [20].

We would argue that a strong motivator for the publication of much information is a desire to be read either as an individual or as someone representing a collective perspective. But while the labor intensity of early publication—whether cave painting or early pamphleteer—acted as natural inhibitors on output, electronic word processing and online publishing have opened the floodgates. Those who had to seek out media to consume, for example, obtain a pamphlet, created *by hand* or printed in relatively modest volumes, now have information delivered directly to them, and in infinitely scalable volumes, via electronic media.

So, within the context of all the above, is it not possible that we are becoming less concerned with information acquisition and husbandry? Does it really matter where the hard copy of a favorite music track *is*, if it can be located and listened to online, where it resides within the huge, growing, but nonetheless amorphous information environment that is digitally and constantly available via hardware and software interfaces that are becoming progressively cheaper and more portable?

Did our ancient ancestors' deep and spiritual connection with and value of a cave painting at least, in part, reflect the time it took to create and the sheer immutability of its platform—a rock face deep inside the earth? Does the ping of a cell phone's "new message" alert hold its own with book, scroll, hide, or rock?

It is worth any writer, whatever his or her subject matter, considering this history thoughtfully. Is the transfer of information from deliverer to recipient no less sacred a task than it was when early human beings still lived in caves or later generations were printing political tracts? That is not to say that information may not sometimes be distorted or trivialized. But, as doctors historically took the Hippocratic Oath, should writers promise to do their best by those they write for? An abstract from the Hippocratic Oath states, "In purity and according to divine law will I carry out my life and my art" [21].

Consider, also, relevant Codes of Ethics and professional conduct for technicians and engineers, for example, the Institute of Electrical and Electronics Engineers (IEEE) Code of Ethics [22] with its emphasis to honesty and *realism* in stating claims or estimates and commitment to improving understanding of technology. Other examples would include the Ethical Principles articulated by the Society for Technical Communication (STC)— "conciseness, clarity, coherence, and creativity" [23]—and the Code of Ethics for the National Society of Professional Engineers (NSPE) [24].

So, an essential consideration for any author must be "delivery context," and how both that context and the information it carries are perceived by the recipient. It does not matter how valuable an author thinks the information is—will the recipient think the same way? How will authors, editors, and deliverers optimize their delivery and information construction to make both consumption easy and value evident?

Within the confines of this chapter, we specifically consider online channels such as social networks, websites, e-mail, and text messaging because of the many ways that these channels have influenced our consumption of information and written language. Most of us understand and practice many of the scanning and skimming techniques that are employed to sift and filter written information. Readers apply these techniques to written text both online and offline. For example, Ulijn and Strother [25] discussed partial parallel processing as a coping model, using text, syntax, and lexis differently, depending on the reader's familiarity with the material or the language in which it is written. Pugh [26] observed that in search mode, readers may be trying to locate information without knowing the precise form it will appear in; therefore, they visually scan the text. He observes that "the periods of close attention to the text tend to be more frequent and of longer duration and, since information is more deeply embedded in the text, there is more observance of the way in which the author structures his subject matter and, hence, the linearity and sequencing."

Online, there are indications that people may be using a subset or refinement of these skills that pose greater or different demands on the author. Nielsen [27] discerned an F-shaped reading pattern for Web pages when these patterns were viewed as a *heat map* using eye-tracking technology. This technology measures the location and duration of the time spent looking at content areas on a screen, and represents this using a color coding that ranges from gray for no time, to blue for the least time, and to red for the most time. In the black-and-white version (Figure 4.1), the colors are not visible, but the F-pattern shows that much content is left unscanned. This has implications for both the placement of the information as well as the importance of brevity. Nielsen [28] estimated typical printed text *consumption* at 28% or only 20% of the words on an average Web page. Nielsen calculated that Web users read 50% of the information on pages with 111 words or less. They dwelled on pages with more words for less average time—i.e., 4.4 additional seconds for each additional 100 words.

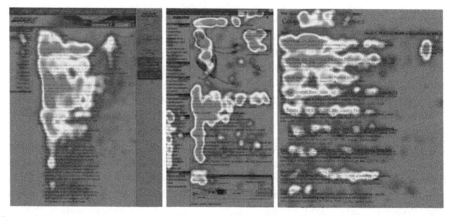

Figure 4.1 Online, people may be using a subset or refinement of scanning and skimming skills that pose greater or different demands on the author. *Heat maps* from eye tracking studies reveal an F-pattern that leaves much content unscanned [23].

4.5 Helping the Reader: Techniques for the Information Shaman

It can be argued that part of a cave painting's potency is how apparently simple early drawings convey richness of meaning. Online information can be made more *consumable* by employing such simplicity and deploying "evolving online best practice," designed to make it easier for readers to pick out key words and phrases when reading quickly. This includes the use of clear language and structuring, including shorter, simpler sentences and language, and the use of information-bearing headings and subheadings. The reader's task may be further simplified if the author/developer/writer places important words and phrases in bold text and exploits "above the fold" screen space on Web pages and within e-mail client preview panes. "Above the fold" is a broadsheet newspaper term used traditionally to describe information above the point at which a broadsheet newspaper is folded in two. It is now employed to describe the portion of information in view on a computer screen before any scrolling occurs.

Online content best practice also advocates for the use of inverted pyramid content structuring [29]. This inverts traditional academic writing style by placing conclusions or results at the beginning.

Authors can also use information placement and visual cues to increase the speed at which readers scan and skim, so that they can quickly assess information (albeit at a surface level) before deciding what to assess in more detail (and take greater time over). Screen icons are the pictograms of high-speed online communication, for example, a graphic of a small envelope indicates e-mail functionality and a line drawing of an old-style telephone draws attention to a contact number. The decoding associated with reading text/information is often being replaced (or at least underpinned) by interpretation of pictograms and icons. Authors need to consider the employment of such visual cues, as well as graphic and grammatical devices such as bullet points and numbered lists, to enhance the readability of their work. To do this effectively requires authors to stay abreast of online technology and understand the increasing visual design capabilities, requirements, and limitations in the online environment.

Note that reading is redundant in this scanning and skimming process. In the case of e-mail, recipients may make decisions about which e-mail to open/read through assessment of "Subject" and "From" (sender) fields. The importance of the From field, particularly where information is unrequested, reflects the importance of the source in ascribing value to information. In the absence of a From field that a recipient recognizes (such as an organization or an individual), the recipient may decide not even to open, much less read, the e-mail or any attachments sent with it. Schade and Nielsen [30] pointed out that "When users were looking for a particular message, they looked to the Sender line to identify the message, then used the Subject line as a way to confirm it was the correct message. Some users sorted their inboxes by Sender to locate a specific newsletter."

The importance of From, i.e., *who* is the sender, also reflects growing fears about spam, e-mail communications that may be fraudulent or contain computer viruses. Encouraging recipients to add Sender e-mail addresses to the contact list contained within their e-mail software is one way to avoid e-mail communications being caught in spam filters. The growth in the use of smart/android cell phones, for example, Blackberry® and iPhone, places even greater demands on authors. This reflects the fact that mobile communication may come with restrictions—cell phone screen size is limited—and, potentially, distractions, for example, the recipient may be on the subway, at a social gathering, or engaged in any number of other activities, rather than sitting at a computer. Traditional laptops impose some restrictions on mobility (e.g., size or weight) while smaller netbooks and tablets, less so.

Authors may have no certainty about how information will be *received* (a developing division between reception mode and delivery mechanism). A letter posted to a recipient will still be a letter, in terms of content, dimensions, and physical nature, when it is received. By contrast, a recipient may be viewing electronic information on a phone, a laptop, a visual display unit (VDU), or a TV screen and be at a desk or on a sofa. Based on the above uncertainties, authors have to recognize that there is an increasing need to pare down (as distinct from condense and distill) information for electronic consumption. Paring down requires further evaluation of what elements of the information are essential (and removing the less essential ones), as opposed to condensing and distilling, which attempts to reduce all the information in a given communication into less space. But will this redaction/reduction create information in a form that is both easier to consume and easier to disregard?

While most of the general population would never have gained entry to, say, a mediaeval library or have been able to read its contents, they may well have wondered at and even been respectful of the power and position (in terms of construction and purpose) such a library commanded. Battles [31] brought this to life by stating the following: "In the library, the reader is wakened from the dream of communion with a single book, startled into recognition of the world's materiality by the sheer number of bound volumes; by the sound of pages turning, covers rubbing; by the rank smell of books gathered together in vast numbers." The library buildings of all civilizations, both ancient and modern, are built (in both size and materials) in a way that communicates value. Book burning, as a form of protest, has a potency that electronic file deletion cannot hope to match.

How is then value communicated by the many modern forms and delivery mechanisms for information that exists primarily in cyberspace rather than in a physical form? At its most basic, attention to detail can communicate value. As previously discussed, good information-bearing e-mail Subject fields should communicate what an e-mail contains and why that is of value to the recipient. Structuring of all online communication should focus on ensuring that the most important information is at the top (inverted pyramid structure). Again, this is the most important information from the perspective of the recipient. It is the nature of authorship to confuse value with effort. Just because the author wrote something, agonized over the research, and battled through a lengthy review process does not mean that the reader will automatically want to consume what has been written. What is in it for the reader? What benefit will the reader acquire from adding the author's information to the load he or she already has to deal with? Basics such as correct spelling, consistency of nomenclature, and clarity of writing style are critical. If an author does not value a piece of written communication enough to check its spelling, even hand-embossed velum will do little to communicate value.

4.6 The Magic of Hypertext Techniques: Journeys at the Speed of Thought

In this digital age, the author must also be aware that information will not be consumed in the order it was conceived. Elements may be extracted without knowledge and certainly without formal permission, finding their way onto social media platforms, blogs, and websites, via content management systems, and so forth. This process is sometimes referred to as content *atomization* [32]. Readers accessing information online will also create their own paths through it, exploring it in more associative ways, much as thoughts grow, connect, and dissipate within the human brain. We create these associative information pathways in digital media by exploiting hyperlinking, which is an essential element within all online content. These links take us from one Web page to another and from one website to another without the need to follow any predetermined route. We can randomly flick through the pages of a book, but the pages of the book do not reorder themselves. The Internet, by contrast, *appears* to reorder itself around our online thoughts and actions as we research a topic of interest by creating (linking) and following our own associative information pathway.

Foutz [33] advocated hypertext as an enabling tool for concurrent engineering teams, as a way of building paths through multiple material sources and locations, reducing "cognitive load." It also frees teams from physically sharing information and enables them to explore information via multiple pathways. Forster and Next [34] demonstrated increased efficiency through using hypertext in the preparation of a design brief. Farkas [35] argued that the *essence* of hypertext and hypermedia (screen-located content that employs hyperlinking and related dynamic media) is choice, allowing the reader/user to decide where to go next. It is also worth considering how this associative, hyperlinked online environment may expose us to an increased risk of manipulation. When authoring in online environments, we need to be mindful of how we use hyperlinking, in order to avoid creating distraction or introducing bias.

Another way readers/users create their own paths through and around the online environment involves *Search*, the location of content on the Internet via key words and phrases. Search engine "spiders" (software programs that index Web content) crawl the Web-sifting content, which is then catalogued by the search engines and matched to the key words and phrases individuals search with. It is possible to manipulate what is a quasi-mechanical process, despite the best efforts of search engines such as Google. Google's algorithms, used by its search engine spiders, are designed and constantly refined to differentiate genuinely relevant content from content that has been optimized but is, in actuality, less relevant. The careful seeding of articles and the creation of multiple pages containing content with key words and phrases that are popular search terms, plus the use of meta tagging—i.e., using key words and descriptions that sit beneath each page and are designed to be read by search bots and aid content aggregation, retrieval, and ordering (content management)—can push websites up the search rankings. The search ranking may therefore not reflect content per se but content *optimization*. An author's choice of tags and writing of meta descriptions may be deemed cynical if the primary objective is simply to enhance page rankings.

So search engine optimization, making content more findable, may draw readers to content that is more findable but, possibly, less relevant. This is unlike a physical library where the content of books is fixed and cannot be "optimized" or even rewritten without time, effort, and high cost. In a library, we must rely on indexing, coupled with our value judgment, prior knowledge, and research, plus an assessment of a book's title and, say, an abstract or flyleaf blurb. We would not be moved by how near the front door a particular book is. Yet, in online Search ranking, nearness to the top of the first page of a Search result returned by a Search engine is an important factor. So, for example, an engineering company could change its Web content to tap into current debate around, say, a new component design, and achieve a higher page ranking than a firm with a better component design but with less-optimized Web pages. As authors, we need to be aware that we may impart bias simply through ordering and optimizing information. A good question is, "I can get this Web page to the top of the Google search rankings—but does it belong there?" If the answer is "no," then what needs to change to make it worth the ranking?

At the same time, authors are working with human nature. Schema Theory [36, 37] can be most simply described as the organization of knowledge into units or schemata. We develop these schemata long before we engage in formal education, and they are the basic building blocks we employ for understanding the world. Our personal schemata help us discern information that is relevant to us. These cognitive processing strategies are also influenced by the use of the Internet as a way of assessing and consuming information. Bellman and Rossiter [38] argued that consumers have schema for product and marketing information consumption, so Web schemata are developed by consumers to deal with the volume and idiosyncratic presentation of multiple pages of information within a website. "However, whereas moderately atypical executions in traditional media need not impair information processing and may even enhance it, a website that does not match general expectations may significantly reduce the ability to navigate the site and search for information effectively" [38]. The Bellman and Rossiter website schema is defined as the consumer's set of beliefs about information locations and routes to those locations for a specific website. The closer a website comes to these beliefs, the better the consumer experience, resulting in more positive feelings about a brand or a product [39, 40].

On a practical level, this is a strong argument for keeping the reader/user the central focus of all decisions relating to Web design and information structuring. This may seem like stating the obvious but it is easy to get caught up in the excitement of harnessing the latest technological developments or choosing a very striking design in an egocentric, rather than sociocentric way. Schema Theory and the Bellman and Rossiter [38] website schema studies underpinned the personal preferences, assumptions, and habits that information recipients bring to information and its assessment. What is less researched and understood is how our fundamental attitudes to, and engagement with, information may have shifted as the amount of information that is instantly available has increased. Does the term "information overload" and its common usage denote a widely held, or widely understood, acceptance of overload as the status quo—i.e., perceptive or actual excess?

4.7 Conclusion: The Responsibilities of the Information Shaman

While it can be argued that we have always scanned and skimmed text and that, for example, the e-mail From and Subject line assessment is remarkably similar in mechanism to traditional book spine scanning in libraries, the *real* shift appears to be away from deciding what to read and toward deciding what *not* to read. This may involve a version of selective exposure, where we scan and skim for content elements that support our position.

Online communication may also be creating changes at a far more profound and intriguing level. Small and Vorgan [41] observed that the neural networks of people who have been brought up with computers are different from those who came in contact with computers at an older age. "Although initially transient and instantaneous, enough repetition of any stimulus—whether its operating a new technological device, or simply making a change in one's jogging route—will lay down a corresponding set of neural network pathways in your brain that can become permanent."

As authors, we must give this careful thought. Our audiences are evolving, so we need to evolve too (and constantly). How will we deliver information in the future to harness these changes and aid fast, effective information consumption and ease potential overload? In rapidly evolving fields such as engineering, this challenge may be greater as we inform audiences about new concepts, breakthroughs, and discoveries while employing new terms, descriptions, and theories. Who will receive this information? When will they access it? What will they do with it? What do I want them to do with it? What about their receiving mode, abilities, and the delivery mechanism itself have I not considered or not considered in enough detail? If words are to carry meaning and value, rather than simply "load," we must treat them respectfully. Ultimately, each new communication of information needs to be considered with the care and thought that an early shaman might have demonstrated when embarking on a new cave painting if it is to inform and augur.

REFERENCES

[1] P. Meggs and A. W. Purvis, *A History of Graphic Design*. Chichester, U.K.: Wiley, 2006.

[2] D. S. Whitley, *Cave Paintings and the Human Spirit: The Origin of Creativity and Belief*. New York: Prometheus, 2009.

[3] H. J. Graff, *Literacy and Social Development in the West*. Cambridge, MA: Cambridge Univ. Press, 1981.

[4] G. Baldasty, *The Commercialization of News in the Nineteenth Century*. Madison, WI: Univ. Wisconsin Press, 1992.

[5] TFC Recycling. (2011, Aug.) [Online]. Available: http://www.tfcrecycling.com/aboutus

[6] D. K. Fremon, *The Watergate Scandal in American History*. Berkeley Heights, NJ: Enslow, 1998.

[7] T. D. Brookes and E. Etkina, "Using conceptual metaphor and functional grammar to explore how language used in physics affects student learning," *Physical Rev. Special Topics Physics Education Res.*, vol. 3, no. 1, pp. 1–16, 2007.

[8] D. G. Andriessen and C. Bratianu, "Knowledge as energy: A metaphorical analysis," in *Proc. 9th Eur. Conf. Know. Manag.*, 2008, pp. 75–82.

[9] Agile Methodology. (2011). *Understanding Agile Methodology* [Online]. Available: http://agilemethodology.org Sep. 9, 2011 (last date accessed).

[10] D. G. Andriessen and M. Van den Boom, "In search of alternative metaphors for knowledge: Inspiration from symbolism," *Electron. J. Know. Manag.*, vol. 7, no. 4, pp. 397–404, 2009.

[11] H. M. Trice and J. M. Beyer, *The Cultures of Work Organizations*. Englewood Cliffs, NJ: Prentice-Hall, 1993.

[12] W. G. Vincenti, *What Engineers Know and How They Know It: Analytical Studies from Aeronautical History*. Baltimore, MD: Johns Hopkins Univ. Press, 1993.

[13] R. Cullen, *Empowering Patients Through Health Information Literacy Training*. Bingley, England: Emerald Group Publishing Ltd., 2005.

[14] R. R. Faden, *et al.*, *A History and Theory of Informed Consent*, Oxford, U.K.: Oxford Univ. Press, 1986.

[15] C. Faulder, *Whose Body Is It? The Troubled Issue of Informed Consent*. London, U.K.: Virago, 1985.

[16] L. Donaldson, *The Expert Patient: A New Approach to Chronic Disease Management for the 21st Century*. London, U.K.: Stationery Office, 2001.

[17] J. Shaw and M. Baker, " 'Expert patient'—dream or nightmare?" *British Med. J.*, vol. 328, no. 7442, pp. 723–724, 2004.

[18] E. Claude and C. E. Shannon, "A mathematical theory of communication," *Bell Syst. Tech. J.*, vol. 27, pp. 623–656, Jul.–Oct. 1948.

[19] J. Nielsen. (2005). *Lower-Literacy Users, Alertbox* [Online]. Available: http://www.useit.com/alertbox/20050314.html Aug. 26, 2011 (last date accessed).

[20] Technorati, *State of the Blogosphere*. San Francisco, CA: Technorati, Inc., 2009.

[21] *The Hippocratic Oath*, U.S. National Library of Medicine, National Institutes of Health, 2011 [Online]. Available: http://www.nlm.nih.gov/hmd/greek/greek_oath.html Aug. 26, 2011 (last date accessed).

[22] IEEE Policies. *Section 7—Professional Activities (Part A—IEEE Policies), 7.8 IEEE Code of Ethics* [Online]. Available: http://www.ieee.org/about/corporate/governance/p7-8.html Nov. 11, 2011 (last date accessed).

[23] Society for Technical Communication. *Ethical Principles* [Online]. Available: http://www.stc.org/about-stc/the-profession-all-about-technical-communication/ethical-principles Nov. 11, 2011 (last date accessed).

[24] National Society of Professional Engineers. *Code of Ethics* [Online]. Available: http://www.nspe.org/Ethics/CodeofEthics/index.html Nov. 11, 2011 (last date accessed).

[25] J. M. Ulijn and J. B. Strother, *Communicating in Business and Technology: From Psycho-linguistic Theory to International Practice.* London, U.K.: Peter Lang Publishing, 1995.

[26] A. K. Pugh, *Silent Reading.* London, U.K.: Heinemann, 1978.

[27] J. Nielsen, "F-shaped pattern for reading Web content," *Alertbox,* [Online] Available: http://www.useit.com/alertbox/reading_pattern.html August 26, 2011 (last date accessed). 2006.

[28] J. Nielsen. (2008). *How Little Do Users Read? Alertbox* [Online]. Available: http://www.useit.com/alertbox/percent-text-read.html Aug. 26, 2011 (last date accessed).

[29] J. Redish, *Letting Go of the Words, Writing Web Content That Works.* San Francisco, CA: Morgan Kaufman (Elsevier), 2007.

[30] A. Schade and J. Nielsen, *Email Newsletter, Usability 165 Guidelines for Newsletter Subscription, Content, Account Maintenance and RSS News Feeds Based on Usability Studies,* 3rd ed. Fremont, CA: Nielsen Norman Group, 2006.

[31] M. Battles, *Library: An Unquiet History,* New York: W. W. Norton & Company, 2003.

[32] D. Chaffey. (2007). *What is atomisation (Web 2.0)?* [Online]. Available: http://www.davechaffey.com/Internet-Marketing/C8-Communications Aug. 26, 2011 (last date accessed).

[33] J. Foutz. (2001). *Hypertext as an Enabling Tool for Concurrent Engineering Teams* [Online]. Available: http://www.smpstech.com/papers/fout93b.htm Aug. 26, 2011 (last date accessed).

[34] J. K. W. Forster and P. L. Nest, "An engineering application for hypertext," *Instruct. Sci.,* vol. 21, nos. 1–3, pp. 199–208. 1992.

[35] D. K. Farkas, "Hypertext and hypermedia," in *Berkshire Encyclopaedia of Human-Computer Interaction.* Great Barrington, MA: Berkshire Publishing, 2004.

[36] J. Piaget, *The Child's Conception of the World.* Towota, NJ: Littlefield Adams, 1926.

[37] R. C. Anderson and P. D. Pearson, "A schema-theoretic view of basic processes in reading comprehension," in *Handbook of Reading Research,* P. D. Pearson, Ed. New York: Longman, 1984.

[38] S. Bellman and J. R. Rossiter, "The website schema," *J. Interact. Advertising,* vol. 21, no. 11, pp. 945–960, 2004.

[39] J. R. Coyle and S. J. Gould, "How consumers generate clickstreams through web sites: An empirical investigation of hypertext, schema, and mapping theoretical explanations," *J. Interact. Advertising,* vol. 2, no. 2. 2002.

[40] D. L. Hoffman, and T. P. Novak, "Marketing in hypermedia computer-mediated environments: Conceptual foundations," *J. Marketing,* vol. 60, no. 3, pp. 50–68, 1996.

[41] G. Small and G. Vorgan, *iBrain: Surviving the Technological Alteration of the Modern Mind,* New York: Harper Collins, 2008.

PRACTICAL INSIGHTS FROM
THE LIMBURG MEDIA GROUP

Mecom Group plc ("Mecom") is a European content and consumer business that owns over 250 printed titles and over 200 websites in its three divisions, with substantial operations in The Netherlands, Denmark, and Poland. The Group generates a readership of 24 million per week and attracts 22 million unique website users per month. The Group's Dutch division is composed of The Limburg Media Group (LMG) and Wegener.

The Limburg Media Group (LMG) is the leading regional newspaper business in the Dutch province of Limburg. The group owns a printing house and is the publisher of two daily newspapers: *Dagblad de Limburger* and *Limburgs Dagblad*. LMG also owns a regional free sheet and a door-to-door title. It services 12 hyperlocal and regional websites. LMG focuses its business in the Limburg province with a unique number readership in print and online of 70%. In 2011, LMG launched *Perspectief*, a paid-for content website, as well as applications for the iPad and mobile device market. LMG has 325 employees, of which 75 are in the printing house, 115 are in journalism, 60 are in sales and marketing, and 75 are additional staff.

Information Overload: An International Challenge for Professional Engineers and Technical Communicators, First Edition. Edited by Judith B. Strother, Jan Ulijn, Zohra Fazal.
© 2012 Institute of Electrical and Electronics Engineers. Published 2012 by John Wiley & Sons, Inc.

Newspaper Position in The Netherlands

Many of the challenges of this news organization are concerned with refining the content of the publications. As with all news endeavors, these journalists must decide what the definition of news is: what is *nice* to know versus what is *necessary* to know? How do you balance the ethics of telling a story objectively with the demands of advertisers and other marketing genres? As Mr. Thijssen says, "Managing the independence of the news is both delicate and tricky."

Since LMG's publications are regional newspapers, they are not the main source of international news for their readership. Instead, the journalists interpret the international and national news for local interests. For example, if there is a new area of crisis in the world, what effect will that have on the gas prices in the Limburg region? If prices go up, what effect might that have on transportation costs for local grocery products? Might there be an effect on the employment opportunities in the region?

The paid newspaper system in The Netherlands is based on subscribers, which makes the setting and mission totally different from many other countries. Trustworthy, independent information for the region, not screaming headlines to attract attention, has been the editorial compass for decades. The confidence readers have in the LMG products is the reason for its success. Similar to all paper media, LMG encounters problems with diminishing subscribers, but LMG still has 170,000 subscribers, which is close to 60% of all households in Limburg. For many years, the balance between supplying independent information and demands of advertisers has been hard to maintain. This is related to the shift in income. For example, 10 years ago, 60% of the income was generated by advertising and 40% by subscribers. In 2010, the balance was more than turned around into 30% ads and 70% subscribers. As always, "whoever pays the most gets the best seats," and that is clearly in favor of independent journalism at LMG. This shift in market position is part of a shift in the business model that LMG intends to implement in the coming years.

Managing Information Overload Using an Evolutionary Approach

How is information overload controlled in this dynamic, ever-changing environment? Mr. Thijssen says that the key is streamlining the process itself. The "string list," which is developed by the Editor-in-Chief, determines what news goes to whom for development. This is the primary source of news stories. In addition, each journalist can present a news item that he or she wishes to write about. Controlling this constant flow is essential if information overload is to be avoided. In this era, news does not enter by the postman any more; instead digital sources are used. While the fact that the majority of news tips today come electronically, mainly through e-mail, has simplified the process of moving the information through the system, managing such a complex information network is still a challenge.

In the past decade, an evolution has occurred toward processing the right information to the right internal address. Today, 95% of the information is offered by e-mail. On a daily basis, out of over 1200 items, 300 news items are selected and directed to various platforms. For instance, breaking news is directly published online and on mobile platforms, while background information and reflective items are published in the regional newspapers.

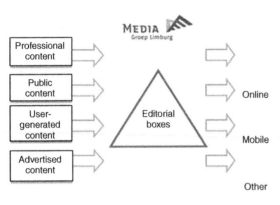

Professional content

Public content

User-generated content

Advertised content

MEDIA
Groep Limburg

Editorial boxes

Online

Mobile

Other

Figure 1 Managing the flow of information by editorial boxes at LMG. Sources of news, information windows, are instructed on how to route items to the right editorial box. For every editorial box, rules have been defined related to the specific formula or goal of the information platform. Currently, 15 editorial boxes are operated. A journalist has access to one or more editorial boxes. This logistical highway organizes the necessary information flow in a way that every journalist directly is able to intercept or launch relevant news items.

In order to manage the process of selection effectively, a fast and tight logistic is required. E-mail software makes this possible, as the sources of news, information windows, are instructed on how to route items to the right editorial box. Currently, 15 editorial boxes are operated. A journalist has access to one or more editorial boxes, depending on his or her discipline. This logistical highway organizes the necessary information flow in a way that every journalist is able to directly intercept or launch relevant news items.

In the evolutionary development of this editorial process, for every editorial box, rules have been defined related to the specific formula or goal of the information platform. Such criteria are guidelines for the journalists to decide what is news or what is not. This initial selection by an individual journalist is followed by a coordinated final decision by the editor-in-chief to prioritize the news stories and decide what will make that day's press. This process takes place three times a day. The coordination includes, for instance, a discussion between the various disciplines involved. Figure 1 illustrates how available and required information is managed by the 15 editorial boxes.

A Revolutionary Perspective

The organization is constantly working to improve its information flow and the prioritization of the information itself. Much of the process has changed through a natural evolution, with the development of the technology. However, now, a revolution is taking place in parallel with this natural evolution. Instead of journalism, marketing, and advertising being separate departments with distinct tasks and operations, they are now unified.

By nature, the role of a publisher is to bring news. LMG, like all newspaper operations, is a receiver as well as a producer of vast amounts of information; therefore, information overload is a daily challenge for its operations. For the Dutch paid-for newspaper, the selection of news is in the hands of an independent editorial board. The news is anchored in reliability and the orientation to bring news that is relevant for the reader. However, with the emergence of a large variety of information sources in addition to a newspaper, marketing and sales have become more important. LMG needed to reflect on a strategy with which a new balance can be found between relevant content and sales interests. In a current project, LMG intends to connect journalistic and sales elements. This strategy required a rather complex decision-making structure. In order to support this strategy, an information structure of a revolutionary kind has been developed. While keeping the editorial independence, journalists also decide on marketing matters. In reverse, the sales department can take initiatives with respect to editorial independency.

Most activities and energy go to the correct setup and technical support of the new strategy perspective, but the fit between the goals of the new strategy with the capabilities of human resources is not achieved without effort. The new organization is distinguished by a multimedia approach, multidisciplinary way of organizing work, innovative orientation of stakeholders, and intensive internal and external networking. The HR strategy is focused on such new key values and is operationalized by an output-oriented approach. Employees and other stakeholders are facilitated and encouraged to develop their talents within the new perspective of the Limburg Media Group. With such a large workforce depending on frequent interaction, the interpersonal aspects of the organization are critical. Without good internal communication, overload can occur at many levels. The final way Limburg Media Group helps to control information overload is through expectation management. While it is essential to manage the process itself, Mr. Thijssen emphasized that it is essential to make good choices—to focus on the outcome or results more than to get lost in the details of the process.

This revolutionary way of working, which was implemented in August 2010, was developed with the help of a government subsidy through the Natural Fund for Journalism, with the goal that LMG's successful results can be transposed to other regions. Since the evolution of the information flow process occurs in parallel with the revolution in the way the previously independent departments are operating, the daily operations are indeed multidimensional and are helping to address the ever-present issue of information overload.

Contributors to Practical Insights from The Limburg Media Group

- Hans Thijssen, Human Resources Manager
- Huub Paulissen, Editor-in-Chief

5

THE INFLUENCE OF CULTURE ON INFORMATION OVERLOAD

Jan M. Ulijn and Judith B. Strother

> *This dominance of technology over culture is an illusion. The software of the machines may be globalized, but the software of the minds that use them is not.*
>
> Geert Hofstede
> Mechanical Engineer and Social Psychologist

ABSTRACT

Too often, all communication genres are analyzed without recognizing the unavoidable impact of a communicator's cultural framework on producing, receiving, or attempting to comprehend the document. Differing cultural preferences, varying rhetorical styles and discourse patterns, in addition to linguistic elements, can easily cause information overload. This chapter analyzes the impact of culture on information overload, primarily using Hall's high context versus low context model. We demonstrate that high and low context cultures correlate with typical discourse styles, using Kaplan's model as an example. Hinds' theory of Reader versus Writer Responsibility for comprehension provides further insights into possible causes of cross-cultural information overload. A shared professional and/or corporate culture may mitigate the cross-cultural issues in some cases, although it does not eliminate them. Cases from specific cultures—Latin American, Japanese, Chinese, and others—are used to demonstrate the issues discussed in this chapter.

With so many textual elements acting as potential problems for international communicators, translating documents from one language to another is often not enough. Localization, or making a

Information Overload: An International Challenge for Professional Engineers and Technical Communicators, First Edition. Edited by Judith B. Strother, Jan Ulijn, Zohra Fazal.

product linguistically and culturally appropriate for the target audience, makes an important contribution to reducing the cognitive load. The significant conclusion is that the major cause of information overload in cross-cultural situations is having to process information in different discourse patterns from different cultural frameworks. This area is in need of solid empirical research to determine the nature and extent of the load, which will inform practices to reduce it.

5.1 Introduction

In so much of our discussion of information overload, the focus is on the technology and what it has brought us—e.g., too many e-mails, 24/7 access to information on the Internet, and almost compulsive connections to our phones and all manner of social networks. In this chapter, we remind ourselves that to really understand what is causing information overload, we cannot neglect the cultural element—what Hofstede so insightfully called *the software of the mind* [1]—in all forms of communication.

Many facets of intercultural communication can affect information overload. Perhaps, the most noticeable occurs when the communicators are using different languages. Just the cognitive effort to communicate in a second or additional language adds to the possible information overload caused by the volume and/or complexity of an individual message. In addition to Hofstede's work [1], there have been a number of studies about the various aspects of culture and their effects on the communicators and the communication events. (See, e.g., the work of Trompenaars [2], Triandis [3], Smith *et al.* [4], and Schartz [5].)

With the recognition of the many interlinguistic issues and the resulting cognitive effects, this chapter focuses on three important cultural factors that can affect information overload—varying discourse patterns, high versus low context cultures, and the related issue of reader-centered versus writer-centered text. We also look at the impact of shared professional culture as a potential mitigating factor in information overload for intercultural engineers and technical communicators.

Almost all professional communicators face linguistic and intercultural challenges as they deal with information. These challenges may result from work with multinational companies, work in different countries, or work in a diverse workplace within one's own country. The latter may apply especially in areas where groups of cultural minorities live and work, for example, in parts of the United States, where Hispanic or Asian populations exist, or in parts of Europe, where Turkish and Moroccan populations are prevalent. However, most countries reflect the result of today's increasingly mobile population, with the result that the diverse workplace is no longer limited to areas with concentrations of ethnic populations.

The proliferation of the Internet is exponentially increasing the frequency of cross-cultural business communication. Because the Internet now connects people around the world, a number of researchers have hypothesized that all forms of electronic communication could somewhat level or equalize discourse styles, thereby decreasing intercultural misunderstandings [6]. However, McFadden and Richard [7, 8] conducted two studies of Chinese and American business associates exchanging e-mails within a corporate setting. They found definite stylistic differences and discovered fairly strong reactions to differences in style, leading to a greater desire to do business with someone who exhibited similar rhetorical patterns and politeness conventions in the e-mail exchanges. As Hofstede [1] stated and other studies have clearly demonstrated, the Internet does not necessarily act to reduce cultural differences and in some instances may help solidify and exaggerate cultural values and communication norms [6, 9].

Brewer [10] conducted a multinational study of the causes of miscommunication in virtual workplaces with participants from Sweden, the United States, France, the

Philippines, Singapore, Canada, and Australia. She found that culture was specifically identified as a cause of miscommunication. In addition, a number of other elements also revealed underlying cultural influences. For example, under the broad heading of Information Sharing, "a lack of clear detail, incorrect assumptions about receiver knowledge, disparity of information, unnecessary information, volume of correspondence, and missing information" were listed [10]. Lack of detail provided in a communication event caused a number of issues, from frustration about knowing how to complete a problem to causing extra time and additional costs to track down the information needed to complete a project. There is no doubt that these kinds of problems could contribute to the information overload issue. It would also be logical to conclude from the above study that high versus low context cultural backgrounds (discussed in Section 5.4), which have a strong influence on the amount of detail provided, are a contributing factor to these communication problems and therefore to information overload.

5.2 Levels of Culture

As many researchers have observed, there are multiple levels of cultures and subcultures—from supranational cultures (regional, ethnic, religious, linguistic) to multiple kinds of group cultures or communities of learning of specific professionals, such as engineers, economists, lawyers, and medical doctors. Karahanna *et al.* [11] visualized the interrelated levels of culture as shown in Figure 5.1. In their efforts to integrate the different levels of culture and individual behavior, they propose an overall perspective for any global

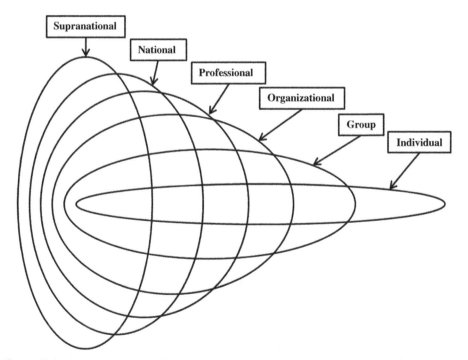

Figure 5.1 Interrelated levels of culture.There are multiple interrelated levels of cultures and subcultures—from supranational (regional, ethnic, religious, linguistic) cultures to multiple kinds of group cultures. This figure helps to illustrate the complexities of culture and it can contribute to the challenges of information overload [7].

information receiver, which, for the current audience, refers to engineers and technical writers (group level) within the scope of their firms (organizational level). The individual level deals more with the psychological level rather than any specific cultural level, an analysis which is beyond the scope of this chapter [12, 13].

Figure 5.1 helps to illustrate the complexities of culture and how challenging it is to discuss its effects on information overload in such a short chapter. However, the contribution of culture to all aspects of communication cannot be ignored. In our discussion, we limit this topic to the main elements of discourse patterns and national or regional cultures.

5.3 Cultural Patterns of Discourse Organization

There is no doubt that in addition to individual characteristics that influence the way text is organized and processed, a person's ethnic and national culture and the related language influence the way he or she deals with information. One of the earliest studies in this area was by Kaplan [14], who analyzed compositions written in English by students with different cultural backgrounds, such as French, Spanish, and Russian. These compositions, written on the same subject, had paragraph structures that diverged in a systematic way from typical linear paragraph structure commonly used in countries such as the United States, the United Kingdom, and The Netherlands, and therefore, reflected a different line of thought. Kaplan then correlated these different text structures with the historical typology of language

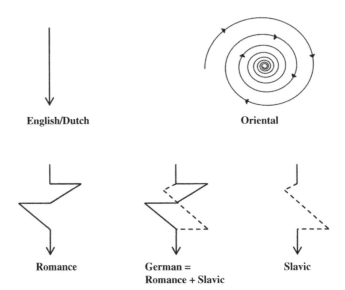

Figure 5.2 Examples of cultural styles of discourse organization. The English (U.S. and U.K.) and Dutch patterns of thought are linear, focused, direct, and monochronic, handling only one thing at a time. The Romance approach (e.g., French, Spanish, and Italian) is polychronic, discussing several things at the same time. In the Slavic culture, the rather long digressions seem to be irrelevant to those unfamiliar with this pattern of discourse. German discourse is a combination of the Romance and Slavic approaches, while the Asian cultures (e.g., Japan, China, and Korea) have an indirect approach, moving in a circular pattern as the writer approaches the subject indirectly [10].

families, such as Romance, Germanic, and Slavic. Figure 5.2 presents a visualization of the differences in styles of organization for some cultural groups.

The English (U.S. and U.K.) and Dutch patterns of thought are linear, focused, and direct without digressions, and they are monochronic, handling only one thing at a time. (From now on, in this chapter, this style is referred to as a linear textual pattern.) The Romance approach (e.g., French, Spanish, and Italian) allows for digressions. These "side paths" are clear and fit into a rational line of argument, which is polychronic, discussing several things at the same time. In the Slavic culture, the rather long digressions seem to be irrelevant to the central topic to those unfamiliar with this pattern of discourse. German discourse is a combination of the Romance and Slavic approaches and accepts both functional and nonfunctional digressions. In contrast, the Asian cultures (e.g., Japanese, Chinese, and Korean) have an indirect approach, moving in a circular pattern as the writer approaches the subject indirectly [14, 15] (referred to, hereafter, as an indirect discourse organization or textual pattern).

Understanding the major storytelling patterns of other cultures can inform how the reader can unpack technical information as well. The challenge comes to a reader who is unfamiliar with the cultural pattern of a text he or she is trying to understand. For example, an Irish engineer (who is familiar with linear writing) may be trying to find the main points from a complex technical report written by the company's Vietnamese subcontractor (who is more likely to use indirect structures). Because the Irish engineer has to cope with an unfamiliar discourse structure and possible second language errors, in addition to the complex technical content, he or she can easily experience information overload. We need to understand how various communication barriers can cause information overload, as defined in Chapters 1 and 2. In the following sections, we examine how differing discourse patterns as well as high and low context cultural patterns can influence the load of handling various forms of communication.

5.4 High Context Versus Low Context

In addition to contrasting patterns of discourse structure, another important way to analyze cultural differences in information structure develops from the work of Edward Hall [16, 17] and his concepts of high context (HC) and low context (LC) cultures. His early work was with Native Americans in the United States, and he later expanded his analysis to European companies in France and Germany using the same anthropological methods.

The theory of high and low context of information relates to linguistic and cultural differences, specifically to implicit versus explicit ways of communicating. Implicit corresponds here to the indirect discourse organization mentioned above; explicit reflects rather linear textual patterns. Those from high context cultures know implicit sources within the culture itself because they are more homogeneous in sharing nontext-based common knowledge. Communicators from these cultures tend to prefer indirect patterns of organization in text, fewer details, and higher levels of politeness and ambiguity. While these people may need some more details when dealing with new or unfamiliar material, they generally prefer communication without detailed explanations. In contrast, people from low context cultures, which are quite heterogeneous, share much less knowledge, so everything must be explained clearly and in much more detail. Communicators from low context cultures usually prefer messages that are direct and concise [18].

Most Asian and South American cultures are high context. In contrast, North America, especially the United States and Canada, as well as Northwest Europe, for example, the United Kingdom, Germany, The Netherlands, and the Nordic countries, are low context. However, it appears that the high–low context determination is not a simple dichotomy. Ulijn [19] suggested a continuum where only the poles are clear: the Swiss Germans are the champions of low context according to Edward Hall, and the Chinese and Japanese lead the high context cultures, but the position of French and British cultures in Europe is not as clear. With that caveat, we present examples of high context cultures (Latin cultures from several South American countries, Mexico, and France, and Asian cultures from China and Japan) and examples of low context cultures (from the United States, The Netherlands, and Germany).

Interestingly, the national cultures involved in the different patterns of text organization (shown in Figure 5.2) correlate with high and low context. For instance, the linear approach from Figure 5.2 reflects a low context culture demanding an explicit message and concrete facts, where the external conditions of the communication are less important. To illustrate, if one is from a low context culture, there may be much more cognitive overload when one tries to understand the real meaning of a communication from a person in a high context culture without benefit of explicit information as in the earlier example of the Irish (low context) engineer trying to cope with a Vietnamese (high context) document.

High context readers from Asian and Latin cultures with the digressions in the text as illustrated in Figure 5.1 can deal with an implicit message. As a case in point, a Chinese reader will see that the circular way of writing about a topic without mentioning it explicitly will lead eventually in the end to a very clear problem description. A western reader, used to the perspective of a linear text structure, expects this problem description to be put up front. One might assume that because of their ability to process diverse information in routine communication experiences, people from high context cultures may not experience information overload as quickly as those from low context cultures. The external conditions, such as the relationship with the writer, are usually more important. Since the oral often overrides the written communication, information may flow quickly; hence, there may be less risk of information overload.

Japanese and Chinese readers might need implicit information from oral sources in order to connect with their communication partners (relationship building). This is very much the tradition of an oral culture, where the spoken word can be trusted, even without written contracts. (See the Japanese example below of how high context communication can be very quick.) A high context Chinese reader who accepts oral information as a routine source of information on which to base business decisions may ignore the written checks and controls that a western low context reader demands. This detailed text may cause information overload for the high context reader, for instance, in the case of intellectual property or engineering specifications, where only the written text counts.

Fussell *et al.* [20] concluded that high context communicators are expected to count more on *affective* trust through relationship building, whereas low context communicators depend more on *cognitive* trust, based upon one's intelligence, competence, and reliability, mainly because these are very task-oriented cultures.

For both high context and low context readers, computer-mediated communication, such as instant messaging, texting, teleconferencing, and video conferencing, is often helpful in reducing intercultural differences, since misunderstandings can be resolved more quickly and rather informally. However, professionals who depend on highly detailed documents (low context), such as attorneys, may not be able to take full advantage of these techniques to resolve complex differences. In addition, those in high context cultures who depend heavily on relationship building may not be able to take full advantage of these short-term solutions.

In studies of annual joint design projects between students from The Netherlands (low context) and Hong Kong (high context), both national and professional cultures led to challenges [21, 22]. The Dutch were fourth year industrial engineering students; in contrast, those from Hong Kong were business and accounting students. Both teams used electronic group support systems, e-mail, and NetMeeting with each other, and they used video conferencing at the beginning and middle of the term. As the Dutch focused on meeting deadlines, they were more distant with each other (within group); however, the Chinese took action only when all uncertainties were resolved through constant interaction. The Dutch suffered more from lack of face-to-face contact and poor interaction with the Chinese, while the Chinese reported time pressure and technical issues to be more problematic. In both cases, increased videoconferencing could have helped resolve their differences, which could have reduced the overall load from both sides [21, 22].

In many cases where differences in cultural framework (high versus low context) and discourse structure exist, localization of content may help at least decrease the probability of information overload caused by intercultural differences.

5.5 Internationalization Versus Localization

When most people think of translation, they often think of only the linguistic elements—taking words from one language and finding equivalents in the other language. However, cultural translation (a key part of localization)—making the discourse structure and overall tone of the text appropriate for the target culture—is critical.

An increasing amount of research is being done on the topic of internationalization (or globalization) and localization of communication products. However, much of the focus, for example, is on information technology, with consideration of character sets and expansion of text due to translation issues. However, the content of the text itself must be localized because of cultural issues among other considerations. Although the Localization Industry Standards Association (LISA) dissolved in Spring 2011, we can use the LISA Educational Initiative Taskforce (LEIT) guidelines to briefly define these key terms [23, 24]. Localization involves taking a product and making it linguistically and culturally appropriate to the target locale (country/region and language), where it will be used and sold. A similar definition for localization is the process of creating or adapting a product to a specific locale, i.e., to the language, cultural context, conventions, and market requirements of a specific target market. Internationalization is the process of generalizing a product so that it can handle multiple languages and cultural conventions without the need to redesign. LISA used the acronym GIL to describe the combination of globalization, internationalization, and localization, referring to all aspects of the process of taking products to an international audience.

Currently, the Globalization and Localization Association (GALA), with a membership of 280 companies from 50 countries, is the world's largest localization organization [25]. GALA stresses that the need for a local mindset is an indispensable part of company's strategy to reach a target market in foreign countries. As the organization purports, "Communicating locally is necessary to succeed globally" [25].

One decision engineers and technical writers have to wrestle with is whether to do a straight translation of text for other countries or whether to localize the text, i.e., do a cultural translation. For efficiency and budgetary reasons, engineers and technical communicators might be tempted to universalize their products and related documents as much as they can, but globalization in this sense is not always an advantage. Recent studies [26–28] indicate that national/ethnic cultural localization might reduce information overload. Localization to a specific national cultural model, in addition to creating more successful business models, can help an organization's members prevent information overload. With regard to international audiences, the high–low context distinction remains an important element in transmitting information.

When document designers localize documents for readers in other cultures, they should consider what text organization will best suit those readers. Let us look at research on three cultures with distinct cultural patterns to help us understand some of the issues.

5.5.1 Latin America

The area covered by the label "Latin America," which includes Central and South America, while very similar in many cultural patterns, also has great diversity. For the sake of our discussion of general cultural differences between Latin America and other cultures, such as the U.S., European, or Asian cultures, we state the caveat that we are generalizing, but include specific examples to demonstrate the points.

In general, the Latin American cultures are high context, which means they do not always need very specific details, especially when dealing with familiar topics. Thatcher's work on adapting information for South American readers (especially in Venezuela, Colombia, Ecuador, Peru, and Bolivia) demonstrated that there were a number of different rhetorical patterns as well as cultural influences that caused misunderstandings and/or challenges for both the Latin American readers as well as for the people from the United States with whom they were communicating [28]. One of the most challenging characteristics was the tradition of orality (conveying most information through oral channels and using an oral-like writing style) by the Latin Americans. This meant that they had a challenging time processing the direct, highly detailed written materials produced routinely by the low context U.S. writers. On the other hand, the U.S. communicators were challenged to understand the very different writing style of their South American counterparts.

This study uncovered a number of other factors that influenced the challenges and misunderstandings, including the purpose technical documents serve in each culture as well as cultural factors such as Hofstede's dimensions of individualism/collectivism and power distance [1]. Both groups were experiencing information overload caused by the extra cognitive load of struggling with different rhetorical structures of the information they had to work with. Thatcher confirmed these same kinds of intercultural issues in his study of joint U.S.–Mexico manufacturing facilities, called *maquilas*. Interestingly, in the four maquilas he studied, he found little hybridity or adaptation between the two cultural styles of communicating [29]. This certainly points out the need for much more research in this area to determine what kinds of strategies are necessary to bridge intercultural and rhetorical differences.

Localization of texts for a specific ethnic culture in a country with a different mainstream culture might be less needed than writing for a national culture with a completely different educational system or diverging information needs. St. Germaine-Madison [30] tested 35 focus groups, which included five Hispanic focus groups (conducted in Spanish), to determine the best approach to communicate information about the dangers of the Human Papilloma Virus (HPV) to the Hispanic population in the

United States. In some cases, members of this group preferred some text that reflected their traditional pattern of organization, a more indirect rather than linear style. However, because of their experiences in the United States, where highly detailed low context texts are common, this audience dealt better with the brochure in a low context style than if it had been in a simple but implicit version. In this case, the low context, more explicit text, spelling out all the details, worked better for Spanish high context speakers located in a low context country than the text with their own more Latin orientation, where fewer details (high context) are more common. For this group, how long they had been in the United States and what kind of educational system they went through may have been influencing factors.

5.5.2 Japan

Nonaka and Takeuchi [31] developed a theory of an ideal knowledge-creating company, based on the best attributes of the Japanese and western cultures and their corresponding typical corporate cultures. They address how each culture has its own communication styles, which often conflict or cause misunderstandings. For example, they mention that the Japanese tend "to overemphasize the use of figurative language and symbolism at the expense of a more analytical approach in documentation." Our audience of technical writers may appreciate the old example of a manual for the first Apple computer introduced in Japan. It started with a fairy tale about an orchard of apple trees and ended up with one apple and one computer being the same [32]. As in this example, too much figurative language and symbolism, which work well in Japan, can also be an indirect cause of information overload as those unfamiliar with this style struggle to understand the main message.

Spyridakis and Fukuoka [33] examined American and Japanese readers' comprehension of and preference for expository text that contains a thesis and is organized either inductively or deductively. The results revealed that while Americans performed equally well with either organizational structure, Japanese readers recalled more information from the inductively organized text. The implications for document designers in English- and Japanese-speaking countries are that, on the one hand, localization of text could prevent information overload for readers from implicit high context traditions. Reader-responsible cultures prefer inductively organized texts, for example, Japan [33], China [26, 34], France [27], and Latin America [28, 35]. On the other hand, if the same cultures are exposed long enough to explicit low context cultures, such as the mainstream Anglo-Germanic one in the United States, such localization may not be as important.

5.5.3 China

Honold [36] conducted a study comparing Chinese and German subjects learning how to use a cellular phone in their home country. Neither group had been trained to use a cell phone in the other country. This study found a clear difference between learning objectives, learning styles, and high context/low context cultural dimensions. The majority of the Chinese users had a definite affinity for depending on relationships for learning about their new cell phones, which is very collectivistic and indicates a high context society. They were quite pragmatic about their learning objectives, feeling that learning only the basic features they needed was enough. For this group, reading the instruction book was a last resort. In contrast, the more individualistic Germans (low context) went directly to the formal instructions as the way to learn.

Another interesting finding in Honold's study was that the Chinese have a much stronger picture orientation, preferring to depend on visuals rather than text for information and instructions. In contrast, the Germans preferred complete and consistent textual information. In this case, it is easy to imagine that information overload could happen for each of these cultural groups if the instruction manual for the cell phone did not meet expectations. For example, the Chinese users could be overloaded with text when they need graphics to better understand the procedure for operating their new phone.

When studying the influence of U.S. and Chinese national cultures on the style of communication and types of knowledge-sharing activities in virtual communities, Siau *et al.* [37] found that Chinese communicators depend more on relationships; thus, they generally had a more closed system of information exchange than the U.S. communicators, who had a more open system of knowledge exchange. In other words, the Chinese participants were more likely to share information with people with whom they already had an establishing relationship, whereas the U.S. participants in the virtual communities were more willing to share information, whether they knew the receiver or not. This is another example of potential information overload (or underload) as members of each group are sending and/or receiving an amount of information considered inappropriate or uncomfortable by members of the other cultural group.

Websites provide additional elements of virtual information exchange. Barnett [9] conducted a longitudinal review of websites from both China and the United States. He reported that U.S. websites were constructed in a significantly lower context structure compared to Chinese websites. Other studies have confirmed this difference in high versus low context in Asian versus American websites [38–40] as well as in other forms of written communication [41–44]. Fujimoto *et al.* [6] found that when individuals from high context cultures communicate with those from low context cultures, they are less satisfied with the communication and are often stressed as a result. As mentioned earlier, McFadden and Richard [7, 8] confirmed this finding.

These cases illustrate that while people from different cultural and linguistic backgrounds are attempting to communicate to create understanding, linguistic and/or cultural interference can contribute to cognitive barriers, therefore leading to the perception of information overload. One might conclude that localization (writing text in either the low context or high context style of the target audience) meets the information load level of readers better than universal texts, which are usually written in a low context style for global audiences.

5.6 The Effect of Professional Culture

Organization or corporate culture deals with the culture of a particular company and can affect the degree of openness of communication, attitudes toward human error, and level of trust between managers and employees. Corporate culture may operate within or across national borders. A professional culture is shared by those who are in the same career field since they share much common knowledge. As cases in point, we provide some examples from the United States and the global aviation sector (airline pilots), Dutch high-tech firms (software engineers, electronic engineers in the military sector, and hydraulic engineers), German and Chinese mechanical engineers, as well as Dutch and French facility managers.

For example, think about the professional culture of commercial airline pilots. Most commercial flights have more than one crew member, and in many cases, crew members come from different cultures. These multinational crews must learn to communicate effectively in order to function as a team. Because of the potential disaster of bad decisions or commands based on flawed information, Cockpit Resource Management (CRM) was implemented [45]. CRM training programs do not deal as much with technical expertise as with the cognitive and interpersonal skills necessary for situational awareness, problem solving, and decision making within this high-stress environment where interpersonal and communication skills are critical. When CRM operates as it should, any member of the cockpit flight crew, regardless of rank or status, is encouraged and empowered to speak out when necessary. However, because of cultural characteristics such as traditional hierarchical or power distance issues, questioning a decision made by the person in authority, in this case, the pilot, may be unacceptable. For example, cockpit voice recordings of various air disasters tragically reveal cases where the first officer and flight engineer attempt to bring critical information to the captain's attention, but they do so in an indirect and ineffective way. By the time the captain understands what is being said, it is too late to avert the disaster. In some cases, language issues are complicating factors, as was true in the case of the crash of Avianca Flight 52. The flight had originated in Colombia with New York as the destination. Because of delays in the flight and being put in a holding pattern above New York, the plane was running very low on fuel. The pilots were handicapped in communicating with the air traffic controllers because of their limited English. The language issue was compounded because the crew did not know, or at least did not use, standard ICAO phraseology, causing the air traffic controllers to misunderstand the urgency of their situation. The National Traffic Safety Board (NTSB) summarized that in addition to poor weather and fuel mismanagement, the cause of the accident was "the failure of the flight crew . . . to communicate an emergency fuel situation to air traffic control before fuel exhaustion . . . and the lack of standardized understandable terminology for pilots and controllers for minimum and emergency fuel states" [46].

Thus, although CRM, aviation's "safety culture," is an essential part of aviation's professional culture, its implementation is still influenced by national culture. Helmreich *et al.* [47] found that while professional and organizational cultures do have an impact on flight operations, national culture has a much stronger influence.

In today's mobile world, the national, professional, and organizational cultural values are interconnected in complex ways. Following our aviation example, an Asian pilot may have been trained in a western country, such as the United States. However, when that pilot returns to Asia to fly for a commercial airline, the home country's culture of practices may conflict with the western teachings, and thereby affect the cockpit culture for that pilot and the crew. This is just one example of a professional group that may have overlapping and sometimes contradictory cultural values [48].

With a different perspective, in their study of professional communication in a global business context, where English is spoken as a *lingua franca*, Louhiala-Salminen and Kankaanranta [49] made a number of important observations about the influence of culture on the communication process. Among other things, they found that there was a form of solidarity among those speaking Business English as a *lingua franca*, which created somewhat of a professional culture beyond just the business knowledge shared by the participants. Their participants seemed aware of varying discourse patterns and other cultural influences on their cross-cultural interactions. These researchers concluded that "multicultural competence stems from the acknowledgement of factors related to national, corporate, and/or professional cultures as fundamentals of any communicative event, and enables the flexibility and tolerance needed for GCC [Global Communicative Competence] to succeed" [49]. This result helps to confirm the idea that a professional culture could reduce information overload somewhat because of shared backgrounds and knowledge as well as the participants' awareness of the need to adapt to differing culturally influenced communication styles.

What effect do professional cultures have on information overload? When people share a professional culture, they might avoid some information overload because of their common knowledge, which creates a kind of high context, as discussed earlier. Hall and Weaver [50] also suggested that people who are attracted to a certain profession share patterns of cognitive learning skills and styles. These cognitive patterns, as well as other facets of a particular culture—such as values, beliefs, attitudes, and customs—are reinforced through common educational backgrounds and workplace behaviors and are passed down to the neophytes in a given profession and/or organization [51]. (See [52] for a discussion of a professional culture's influence in the field of midwifery; also see Chapter 4.)

In the context of five Dutch high-tech firms involving the professional cultures of software engineers, electronic engineers in the military sector, and hydraulic engineers, Van Luxemburg *et al.* [53] studied the use of computer-mediated communication (CMC) in the design process between supplier and customer. Based on their findings, we can infer that the suppliers and customers have reduced information overload because they share a similar professional culture.

However, we cannot overgeneralize. Wang and Wang [26] conducted a user study with German and Chinese mechanics using their own service manuals for the same injection system of Mercedes-Benz. This study was unique in linking the professional culture of German and Chinese mechanical engineers to the necessary audience analysis and usability testing Mercedes-Benz had to do to export its products to China. The study was done in each country using two manuals with the two different styles in each country. The Chinese version had more implicit text organization (more graphics and less text) and the German one relied on explicit patterns: more text and details and fewer graphics.

The Mercedes-Benz Chinese manuals reflect the *chi-cheng-juan-he* structure [54, 55] given below.

- The beginning, *chi*, would contextualize but would not correlate directly to the English or German topic sentence in a western linear prose structure.
- Next, the *chi* and *cheng* elements would be very general.
- The *juan* element would provide a digression (such as the *ten* in Japanese). In the *juan*, there is a connection to the main topic, but since it does not seem directly related, it is confusing for western readers.
- Finally, the *he* element would include all the details and the conclusion in a very inductive style.

In the automotive example, this means that the Chinese writer would come to the point of troubleshooting much later in the text, which would not seem as logical for the western reader. On the other hand, a German version would have these details up front, which could seem confusing to a Chinese reader. Those findings confirm the high–low context distinction between these two cultures, evidenced earlier by Ulijn and St. Amant [34], who also used German and Chinese subjects in a study of the interpretation of oral negotiation discourse. Therefore, while technical communicators can use the common professional culture to reduce information overload, they should be prepared for different national cultural styles, in this case between Germany and China. In many cases, the engineers may have not been trained in the same way in their respective countries, and they probably have different communication styles, especially relating to discourse structure.

Another example of the need for localization comes from a French client purchasing a Dutch coffee maker. The French client wanted the Dutch coffee maker's manual localized through cultural translation. In a simple organization pattern of (1) technical data, (2) operation, (3) maintenance, and (4) troubleshooting, the Dutch writers preferred to put the operation up front; however, the French are more used to having the technical data up front. When the two versions of the manual were tested with both Dutch and French readers, for both cultures, the readers were able to complete a reading-to-do task faster in their own cultural structure [56]. In this case, the Dutch and French share a fairly limited and multilayered professional culture, which ranges from facility managers, who have to operate professional coffee makers in restaurants, to technicians, who have to serve as troubleshooters if the apparatus fails to do its work.

For all of these examples, each group was able to better handle text that was written in the discourse style that the readers were familiar with. This implies that corporations must use this knowledge when they are writing documents for international audiences.

5.7 Japan and U.S. Discourse Structures

Section 5.5.2 outlines the way Japanese see sharing and increasing professional knowledge. While this probably has consequences for the development of professional cultures within Japan, in this chapter, we limit our discussion to the consequences for the technical communicator, in this case, for writers from the United States producing text for Japanese audiences. Using Spyridakis and Fukuoka's results [33] (see Section 5.5.2), it is easy to see that the indirect, circular, high context, inductive, implicit style of Japanese communicators contrasts directly with the linear, low context, deductive, explicit style reflected in American English.

Perhaps, Hinds' theory about Reader versus Writer Responsibility [55] can help explain text differences here so that a technical communicator could act as a bridge builder between two cultures. Hinds points out that a key issue in crafting any text, especially for those used to the English pattern of organization, is unity. He emphasizes that transitions and landmarks are essential and help the reader follow the pattern of organization of information presented in the text. Since these transitions and landmarks may be scarce in, for example, Japanese text, increased cognitive load may result on the part of English readers from text that is organized according to Japanese rhetorical organization patterns.

The example of the Apple computer manual introduced in Japan, mentioned earlier, illustrates this indirect, quasi-inductive expository style. For western readers and clients, this kind of indirect story may give the impression of lack of focus. Hinds [55, 56] made this point of focus, along with unity and coherence, explicit in his translation experiment. A reader-responsible culture, such as the Japanese one, will look in an American text for focus, unity, and coherence in a Japanese way, which is through the *ki-shoo-ten-ketsu* discourse structure. In this style, *ki* introduces the topic, *shoo* develops the topic, *ten* forms an abrupt transition or a vaguely related point, and *ketsu* concludes the topic. A U.S. reader, coming from a writer-responsible culture, might blame the Japanese writer for causing miscommunication by using so much indirect language, even if the text is translated literally into English.

Consider the example of a popular Japanese newspaper column, "Tensei Jingo," which reported on a car accident in a tunnel in northern Japan. The article was both literally and culturally translated into English. The literal translation kept the traditional Japanese discourse structure, in which the *ten* part had considerable digression, with percentages about car accidents during the last 10 years in Japan. Both Japanese and English readers were asked to evaluate the two versions with regard to focus, unity, and coherence, and not, surprisingly, each group preferred its own discourse structure. For the native English readers, the *ten* part of the Japanese version, which had a connection but not a direct association with the rest of the text, simply did not meet their western expectations of focus, unity, and coherence of a text. This could have increased cognitive load required to process this different style of text certainly contributed to information overload. Other studies have confirmed that comprehension is definitely affected by rhetorical patterns and discourse markers (e.g., see [54, 58, 59] for several languages, including Chinese).

In another study, Maitra and Goswami [60] asked American readers who had been trained in document design to analyze Japanese annual reports that had been translated into English. The Japanese reports had been designed carefully to reflect the cultural framework of valuing aesthetics and ambiguity. However, the American readers used their own perceptual framework when reading these reports. This meant they expected a clear logical line of reasoning and images that were purposeful and conveyed specific information to support the text. This group of readers felt the images lacked function and were too "flashy" whereas the Japanese readers believed the images conveyed important cultural values. (For additional information on visual communication, see [61] and Chapter 6.) The above might suggest that the Japanese version is illogical, nonpurposeful, and difficult to understand. Although this may be the perception of the western low context reader, it certainly is not the perception of the Japanese.

5.8   Cultural Issues in Reader Versus Writer Responsibility

Hinds' [55, 56] landmark typology dealing with reader versus writer responsibility within different cultures purports "that in some languages, such as English, the person primarily responsible for effective communication is the speaker, while in other languages, such as Japanese, the person primarily responsible for effective communication is the listener." For English speakers, this means that the speaker or writer has the responsibility to clearly communicate his or her message explicitly. In contrast, for Japanese speakers, the listener or reader is responsible for decoding and understanding what the author's message is. This reader versus writer responsibility can be correlated with high versus low context cultures.

Thus, any breakdown in intercultural communication would be analyzed very differently depending on the national culture of both the sender and the receiver. Consider the following example: a woman from the United States was taking a taxi to the Ginza Tokyo Hotel in Japan. The taxi driver mistakenly took her to the Ginza Dalichi Hotel. She said, "I'm sorry, I should have spoken more clearly." The taxi driver responded, "No, I should have listened more carefully" [55]. Although the politeness conventions of the Japanese culture probably influenced the taxi driver's response, it is clear that the driver was taking responsibility for his perceived role in the conversation.

Communicators from high context cultures, such as the Japanese client/reader, who consults product information or the Japanese taxi driver in the above example, tend to feel more responsible for digesting the text themselves rather than depending on the communicator to make all information explicit. On the other hand, people from low context cultures, such as the U.S. female taxi passenger or a German engineer reading product information, feel the ones originating the communication (speakers or writers) are definitely responsible for conveying all the necessary details. In cases where international professional communicators do not provide enough contextual information, these people from low context cultures may be more likely to feel information underload.

When a technical professional from a low context culture, such as the United States, writes a highly detailed text (a marketing message or user manual for a new American product), people from different cultures could have dramatically different reactions to those texts. If those texts were to be read by a Japanese reader (high context culture), who had difficulty understanding, that reader may respond, *I am experiencing information overload. I must not be intelligent enough to understand this* (reader responsibility) *and therefore I will not buy it.* A U.S. reader facing the same situation would blame the writer for any information overload and may respond, *I am entitled to explicit, clear information without overload or underload* (writer responsibility) *and without that, I will not buy it.* The (negative) marketing result might be the same for both people: *If the product information does not meet my cultural expectations, I will not buy that product.*

The same issue extends to the situation of the readers of, for example, German product information if they are not fluent enough in that language. If they do not understand the information, they will not blame the writer, but consider themselves to be not intelligent or not educated enough, and hence, may not buy the product, a case of reader responsibility. In contrast, a low context American reader/client may sue the product information provider, often a technical writer and/or engineer, for not providing clear enough information (writer responsibility).

5.9 Implications for Engineers and Technical Communicators and Their Corporations

What are the consequences for the writing process of an engineer or a technical communicator working on a global scale? In all cases, international professional communicators will face intercultural differences, which result not only from national but also from professional and corporate sources. The high context of the same nation, same ethnic group, the same profession, or the same corporate environment might reduce information overload because, in those cases, not everything has to be spelled out as explicit information. According to a Dutch proverb, *a good listener needs only half a word.*

However, in most cases, localization of information can certainly help a corporation serve its clients better. For example, as mentioned with high versus low context cultures and various discourse patterns, producing text in a form that readers are familiar with reduces cognitive load and therefore may decrease the amount of information overload. Too often, companies are tempted to take their documents and make a direct, linear translation from one language into another. However, the above-cited examples of culturally diverse text organizations demonstrate that cultural translation—making the appropriate cultural adaptations to the original text—is essential.

It is well established, as shown by the research results presented in this chapter, that there are culturally different discourse patterns that are evidenced when producing or processing text. It is also logical to think that processing text from a culture that uses a different discourse pattern creates a heavier cognitive load than processing text in familiar discourse patterns and that increased cognitive load can contribute to information overload. Therefore, technical writers must understand that it is very difficult to globalize all text. As demonstrated through this chapter's examples, some localization is critical for linguistic and cultural regions that have contrasting rhetorical patterns and cultural contexts. However, as noted in the Latin American study of Hispanics living in the United States, sometimes, a more specific kind of localization (e.g., adapting to patterns of organization taught in a certain educational system) may be even more important than national or ethnic localization.

For knowledge creation, companies must "have the organizational capability to acquire, accumulate, exploit, and create new knowledge continuously and dynamically, and to recategorize and recontextualize it strategically for use by others in the organization . . . " [31]. For example, consider a Dutch technical information firm, Tedopres, which specializes in troubleshooting software for the aviation industry. In this case, 15 Indonesian engineers had to cooperate with 15 Dutch technical writers who had the task of localizing the software documentation for international clients through both linguistic and cultural translation. Engineers know their product very well, but the organization often needs to employ technical writers to prepare the content, in this case, the product knowledge, in such a way that their clients can use it successfully. To do so, the engineers (Indonesian) and technical writers (Dutch) within the organization had to find an effective way to communicate with each other so the technical writers could understand the product information well enough to make the cultural translation [12]. It took six months of training for the Indonesian software engineers and the Dutch technical writers to be able to communicate effectively enough to prepare the required aviation software documentation.

This implies that strong communication systems must be in place to control the quantity as well as quality of information so that information overload is not a constant threat both up and down the chain of command. Nonaka and Takeuchi " . . . believe that the future belongs to companies that can take the best of the East and the West and start building a universal model to create new knowledge within their organizations. Nationalities will be of no relevance. . . ." [31]. Needless to say, this is also true for cultures from the North and South, where differences can be just as dramatic. Although this may seem extreme, it is certainly true that to reduce intercultural interference in communication, all communicators will have to find ways to adapt—meet in the middle—to reduce both cognitive and information overload. Audience analysis has always been an essential tool for the technical writer and, as this chapter has demonstrated, analyzing the reader's cultural framework is an essential part of this analysis.

Technical communicators must increase feedback opportunities to fully address issues leading to miscommunication. An open dialog can at least begin to address information overload issues among all parties of cross-cultural communication events. Computer-mediated communication (CMC), such as videoconferencing and teleconferencing and instant messaging, can facilitate this process in real time.

It is essential to train all technical communicators and writers about cultural elements, such as reader expectations of discourse organization as well as amount and kind of textual detail expected and needed to process a text without excessive cognitive load. It is also important to encourage better communication and teamwork between engineers, the experts on the technical information, and those who must write technical documents and make them readable for a multicultural audience.

5.10 Conclusion

This chapter barely scratches the surface of the impact of culture on information overload as it is relates to the way people receive and decode texts and other forms of information from a variety of national rhetorical styles, cultural preferences, and professional and corporate cultural styles. It might, however, give practicing engineers and technical and professional communicators an awareness of how to decrease the cognitive load and therefore to reduce potential information overload.

Discourse patterns and high versus low context cultural frameworks give solid insight into elements necessary to reduce the effort it takes to process cross-cultural oral and written discourse. Perhaps, the most important conclusion is that these patterns *within* a culture are not so important. It is the extra cognitive load caused by having to process information in *different* discourse patterns from *different* cultural frameworks that is a major factor in information overload. The examples given in this chapter clearly demonstrate that corporations must localize their product information and that localization must include thorough audience analysis, consideration of cultural discourse structural preferences, as well as careful usability testing. To produce successful documents, engineers and technical managers must attempt to match the expected discourse structure and therefore the cognitive load of the target reader to truly reduce information overload.

REFERENCES

[1] G. Hofstede, *et al.*, *Cultures and Organizations: Software of the Mind*, 3rd ed. New York: McGraw-Hill, 2010.

[2] F. Trompenaars, *Riding the Waves of Culture: Understanding Cultural Diversity in Business*. London, U.K.: Nicholas Brealey, 1994.

[3] H. Triandis, *Culture and Social Behavior*. New York: McGraw-Hill, 1994.

[4] P. B. Smith, *et al.*, *Understanding Social Psychology Across Cultures: Living and Working in a Changing World*. London, U.K.: Sage, 2006.

[5] D. G. Schwartz, *et al.*, *Internet-Based Organizational Memory and Knowledge Management*. Hershey, PA: Idea Group Publishing, 2000.

[6] Y. Fujimoto, *et al.*, "The global village: Online cross-cultural communication and HRM," *Cross-Cultural Manag.: Int. J.*, vol. 14, no. 1, pp. 7–22, 2007.

[7] M. McFadden and E. M. Richard, "Cross-cultural differences in business request emails," *presented at the 25th Annual Society for Industrial and Organizational Psychology*, Atlanta, GA, Apr. 2010.

[8] M. McFadden and E. M. Richard, "Cultural and individual level differences in business request emails," *J. Appl. Psychol.*, unpublished.

[9] G. A. Barnett, "A longitudinal analysis of the international telecommunications network: 1978–1996," *Am. Behavioral Scientist*, vol. 44, no. 10, pp. 1638–1655, 2001.

[10] P. E. Brewer, "Miscommunication in international virtual workplaces," *IEEE Trans. Prof. Commun.*, vol. 53, no. 4, pp. 1–17, Dec. 2010.

[11] E. Karahanna, *et al.*, "Levels of culture and individual behavior: An integrative perspective," *J. Global Inform. Manag.*, vol. 3, no. 2, pp. 1–20, Apr.–Jun., 2005.

[12] J. Ulijn and M. Weggeman, "Towards an innovation culture: What are its national, corporate, marketing and engineering aspects, some experimental evidence," in *Handbook of Organisational Culture and Climate*, C. Cooper, S. Cartwright, and C. Early, Eds. London, U.K.: Wiley, pp. 487–517, 2000.

[13] J. Ulijn and T. Brown, "Innovation, entrepreneurship and culture, a matter of interaction between technology, progress and economic growth," in *Entrepreneurship, Innovation and Culture: The Interaction Between Technology, Progress and Economic Growth*, T. Brown and J. Ulijn, Eds. Cheltenham, U.K.: Edward Elgar, pp. 1–38, 2004.

[14] R. B. Kaplan, "Cultural thought patterns in intercultural education," *Language Learning*, vol. 16, no. 1, pp. 1–20, 1966.

[15] J. Ulijn and J. Strother, *Communicating in Business and Technology: From Psycholinguistic Theory to International Practice*. New York/Frankfurt, Germany: Lang, 1995.

[16] E. T. Hall, *Beyond Culture*. New York: Doubleday, 1976.

[17] E. T. Hall, "Three domains of culture and the triune brain," in *The Cultural Context in Business Communication*, S. Niemeier, C. P. Campbell, and R. Dirven, Eds. Amsterdam, The Netherlands: John Benjamins, pp. 11–30, 1998.

[18] K. Leung, *et al.*, "Culture and international business: Recent advances and their implications for future research," *J. Int. Business Studies*, vol. 36, no. 4, pp. 357–378, 2005.

[19] J. Ulijn, "Is time still money? Why a new construct of innovation culture in an East-West setting would need inclusion of the concept of time?" in *Strategie Maakt het Verschil (Strategy Makes the Difference)*, G. J. Melker, W. Have, N. Filipovic, and F.van Eenennaam, Eds. Amsterdam, The Netherlands: Mediawerf, pp. 88–109, 2009.

[20] S. R. Fussell, *et al.*, "Global culture and computer mediated communication," in *Handbook of Research on Computer Mediated Communication*, S. Kelsey and K.St. Amant, Eds. New York: Information Science Reference, vol. 2., pp. 901–916, 2008.

[21] D. Vogel, *et al.*, "Exploratory research on the role of national and professional cultures in a distributed learning project," *IEEE Trans. Prof. Commun.* vol. 44, no. 2, pp. 114–125, Jun. 2001.

[22] A. F. Rutkowski, *et al.*, "E-collaboration, the reality of virtuality," *IEEE Trans. Prof. Commun.*, vol. 45, no. 4, pp. 219–213, Dec. 2002.

[23] Localisation Industry Standards Association. (2010) [Online]. Available: http://lisa.org/leit/terminology.html (Website no longer active.)

[24] Localization Institute Terminology. (2011) [Online]. Available: http://localizationinstitute.com/switchboard.cfm?page=terminology Aug. 22, 2011 (last date accessed).

[25] Globalization and Localization Association. (2011) [Online]. Available: http://www.gala-global.org/about-association Dec. 22, 2011 (last date accessed).

[26] Y. Wang and D. Wang, "Cultural context in technical communication: A study of Chinese and German automobile literature," *Tech. Commun.*, vol. 56, no. 1, pp. 39–50, 2009.

[27] J. Ulijn, "Translating the culture of technical documents: Some experimental evidence," in *International Dimensions of Technical Communication*, D. Andrews, Ed., Washington, DC: STC Press, pp. 69–86, 1996.

[28] B. Thatcher, "Cultural and rhetorical adaptations for South American audiences," *Tech. Commun.*, vol. 46, no. 2, pp. 177–195, 1999.

[29] B. Thatcher, "Intercultural rhetoric, technology transfer, and writing in the U.S.-Mexico border maquilas," *Tech. Commun. Quar.*, vol. 15, no. 3, pp. 383–405, 2006.

[30] N. St. Germaine-Madison, "Localizing medical information for US Spanish speakers: The CDC campaign to increase public awareness about HPV," *Tech. Commun.*, vol. 56, no. 3, pp. 235–247, 2009.

[31] I. Nonaka and H. Takeuchi, *The Knowledge-Creating Company: How Japanese Companies Create the Dynamics of Innovation*. New York: Oxford Univ. Press, 1995.

[32] J. Ulijn, private communication.

[33] J. H. Spyridakis and W. Fukuoka, "The effect of inductively versus deductively organized text on American and Japanese readers," *IEEE Trans. Prof. Commun.*, vol. 45, no. 2, pp. 99–114, Jun. 2002.

[34] J. Ulijn and K. St. Amant "Mutual intercultural perception: How does it affect technical communication? Some data from China, The Netherlands, Germany, France and Italy," *Tech. Commun.*, vol. 47, no. 2, pp. 220–237, 2000.

[35] S. Koeszegi, *et al.*, "National cultural differences in the use and perception of internet-based NSS—Does High or Low Context matter?" *Int. Negotiation*, vol. 9, no. 1, pp. 79–109, 2004.

[36] P. Honold, "Learning how to use a cellular phone: Comparison between German and Chinese users," *Tech. Commun.*, vol. 46, no. 2, pp. 196–205 May 1999.

[37] K. Siau, *et al.*, "Effects of national culture on types of knowledge sharing in virtual communities," *IEEE Trans. Prof. Commun.*, vol. 53, no. 3, pp. 278–292, Sep. 2010.

[38] N. Singh, *et al.*, "Cultural adaptation on the Web: A Study of American companies' domestic and Chinese websites," *J. Global Inform. Manag.*, vol. 11, no. 3, pp. 63–80, 2003.

[39] A. Marcus, "International and intercultural user interfaces," in *User Interfaces for All: Concepts, Methods, and Tools*, C. Stephanides, Ed., Mahway, NJ: Lawrence Erlbaum, 2001.

[40] T. Ahmed, *et al.*, "Website design guidelines: High power distance and high-context culture," *Int. J. Cyber Soc. Edu.*, vol. 2, no. 1, pp. 47–60, Jun. 2009.

[41] Y. Chang and Y. Hsu, "Requests on e-mail: A cross-cultural comparison," *RELC J.*, vol. 29, no. 2, pp. 121–151, 1998.

[42] E. Chen, "The development of email literacy: From writing to peers to writing to authority figures," *Language Learn. Tech.*, vol. 10, no. 2, pp. 35–55, 2006.

[43] A. Kirkpatrick, "Information sequencing in Mandarin in letters of request," *Anthropologic. Linguistics*, vol. 33, no. 2, pp. 183–203, 1991.

[44] A. Kirkpatrick, "Information sequencing in modern standard Chinese," *Australian Rev. Appl. Linguistics*, vol. 16, no. 2, pp. 27–60, 1993.

[45] R. L. Helmreich, *et al.*, "The evolution of crew resource management training in commercial aviation," *Int. J. Aviation Psychol.*, vol. 9, no. 1, pp. 19–32, 1999.

[46] R. L. Helmreich, *et al.*, "Models of threat, error, and CRM in flight operations," in *Proc. 10th Int. Symp. Aviation Psychol.*, 1999, pp. 677–682.

[47] Flight Safety Foundation. (2011). Accident description, in *Aviation Safety Network* [Online]. Available: http://aviation-safety.net/database/record.php?id=19900125-0 Dec. 22, 2011 (last date accessed).

[48] J. B. Strother, "Cultural adaptation of cybereducation," in *Culture, Communication, and Cyberspace*, K. St. Amant and F. Sapienza, Eds. Amityville, NY: Baywood Publishing, 2011.

[49] L. Louhiala-Salminen and A. Kankanranta, "Professional communication in a global business context: The notion of global communicative competence," *IEEE Trans. Prof. Commun.*, vol. 54, no. 3, pp. 244–262, Sep. 2011.

[50] P. Hall and L. Weaver, "Interdisciplinary education and teamwork: A long and winding road," *Med. Edu.*, vol. 35, no. 9, pp. 867–875, 2001.

[51] P. Hall, "Interprofessional teamwork: Professional cultures as barriers," *J. Interprofessional Care*, vol. 19 (Suppl. 1), pp. 188–196, May 2005.

[52] S. Kelsey and K. St. Amant, Eds., *Handbook of Research on Computer Mediated Communication*. New York: Information Science Reference, 2008.

[53] A. P. D. Van Luxemburg, *et al.*, "Interactive design process including the customer in 6 Dutch SME cases: Traditional and ICT-media compared," contribution to a special issue of the *IEEE J. Prof. Commun.* on The future of ICT-studies and their implications for human interaction and culture in the innovation management process, J. Ulijn, D. Vogel, and T. Bemelmans, Eds., vol. 45, no. 4, pp. 250–264, 2002.

[54] U. Connor, *Contrastive Rhetoric: Cross-Cultural Aspects of Second Language Writing*. Cambridge, U.K.: Cambridge Univ. Press, 1996.

[55] J. Hinds, "Reader versus writer responsibility: A new typology," in *Landmark Essays on ESL Writing*, T. Silva and P. K. Matsuda, Eds. Mahwah, NJ: Erlbaum, 2001.

[56] J. Hinds, "Inductive, deductive, quasi-inductive: Expository writing in Japanese, Korean, Chinese, and Thai," in *Coherence in Writing: Research and Pedagogical Perspectives*, U. Connor and A. M. Johns, Eds. Alexandria, VA: TESOL, 1990, pp. 87–110.

[57] J. Ulijn, "Translating the culture of technical documents: Some experimental evidence," in *International Dimensions of Technical Communication*, D. Andrews, Ed., Washington, DC: STC Press, 1996, pp. 69–86.

[58] X.-Q. Yang, "The influence of discourse organizational patterns on Chinese EFL learners' listening comprehension," *US-China Foreign Language*, vol. 5, no. 3, pp. 22–40, 2007.

[59] U. Connor, "Intercultural rhetoric research: Beyond texts," *J. English Acad. Purposes*, vol. 3, no. 4, pp. 291–304, Oct. 2004.

[60] K. Maitra and D. Goswami, "Responses of American readers to visual aspects of a mid-sized Japanese company's annual report: A case study," *IEEE Trans. Prof. Commun.*, vol. 38, no. 4, pp. 197–203, Dec. 1995.

[61] S. Hilligoss and T. Howard, *Visual Communication: A Writer's Guide*, 2nd ed. New York: Longman, 2002.

PRACTICAL INSIGHTS FROM A2Z
GLOBAL LANGUAGES
GLOBALIZATION VERSUS LOCALIZATION

Globalization (G11N—an acronym meaning "globalization," which starts with a "g," has 11 letters and ends with an "n") means different things to different people. It could refer to taking a product or information out to the big wide world—in other words, trying to make the product or information suitable to any market (something we typically call Internationalization—I18N). The same principle applies to working with culturally diverse folks all over the world in unison. As any user might have difficulty with new subject matter in a presentation, he or she would definitely experience information overload if forced to consume the material while burdened with an approach that was entirely foreign. (For the theories behind these issues, see Chapter 5.) To communicate at such a global or international level requires patience, flexibility, and understanding. The following text includes examples that I have encountered in my business, in which we help clients localize or internationalize their information.

Flexibility is probably the key feature to understand and accept the cultural habits of other participants who share similar goals although they differ in their approaches to achieving those goals. For example, the Japanese tend to organize their text in a more circular or indirect manner while Americans are taught to think sequentially. I once was asked by an American in Tokyo if I could translate a very large training manual into Japanese. As I reviewed the document, I noticed that it was created with very beautiful slides in a very easy-to-follow sequential order. I pointed out that the Japanese audience would be very confused by this type of linear presentation. If the American author wanted to localize the manual, he should have recreated the entire presentation with a Japanese person participating in the rewriting.

Information Overload: An International Challenge for Professional Engineers and Technical Communicators, First Edition. Edited by Judith B. Strother, Jan Ulijn, Zohra Fazal.
© 2012 Institute of Electrical and Electronics Engineers. Published 2012 by John Wiley & Sons, Inc.

Cultural differences have the strongest impact when one is trying to be accepted by another culture. For example, it can be extremely challenging for those from the United States to adapt to foreign cultures, especially since they are accustomed to an open spirit and a drive that is often seen as aggressive or even obnoxious in many cultures. Consider the following examples of someone from the United States attempting to understand someone from another culture.

- How one negotiates has been a rude awakening to many from the United States who are trying to understand those from Japan. As they present their cases and get a persistent response of "yes" in Japan, they may not realize that the "yes" merely means "I understood you" and not that the beholder agrees with the message being delivered.
- Similarly, imagine someone from the United States leading a "Train the Trainers" meeting in Indonesia. When asking the potential trainers for comments, he or she will get virtually no feedback, as speaking out as individuals would be embarrassing to the Indonesians. They respond better once they have a consensus of all the participants. This means they must get together, discuss the training methods proposed, and reach an agreement before they respond.

Now this does not mean that those from the United States must in essence "become Japanese" or "become Indonesian." We can expect that participants from other countries should also be culturally sensitive to Americans. However, I must point out that, no matter what culture one is from, the onus is on the seller to adapt and present his or her information or product so that it will be accepted by the local recipient or buyer. This is called localization (L10N), where one adapts the product/information to the local market, including locale-specific requirements. This effort can be adversely affected if the presenter does not know the local marketing and sales styles; uses offensive language, inappropriate colors, or icons that have no meaning whatsoever in the country; or worse yet . . . tries to use a power cord that does not fit the outlet! While localization is extremely important for success, businesses must remember not to go overboard. For example, a debate about whether to call a device a "probe" or a "sensor" kept one company deliberating over a single term for several years! What happened to the "Time to Market" principles? Make an educated decision and move forward, or consult with a localization company if need be.

Even when everyone is working in the same language—English—depending on the verbiage and order of words, readers from different cultures may perceive information differently. For example, the users or workers from one culture perceive that they are responsible for their own safety in the workplace whereas people from another culture may perceive the science behind the process, the equipment itself, or the document to be responsible for safety concerns.

When content is not localized for a particular culture, miscommunication or misunderstandings often result. The following are a few examples of noninternationalized content for a Standards of Business Conduct document written by an American author for an international audience in 23 countries.

- An employee who was gay asked his boss what he needed to do to get a promotion. The boss took him aside and said, "You need to play down the 'gay' thing. If I promote you and your new supervisor finds out you are gay, it will reflect poorly on me." (The lesson: Find out whether this kind of reference to being gay is a nonissue or could mean a beheading.)

- A broker approaches you and complains that things have not been going well for him recently—at work, in his financial situation, and in his personal life. He tells you that one of these days, someone will catch him on the wrong day and they would "pay" for it. A few days later he was seen throwing his calculator on the floor while screaming at a colleague on the phone. (The lesson: Find out the limits of acceptable behavior. Some behavior may be so offensive that it simply would not be allowed.)
- An American girl joined a Japanese company and found it had no softball team, so she decided to start a company softball team. The poor girls in Japan had no idea what softball even was. (The lesson: Be sure your audience will understand any cultural references. For example, avoid references to sports that may not exist everywhere in the world.)

As the examples given above show, it is important for anyone who is trying to work in a different culture to be conscious of word choice. Additionally, common images to avoid are symbols such as birthday cakes, mail boxes, phone booths, and the like. It is also important to learn about such cultural issues as gender roles. For example, Arabic countries have a whole host of rules for women that do not apply to men. The international communicator must be sure to know them before he or she graphically uses a woman's picture, especially uncovered, or even refers to a woman doing certain work that would be unheard of in an Arab country.

Even software strings and dialog boxes require careful thought. Keep in mind that in software strings, while the dialogs might automatically resize, text space will expand in some languages (Spanish and French are examples), whereas it will contract in others (Japanese is an example). In the Japanese culture, the style of technical communication is much more graphic and filled with cartoons. This is unheard of in the United States, as well as most western cultures. We might surmise that the Asians would be overwhelmed with large volumes of text to convey something that a simple picture might convey better.

Internationalizing your text requires study and understanding of how people in the target cultures think and how they remit information. If a text is not internationalized adequately, then the localization of that text is virtually impossible without extensive rewriting. In order to keep the reader engaged, make sure to incorporate locale-specific information gathered from the local marketplace. Any irregularities may cause the reader to feel overwhelmed. He or she is trying to understand the document, but the method of delivery or the subject matter being used to make the point may hinder the process, causing the perception of information overload and hindering the overall purpose of the document.

Contributor to Practical Insights from A2Z Global Languages

- Theodora Landgren, Director

6

EFFECT OF COLOR, VISUAL FORM, AND TEXTUAL INFORMATION ON INFORMATION OVERLOAD

Noël T. Alton and Alan Manning

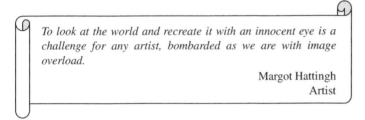

To look at the world and recreate it with an innocent eye is a challenge for any artist, bombarded as we are with image overload.

Margot Hattingh
Artist

ABSTRACT

Technical communicators in both academia and industry need to be made aware that there is no clear-cut boundary between "mere" document design issues and actual information-content issues. Current online menu design tends to overload menus with too many separate entries, too many distinct colors, too many flashing images or signals, too many formal boxes and boundaries, and so forth. The same design overload can be found in printed menus, which include (of course) restaurant menus as well as the tables of contents of books and magazines, along with their organizational layouts, which readers use to navigate to specific information elements. These are all, in essence, menus that can often suffer from parallel difficulties: too many listed entries in one block, too many colors used as separate indicators, and excessively deep nested hierarchies of information. Key information both in print and online is too often buried in a sea of distraction. Therefore, document design issues are information issues. The recent literature on visual design recognizes the need for some kind of "balance" between color/form elements and textual elements thought to convey the essential information, but most discussions have lacked both a coherent theory of information design and any kind of clear quantitative evidence to support specific claims

Information Overload: An International Challenge for Professional Engineers and Technical Communicators, First Edition. Edited by Judith B. Strother, Jan Ulijn, Zohra Fazal.
© 2012 Institute of Electrical and Electronics Engineers. Published 2012 by John Wiley & Sons, Inc.

about how much decorative color/form is too much, on one hand, and how much textual information is too much, on the other hand. This chapter describes a coherent theory of information design as well as empirical evidence in support of that theory and also in support of specific design recommendations.

6.1 Introduction

Key information both in print and online is too often buried in a sea of distraction: too many colors, bullets, menu items, and so on. Therefore, document design issues are information issues. The recent literature on visual design recognizes the need for some kind of "balance" between color/form elements and textual elements thought to convey the "essential" information. However, the boundaries between "mere" decoration—visual contrasts that tell readers where to look—and textual information remain unclear. It therefore remains unclear how much of each of these distinct information elements is too much. In this chapter, we offer a theory of information that clarifies these boundaries and therefore enables a more focused discussion of information overload.

We can process large amounts of information only when it is organized properly. Even a respectable ancient library with just a thousand books in it (let alone a million) was hardly usable if the books were shelved randomly. Any specific book would require more time to locate than most library users would be willing to spend. Thus, an effectively organized index or table of contents determines and always has determined the usability of any large quantity of information.

Any index or table of contents is essentially a menu of information available. If that menu is so poorly organized or, in other words, so poorly *designed* that users cannot readily locate the pointers to the information they need, then they are back to the problem of searching over the whole pile of books, and, indeed, over each page in each book. This image of a vast stack of ancient books, pages, or scrolls to search through still captures the essence of the information overload problem, even now when many of these pages and scrolled text exist only in a virtual, electronic form.

In this volume, each of the other chapters recognizes, implicitly or explicitly, the role of document design in mitigating or exacerbating the information overload problem. Caborn and Cooper (Chapter 4) note that information users are hardwired to evaluate information in a "delivery context," the design of the information implicitly being a key part of its delivery. Bidkar (Chapter 7) includes in the calculation of information load any aspects of communication (including design issues implicitly) that make demands on the cognitive capacity of users and that potentially creates distraction or "noise" that again might increase the time required to process the information.

Other chapters recommend visual-design strategies explicitly for reducing information overload. Mengis and Eppler (Chapter 10) argue that information visualization, in general, can minimize information overload. Hoenkamp (Chapter 8) suggests that search-query efficiency increases if users have, among other things, a visual display showing degrees of similarity between their search queries and those documents returned by those queries.

This chapter, likewise, starts with the understanding that ineffective visual document design adds to the sense of information overload. Conversely, effective visual design can help minimize overload, both the *feel* of overload, which Caborn and Cooper assert is the main factor, as well as *physical* parameters of overload, as measured by time and effort costs of information. Visual design is considered effective to the extent that it reduces mental effort and the time required to locate pointers to desired information in menu structures, data tables, and other relational visuals. In effective visual designs, color and form are deployed with careful consideration of their effects over distinct channels of information, each of which may contribute to information overload if overused or misused.

As Vossen (Chapter 3) points out, we may differentiate between several specific information loads, one for each channel of information processing. Also note that once we accept such a definition, it becomes increasingly unclear where in the chain of information processes we actually should measure this information load: Just at the very beginning (sensory input)? Somewhere during its intermediate inner processing (cognitive through-put)? Or at the very end of the information chain, where decisions are mapped to actions (motor output)?

In our view, it is useful to differentiate at least three discrete channels of information.

- *Emotional.* Elements that promote a feeling-based response in the viewer (e.g., a sense of fun, agitation, or focus).
- *Indicative.* Elements that tell viewers to perform specific actions (e.g., looking, focusing, dividing, and/or contrasting one element from another).
- *Propositional.* Textual information (i.e., asserted statements of fact such as "the test material *did (or didn't)* demonstration superconductivity").

Color and form, when used decoratively, convey a channel of *emotional* information carried by decorative elements, which, if deployed carelessly, can and do interfere with the channel of *propositional* information, which is mainly conveyed by text.

Between emotional and propositional information, there is yet another information channel that color, form, and text influence equally: the channel of *indicative* information. This can be seen in phrases such as "see below," numbered or bulleted list points, or other visual elements that do not assert *did/didn't* propositions but do tell viewers to perform specific actions of looking, focusing, dividing, and/or contrasting one information element from another. Here again, careless deployment of indicative elements overloads viewers with an excessive number of discrete visual actions to perform. As pointed out by Remund and Aikat (Chapter 11), the actions provoked by information are likewise a key factor to consider in appropriate information loads.

This chapter addresses the use and overuse of color as an information carrier, where that information is primarily emotional or action provoking in a way that should be complementary to rather than in conflict with the propositional information of text. We offer engineering professionals and technical communicators specific advice, based on quantitative data, about how to keep emotional, indicative, and propositional information in balance in order to avoid information overload.

Broadly speaking, color and form can convey three kinds of information:

1. evoked *feeling*, in which case we refer to color or form's communicative effect as *decorative*, for example, when the color red (particularly in Asian cultures) is used to evoke feelings of happiness or prosperity, or (particularly in Western cultures) to evoke feelings of passion (a positive reading of red) or act as a warning (a negative reading);
2. provoked *action*, in which color or form serves to (at minimum) focus attention on a specific point or otherwise provoke some physical response, in which case we refer to the communicative effect as *indicative*, for example, when a red traffic light causes drivers to first notice the light and then physically stop;

3. asserted *propositions*, such as when litmus paper, turning red, asserts that "this liquid *is* an acid" or an instrument light, showing green, asserts that "this unit *is* functioning," in which case we refer to color or form's communicative effect as *prepositional*.

These primary functions of color and form line up with the three main categories of visual-design analysis [1,2], which, in turn, are based on the semiotic work of Peirce [3,4]. Throughout his work, Peirce postulated three fundamental categories of experience that can be communicated: "Firstness" = feeling, "Secondness" = action, and "Thirdness" = propositional argument. We should note parenthetically that Peirce's numbering of the categories refers to a triangular "logic of relations" between the categories and *not* any necessary temporal order. We mention Peirce's terms here only to give due credit to the source of our organizing framework, not to consider all logical ramifications of Peirce's ordering of the semiotic categories, which is far beyond the scope of our discussion here.

Propositional meaning requires a single, fixed, and codified system of meaning, which is relatively uncommon for color in Western culture [5]. Color only rarely conveys specific propositional meaning, and only then in highly specialized contexts (for example, when the color of velvet on a graduation gown specifically *asserts* the degree awarded: green = the student's degree *is* a medical doctorate, purple = the student's degree *is* a legal doctorate, etc.).

In this chapter, we focus on color's more widespread decorative- and indicative-information functions, precisely because these information functions are more common in Western culture (and therefore more likely to be sources of information overload). To clarify the difference between propositional information and indicative/decorative information, the color red may be used in a danger sign, but the function of the color is *not* to assert that there *is* danger (which is instead asserted by the word *danger* typically). Rather, the color red in this case functions to physically provoke viewer attention (indicative function = *look at this sign*) and also evoke an agitated feeling (decorative function = *feel concern here*). If these functions are carefully deployed, they may complement one another as well as the propositions asserted by words (such as *danger* or *caution*) or propositions asserted by standard Occupational Safety and Health Administration (OSHA), American National Standards Institute (ANSI), or International Organization for Standards (ISO) pictographs. However, our focus in this chapter is on cases where colors are not deployed carefully and therefore more typically do interfere with one another and with surrounding text or other visual elements that assert propositions. This case results in information overload.

6.2 Previous Studies of Decorative and Indicative Effects

Color-information overload generally emerges in the clash between decorative effects, i.e., what the color makes readers feel, and indicative effects, i.e., where color forces readers to look. There is some awareness in the visual design literature that color has these distinct functions, "to communicate ideas and emotions, to manipulate perception, to create focus, to motivate and influence actions," and so on [6]. However, what is not widely understood is that decorative purpose of color (to evoke emotion) is intrinsically at odds with the indicative purpose of color (to focus, motivate, and influence), and that both of these can disrupt or distract from propositional information (to communicate ideas).

The primary goal of a decorative strategy is to evoke an emotion or, in other words, a felt, aesthetic response, but this in turn requires some kind of aesthetic unity, with all the decorative parts tying together to create one emotional effect which in most cases is also an aesthetically pleasing whole (corresponding with the Peircean idea of "Firstness" or, in other words, *oneness*). This is why, for example, a person's belt is generally supposed to match the color of his or her shoes and why aesthetic unity is identified as an important element in nearly all discussions of visual design [6–9].

In contrast, the primary goal of an indicative strategy is to provoke action, and this inherently requires the creation of divisions or contrast (corresponding with the Peircean idea of "Secondness" or, in other words, *opposition*). This same concept of indication as disruption is articulated by Harris [7], who pointed out that the reason for emphasizing certain pieces of text is to make sure the reader is led to the pertinent information: "To be successful, you need to disrupt the mostly automatic reading process—if only for an instant. You need to make it clear . . . that certain words (or images) are being stressed and therefore deserve special attention."

In general, too many color contrasts compel readers to look in too many places at once, and these same contrasts also disrupt the overall aesthetic that a text should have, which generally should be experienced as a unified gestalt or aesthetic whole. Thus, excessive color contrast results in readers feeling disrupted, chaotic, or conflicted, and simultaneously overworked by the sheer number of distinct areas where they are compelled to focus attention. These effects then obscure whatever propositional information is in the text.

Of course, if propositional content of a text is only of minor importance (or no importance), then deliberately excessive color contrast may serve purposes of distraction and/or misdirection. Fraser and Banks [8] noted that we can often determine the target audience of a newspaper without considering text, but rather just by looking at the color scheme of the publication. They note that "the more garish the front page the further down-market the content."

Also implicit in this observation is the idea of an "up-market" (i.e., more prestigious) aesthetic that shuns multiple and vivid color contrasts, in favor of more understated and unified color schemes [9]. This more "prestigious" aesthetic is more than mere fashion, however, since it tends to enhance the accessibility of propositional content in a text rather than to distract or misdirect readers away from it. This polar opposition between "high" aesthetic unity and "low" clamoring contrast, we ultimately argue, encapsulates the tension between *decorative* color function and *indicative* color function. Here, decoratives are defined as visual elements conveying information as *feeling* and indicatives are defined as visual elements conveying information as *active focus*, usually by means of visual contrast.

In summary then, previous discussions of visual design have generally lacked the following:

1. explicit awareness of the tension between decorative unity and indicative contrast;
2. adequate emphasis on the practical limits of readers' capacity to process indicative contrasts, each of which imposes a burden of visual action;
3. awareness of and emphasis on color schemes as information carriers, these being passed over or dismissed as mere design elements rather than emotional information that may enhance or distract from propositional information.

We now turn to one more example that illustrates critical lack of awareness in each of these three areas; Edward Tufte has made the argument (oft-repeated) against any and all visual elements that do not communicate propositional information (i.e., data):

The purpose of decoration varies—to make the graphic appear more scientific and precise, to enliven the display, to give the designer an opportunity to exercise artistic skills. Regardless of its cause, it is all non-data-ink or redundant data-ink, and it is often chartjunk. Graphics do not become attractive and interesting through the addition of ornamental hatching and false perspective to a few bars. Chartjunk can turn bores into disasters, but it can never rescue a thin data set [10].

For examples of what is generally meant by "chartjunk," see Figure 6.1b and c, discussed later in this chapter. For non-chartjunk versions of the same information, see Figure 6.1a and d.

Our main point is that Tufte [10] has not been clear in his oft-cited "non-data-ink" and "chartjunk" descriptors about the distinction between decoratives (visual elements that create feeling) and indicatives (visual elements that attract focused attention). The difference is an important one, because it can be argued that a visual element such as "hatching" or bright color is not just decorative or ornamental but may also be deployed to call attention to specific data points.In other discussions, Tufte has said that indicative visual elements should not be abandoned altogether, but rather should be used sparingly:

. . . when *everything* (background, structure, content) is emphasized, *nothing* is emphasized; the design is often noise, cluttered and informationally flat . . . . Minimal distinctions reduce visual clutter. Small contrasts work to enrich the overall visual signal by increasing the number of distinctions that can be made with a single image [11].

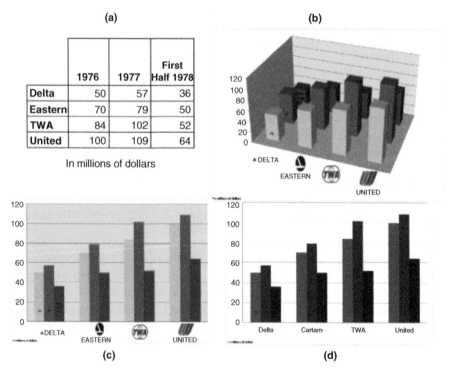

Figure 6.1 Graph design alternatives tested in study two. The graphs vary in design as follows: (1) simple table, (2) brightly colored 3-D graph, (3) brightly colored 2-D graph, and (4) grayscale graph.

These are valid observations, but the critical, unanswered questions here are: (1) what the boundary is between decoration and indication, (2) how much indicative contrast is the minimum necessary, and (3) how much indicative contrast is too much. While most agree that using color in presentations can make them livelier and affect the message being sent, visual design experts typically give advice that pulls the novice designer in different directions. For example, "Most experts agree that your color scheme should include one or two bright colors for emphasis—but to preserve the power of these colors, use them with restraint" [12].

For anyone who has to create a visual representation of information, color contrast is often seen as *the* strategy for getting viewer attention, where "large color differences are sought to make objects as conspicuous as possible . . . in commercial packaging, in advertising, and even in art" [13]. This overemphasis on indicative contrast easily leads to overuse, and to indicative overload resulting from a lack of understanding of viewers' processing limitations.

Another significant part of the problem lies in the pervasive myth that between seven and nine contrasts is acceptable (e.g., seven to nine separate items in a list, in a set of tabs, or as distinct visual blocks on a page, colored or not). Miller's original 1956 article, "The Magical Number Seven, Plus or Minus Two," [14] in fact, identifies seven to nine as the *breakdown point* of memory capacity, the upper limit where errors in recall become *completely* unacceptable in practical terms. Miller's research also identifies a lower, more practical limit, where recall errors are rare, which is three to four [15]. Thus, three to four contrasts is the typical limit that text designers should be imposing upon any one aspect of any given layout. Thus, this three to four contrast limit should apply to the number of contrastive colors on a page, the number of items in any one list, the number of major information divisions on a page, and so on. Otherwise, reader fatigue and errors in attention and recall are highly likely. It follows that it might be possible to have two or three separate information divisions on a page, each with a three or four item list, but again, reader fatigue will become an issue if even these nested-contrast strategies are pushed too far.

Such problems are readily apparent in the design of both restaurant menus and Web pages, as shown in Figure 6.2. In both restaurant menus and Web page designs, we find a similar range of problems introduced by too many competing design elements. If highly technical information is also involved, then these high-contrast design strategies are even more likely to create information overload.

Hays [20] predicted that with the aid of computers and layout software, restaurant menus would become more user friendly. Unfortunately, more than 15 years later, restaurant patrons are too often accosted by overcomplicated designs that detract from the information being presented about the food. Design complications are, in fact, facilitated by computers and software that make color and image elements extremely easy to add and thereby tempt visual creators to overuse color and image in an effort to show off their prowess at using specific software tools, at the expense of clear, uncluttered visual information. Comparable problems also persist in Web page design, in spite of the availability of increasingly sophisticated design and layout tools.

It is not surprising, therefore, that both Web page menus and food menus suffer from the same design difficulties. Most Web pages are a kind of menu, specifically a menu of hyperlinks offering an array of information and service choices. In this sense, Web pages are comparable to Zwicky and Zwicky's ethnolinguistic description of restaurant menus as "a catalog, a sort of list, usually divided according to traditional parts . . . " and "conveying information about [choices], advertising the [choices], and doing so in a relative small space" [21]. We also expect that solutions to the shared problems of menu

Figure 6.2 Samples of restaurant and Web page menus. Restaurant menus and Web page menus both commonly suffer from emotional and indicative overload, typically creating interference with propositional information [16–19].

and Web page design, along with strategies for achieving their shared design goals, would be based on the same principles.

Although principles leading to a solution are not widely agreed upon, the essential difficulty of finding a balance between "too little" and "too much" indicative contrast is widely recognized in the literature. Farkas and Farkas [22] recommended that graphics, including color elements, be limited to those used for grabbing attention, but they also note that their overuse detracts from the information being presented. Karg and Sutherland [23], in turn, said that an effective amount of contrast "results in a pleasing and readable appearance. Too little contrast can make the text virtually unreadable. Too much contrast can make them visually jarring." Note that this observation merely echoes what was said years earlier by Tufte [11], and although correct, fails to identify clearly what constitutes too much or too little color or contrast. Kress and Van Leeuwen [24] noted that color used in business documents can increase attention spans by 80%. However, when used incorrectly, color no longer allows the reader to assimilate the information being presented due to distractions that the color creates. Farkas and Farkas [22] mentioned that it is "very easy to misuse color" and "a key guideline . . . is restraint." Again, the problem lies in questions such as what defines restraint and how much color is too much color.

When referring to Web page design, Lynch and Horton [25] have said that visual design should balance visual sensation and information. This ideal is echoed by Lawrence and Tavakol [26]. Sammons [27] likewise discussed guidelines for color choice: make sure the colors chosen fit the topic being presented, the product or organization being represented, and the mood being targeted. Other commentaries put special emphasis on matching color choices to cultural sensibilities [5,24]. In terms of use, Sammons echoed Farkas and Farkas by saying color should be used sparingly and used to highlight and focus attention to the important information. She goes so far as to suggest limiting the use of colors to only two or three per document.

Studies such as these recognize the general value of balance in visual design and restraint in color choices, but these studies also lack a more precise specification of how that balance is achieved, other than by blindly applying rules like Sammons' three-color limit. We would prefer, instead, some externally motivated principles that can be used to determine the exact tipping point, as it were, between effective and overused colored forms and colored imagery.

We therefore emphasize the following:

1. The distinction between decorative, indicative, and propositional information has been largely overlooked in previous color discussions, very few of which have been informed by actual empirical study.
2. The series of empirical studies that we developed using this distinction measured user response to distinct levels of each information type.
3. This threefold division of information types ultimately proves to be a critical factor in determining how much detail, decorative, indicative, or informative, can successfully be loaded into a menu, which is primarily directed to serve the actions of choice and purchase.
4. This theory of distinct information types lays the foundation for what is effectively a grammar for information design, a grammar that can guide text designers in solving text-design problems in principled ways, no longer just "hacking" at text designs in impressionistic ways.

6.3 Experiments and Results

In this section, we report the results of three empirical studies to identify the actual point of balance between decorative, indicative, and propositional information in visual designs. The first study was a quantitative comparison of several study participants' impressionistic response to different restaurant menu designs. Recall from our discussion of Figure 6.2 that restaurant menus have important design issues in common with more technical information genres such as Web page help menus, and we ultimately find the same design issues relevant in the construction of even the most fundamental of technical visuals such as diagrams and graphs. The second and third studies directly measured how much propositional information was retained from diagrams and graphs with different levels of color and form used decoratively and indicatively.

6.3.1 Study One: Restaurant Menu Design

In our first study (described in more detail in [28]), several restaurant menus were constructed representing a range of design strategies, from no color to a wide palette of color, from no images to several, from plain headers to boxed headers, and so on. The various designs tested are shown in Figure 6.3. A minimum black and white menu was contrasted with muted color designs, with and without images, and against a version with more aggressive, higher-contrast designs, with and without images.

Fifty study participants were surveyed about the menus. They ranged in age from 18 to 61; their average age was 28.6. Responses of participants older than 35 were not statistically distinct from those of younger participants. The survey first asked the participants to look at each menu individually and award points based on a Likert scale:

1 point meant "Dislike [this menu]."
2 points meant "Somewhat Dislike."
3 points meant "No Opinion."
4 points meant "Somewhat Like."
5 points meant "Like [this menu]."

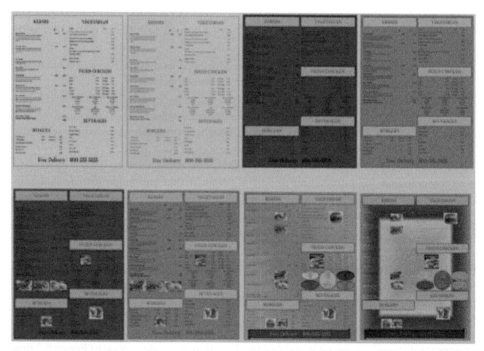

Figure 6.3 Menu design alternatives tested. While these images are black and white, the originals varied as follows: (1) black on white, (2) dark red on cream, (3) gold on blue, and (4) light purple on gray. In (5) and (6), images were added, and in (7) and (8), more images with decorative forms and colors were added [28].

Participants were then asked to compare menu designs in pairs and evaluate each design in the pair as relatively better or worse. Points were awarded to each design based on these responses. Point totals for each menu design are shown in Figure 6.4 .

After the participants completed the quantitative ranking of the menus, they were allowed time to comment on the designs of each of the menus. Many indicated in their open-ended comments that "too much" color and form tended to overload their senses, making it almost impossible to glean any information about the food items themselves. Participants also indicated that colorless presentations could feel bland and uninviting. All participants expressed some preference for color, but the simplified color schemes preferred overall were those that conformed to principles of appropriate decorative purpose as well as indicative restraint, as discussed by Amare and Manning [1, 2], and explicated using the Peircean model of design. In very specific terms, this theory asserts that indicators operate by means of contrast. Contrastive form without indicative purpose constitutes an ill-formed visual in the theory.

In other words, each contrast takes a bit of mental effort to process. Merely decorative elements can be added that create further visual contrast, as for example, when the background color is made darker in Menu 3, but these changes do not further contribute to the practical separation of indicative elements. In that case, effort expended on seeing the decorative elements begins to interfere with separation of the indicatives.

The most striking aspect of these results is that the spare Menu 1 design, *with no design features beyond tabular organization*, tested better (145 points) than most of the alternatives offered. The Menu 2 design, with only minimal color added, tested still better (176 points) than all but one alternative.

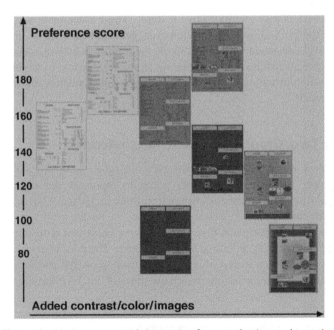

Figure 6.4 Effects of added contrast and decorative features (action and emotion information) on viewer preference. Contrastive elements with nonindicative function significantly lowered preference scores.

Although participants preferred Menus 2, 4, and 6, all of which had some visual interest, the viewers noted that visual interest had to operate well below a definite threshold, i.e., the point where decorative considerations began to interfere with the main indicative purpose of the menu.

Since Menu 2 (176 points) adds decorative interest without adding any contrast, its preference score increased. Menu 4 (159 points) adds decorative color but also some contrast in the form of boxes that indicate main headings; its preference score increased, but not as much. However, Menu 6 includes additional small images, which not only add some further decorative interest but also assist in the visual indication of food categories, so its preference score increased to an optimal level.

All other menu designs take preference in the other direction because they add increasing levels of contrast without serving any added indicative benefit. Menu 3 (86) critically differs from Menu 4 (159) in the saturation level of its color choices. Saturated, complementary colors create contrastive effects without indicative purpose, which severely lowered the preference score of the visual. Menu 5 (132) is somewhat redeemed by the addition of partly indicative images, but the overly contrastive color continues to drag down the score of the visual. Menus 7 (119) and 8 (79) simply continue that downward trend with the further addition of unproductive color contrasts.

As striking as these results are, this first study of menu design only indirectly measured information overload. Participants were asked to imagine how they would feel when using each menu in a food-ordering situation and to assess them accordingly. However, they did not, in fact, have to process information from the menus.

In subsequent studies, we have been measuring information recall more directly, with the degree of enhancement or interference that comes from different visual designs with different kinds of decorative and indicative features. These subsequent studies are, at this writing, still in the preliminary stages. However, general trends in the early accuracy-testing data are still consistent with trends in the menu-design study based on impressionistic responses, except for the tendency among study participants to slightly undervalue uncolored tables of information in impressionistic assessment, relative to their actual value in information recall.

6.3.2 Study Two: Graph Design and Recall Accuracy

In our second, preliminary study, we constructed one table, a colored and an uncolored multiple-bar graph, and· one colored 3-D graph to represent the same information: commissions paid to travel agents by different airlines in 1976, 1977, and the first half of 1978, as shown in Figure 6.1.

Thirty-three study participants (ages 25–62) were presented with one of these four visuals, selected at random by an online survey program. The presented visual was studied for about 1 minute and then irreversibly removed from view. All participants were then presented with the same three multiple-choice questions about the information presented in the visual:

- In what year did United pay more commissions?
- In 1978, which company paid more commissions?
- What accounts for the relatively low amounts in the last year?

We tabulated the total number of correct answers elicited by each visual and divided this figure by the total number of answers (=3 × number of participants looking at that visual). The result is a relative accuracy score, as a percentage, for each visual design, as shown in Figure 6.5.

These information-recall results are generally consistent with the menu-design preferences, with the caveat that impressionistic evaluations may somewhat undervalue plain-table information (i.e., Menu design 1 in Figures 6.3 and 6.4, equivalent to the plain table in Figures 6.1 and 6.5) in terms of the actual utility of the visual. It is also worth noting that the simple 3-D graph is only slightly less effective than the table, although it employs a design strategy sharply condemned as "chartjunk" by Tufte [10,11]. In our analysis, we would say that the 3-D graph is roughly equivalent to Menu design 5 in Figures 6.3 and 6.4, somewhat more decorative and contrastive than necessary, certainly, but still using color and contrast in a mostly effective way, not yet reaching the point of severe information overload.

This preliminary data suggests instead that multiple-bar graphs are more problematic inasmuch as their more linear design makes cross-dimensional comparisons of data more difficult. In other words, color and/or contrast in the multiple-bar graphs are not sufficiently

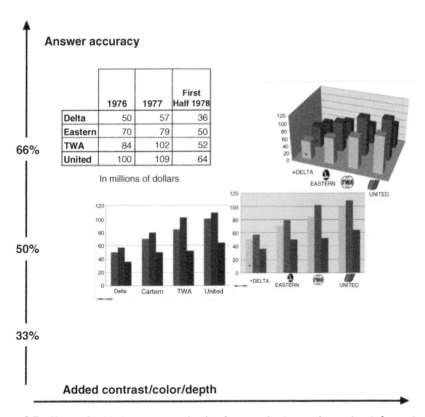

Figure 6.5 Effects of added contrast and color features (action and emotion information) on response accuracy. A simple table is slightly more effective (72%), in terms of immediate-recall accuracy, than a (simple) brightly colored depth graph (66%). Multiple-bar graphs, in comparison, appear less effective (grayscale, 46%; colored, 53%). Cross-dimensional data connections are apparently harder to make in the more linear bar-cluster format.

indicative of the necessary comparisons; therefore, color and contrast here work against the clarity of the visuals.

6.3.3   Study Three: Diagram Design and Recall Accuracy

In our third, preliminary study, we compared a solid "wall" of informational text with three alternative diagrams of the same type of information. The text slide or one of the three diagrams was presented, at random, to 50 participants (ages 20–53), using the same survey framework as in study two. After the visuals were irreversibly removed from view, participants were presented with these multiple-choice questions:

- In what month do Japanese beetles lay eggs?
- In what month do Japanese beetles develop into fully grown larvae?
- In what month do Japanese beetles die?

The relative accuracy of recall for each visual was calculated in the same manner as in study two, and is shown in Figure 6.6.

As initially expected, the solid "wall" of information was least useful in terms of accurate recall of information, even though the text spells out answers exactly. This dramatically illustrates what we mean when we say visual design issues are in fact

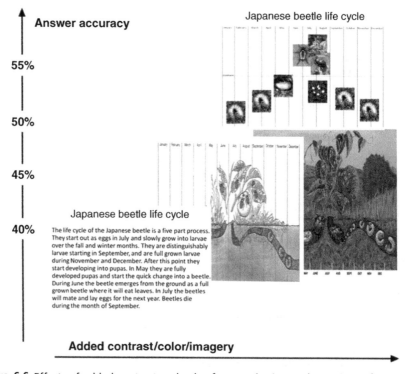

Figure 6.6 Effects of added contrast and color features (action and emotion information) on response accuracy. A solid block of text is less effective (40%), in terms of immediate-recall accuracy, than diagrams with varying degrees of contrast (grayscale, 44%), color (45%), and imagery (contrastive clip images, 55%).

information issues. The diagrammatic presentations of this information show results generally consistent with the menu-design preferences and with graph-design recall accuracy. Here again, the plain grayscale diagram (44% accuracy) is roughly equivalent to the plain Menu 1 in Figures 6.3 and 6.4, or the plain table in Figures 6.1 and 6.5. In this case, however, the colored-drawing version is equivalent to Menu 4 in Figures 6.3 and 6.4, only slightly improved (if at all) with the added color, and the clipart-image version of the diagram is equivalent to the optimal, image-and-light-contrast Menu 6.

In preliminary studies two and three, we did not generate visual designs in which significant information overload occurs, but based on the findings of the menu study, we would expect that preference scores and accuracy of recall would probably fall off quickly if significantly more color or contrastive form were added to either the 3-D graph in study two or the clipart-image graphic in study three.

6.4 Practical Implications for Engineers and Technical Communicators

Text-design problems arise if too many decoratives are used to create emotional effects when the practical purpose of the information is to promote the action of choice selection or to convey information about the choices, as in a menu list [1]. Instead of guiding the viewer, the menus frustrate the viewer because he or she cannot find the desired information amidst an overloaded contrastive clutter of decorative forms, fonts, and images.

Decoratives should be used to create feeling that draws the viewers/readers into the visual and holds their attention. Many TV commercials use decorative elements to entice viewers to buy a particular object. Take, for instance, some recent DeBeers print-advertisement strategies. The only visuals in these are shadows and one piece of shining diamond jewelry. Using no words, only images, these visuals can create within the viewers a felt desire to own the jewelry being presented to them. By using the right decoratives, the overall message is enhanced.

On the other hand, when communicators need to draw particular attention to an item or create action relative to that item, an indicative can be used to accomplish this. A blinking arrow is an indicative. It draws the eye, perhaps causing a viewer to move a mouse and click a link. This is the active effect of indicatives.

However, communicators from all fields, including engineering, need to be more cognizant of the fact that each separate indicative element requires mental effort to see and separately to recognize. Too many indicatives, bullets, arrows, or even images can fatigue viewers, and fatigue is the natural effect of information overload in the most direct sense. Worst of all is the use of indicative-like contrast with only decorative function. Any kind of animation or imagery that does not point to essential information is likely to be very problematic.

For example, the current design trends in networking sites (Myspace[TM] [www.myspace.com], Facebook [www.facebook.com], etc.) tend to overuse indicatives. Each type of newsfeed (item posted on the profile of a user) has its own icon meant to point to the type of item that has been posted. Since each different application has a different icon/bullet point, the eye is no longer attracted to any one in particular. This means that the user (the viewer) of the page is unable to act upon the different indicatives without creating some sort of mental filter against them, which requires further mental effort. Such sites are still usable, but information is easily lost in them.

Balanced design has been frequently referenced as an ideal, but there has been a general lack of precision about what balance really means or how it is to be achieved. Design books such as those we have cited in this chapter [5–9, 22–25, 28], as well as many others, have talked about the key "elements" of design and the way these elements work individually, but there have been few (if any) clearly defined parameters widely recognized, or even mentioned, that identify the useful limits of any given design strategy.

In other words, there has been a general failure to articulate the problem of perceptual or informational overload that is directly caused by the overuse of any given design strategy. This is especially the case with color. As a result, over-cluttered color schemes are still rampant in the visual world, with designs that often obscure and distract from information content. So, for example, Whitbread [9] observed that "Colour should be used mainly for attraction, not for communication, so use it liberally in images, display text, backgrounds and borders, but be particularly conservative in its use in text."

In spite of this caution about integrating color with text, Whitbread also seems to strongly recommend the use of color coding to identify distinct sections in a text, different kinds of documents, or different kinds of information in a document:

> Colour can identify an organisation or the type of document a reader is perusing, or distinguish points of difference between one printed piece and another. Coding enables people to quickly find what they need . . .

Whitbread goes on to say that when a color or series of colors is assigned to each different point or section and those colors are used consistently, the reader makes the relevant association with "a particular style of document or information" [9].

Besides the apparent conflict with advice to use color conservatively with text, Whitbread does not mention any kind of limit on the number of colors/divisions that document users can easily keep track of. Recall that Miller's [14] findings would imply that excessive color coding of text divisions (i.e., more than three or four distinct colors) is very likely to create confusion, with users often forgetting which color goes with which text division. Similar problems are likely when colors are overused in CAD drawings, electrical diagrams, or other complex figures. Three or four color contrasts can be used to effectively indicate contrasts but more are likely to overwork users in terms of visual attention capacity and memory.

If more than six colors are used (i.e., the three primary and three secondary colors), then the additional problem of distinguishing shades of color can arise—whether readers are looking for information in the "green" section that is actually in the "olive" section, for instance. Careless color divisions can make a document seem simultaneously carnival-esque and yet intimidating, with color indicators both hard to remember and difficult to differentiate, as exemplified in Figure 6.7 [29].

A more sound approach to the information in Figure 6.7 is illustrated in Figure 6.8, where the same information is organized in terms of three major divisions, with only three colors (red–gold–blue) used to code those divisions.

The contrast between Figures 6.7 and 6.8 here gives substance to White's [30] observation that:

> Color contrast has the same potential for communicating hierarchy as typeface, type weight and size, or placement contrasts. Random application or changes in color work against the reader's understanding just as do any random changes in design.

Each one of these layers alone is capable of stopping a terrorist attack. In combination their security value is multiplied, creating a much stronger, formidable system. A terrorist who has to overcome multiple security layers in order to carry out an attack is more likely to be pre-empted, deterred, or to fail during the attempt.

20 layers of security

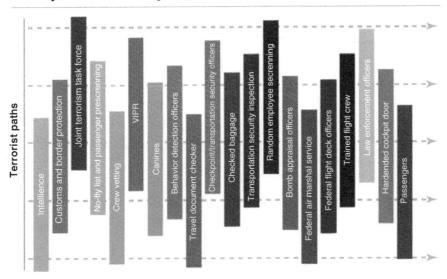

Figure 6.7 Excessive divisions in color coding. Color divisions can make a document ineffective, with color indicators both hard to remember and difficult to differentiate. In this figure, for example, more than three shades each of blue > green > orange > purple > red > olive were used.

The problem lies in specifying precisely why "random applications" of color are problematic, and what "random" really means in this context. Essentially, if color is used to organize text without specific awareness of its limitations, then color is being deployed randomly. Conversely, if color is used without specific awareness of the difference between its decorative and indicative functions, which are inherently in conflict with each other, then color is being used randomly, and information overload is the likely result.

6.5 Conclusion

In contrast to the only vaguely delimited kinds of visual-design advice commonly found in design textbooks and manuals, our 2009 study was directed at providing and validating some specific design criteria that clarify the notion of decorative/indicative balance and identify limits for decorative and indicative elements, the point at which either strategy may begin to distract from other information in the text.

Referring again to Figure 6.4, we would emphasize that as far as indicative menu design is concerned, it turns out that Menu 2, the most minimally decorated visual design, exemplifies the most accessible kind of balance between decorative and indicative purposes. To clarify what we mean by balance in this case, it is evident that the ability to use minimal decoration of this type is within the reach of even a beginning designer—i.e., to create a very muted, low saturation color background, and only subtly

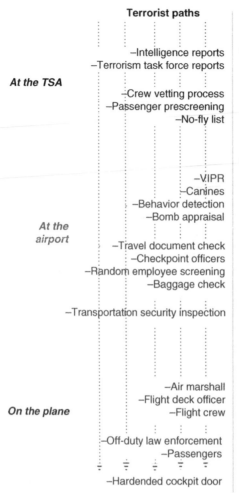

Figure 6.8 Limited but intelligible divisions in color coding. When information is organized in terms of three major color divisions, as opposed to multiple color divisions, the result is more effective. In this figure, for example, red > gold > blue were used.

colored text. A design like Menu 6 is somewhat superior in overall viewer preference, but requires considerably more finesse and understanding to create. Design choices like Menu 2 would thus be the most effective compromise—a balance between ease of creation and beneficial effect. It is also worth emphasizing that Menu 1, i.e., a design without any color, is preferable to those designs in which color is poorly deployed.

There is, of course, a more desirable kind of balance, where a fully competent designer can create an optimal design with the optimal viewer response. This kind of balance, however, requires either a good deal of luck to achieve or a nuanced understanding of the interaction between decorative, indicative, and informative effects.

Our observations about color-overload avoidance, briefly summarized, are as follows.

1. Color schemes must be unified; too many distinct colors in one visual disrupt this fundamental unity requirement, creating a feeling of clash, which translates into emotional-information overload. As a corollary to this unity requirement, we would further note that color schemes are interpreted based on prior associations an audience has made (e.g., red and green associated with the Christmas season in the

United States). If dominant color associations are not consistent with other information in the visual (e.g., bright reds and greens in a non-Christmas-oriented Web page), this again translates into information overload.

2. Color schemes primarily communicate emotion, but color can also indicate, drawing attention to a particular visual area (e.g., a warning note in red letters). However, this indicative function does not cancel the dominant emotional effect (e.g., red creating a general feeling of agitation). As such, vividly distinct colors, if they disrupt decorative emotional unity, must convey specific, necessary, and easily recognizable indicative information. More than three or four indicative contrasts cannot be easily processed by readers as indicatives. Contrastive colors not instantly recognized as indicative result in a feeling of clash and, again, information overload.

3. Given these principles, we have a reliable heuristic for determining when color, visual form, and textual information create a synergistic whole, and when they conflict and create information overload. With this heuristic, visual designers will be better able to resist current information-design trends that tend to overuse decoratives and indicatives, especially colorful and merely decorative imagery, at the expense of propositional information.

REFERENCES

[1] N. Amare and A. Manning, "Back to the future: A usability model of hypertext based on the semiotics of C.S. Peirce," in *Proc. Intern. Professional Commun. Conf.*, Saratoga, NY, 2006, pp. 47–56.

[2] N. Amare and A. Manning, "A Language for visuals: Design, purpose, usability," in *Proc. Intern. Professional Commun. Conf.*, Montreal, QC, Canada, Jul. 2008, pp. 1–9.

[3] C. S. Peirce, *Collected Papers*, vol. VI, C. Hartshorne and P. Weiss, Eds. Cambridge, MA: Harvard Univ. Press, 1935.

[4] C. S. Peirce, *Collected Papers*, vol. VIII, C. Hartshorne and P. Weiss, Eds. Cambridge, MA: Harvard Univ. Press, 1958.

[5] L. K. Peterson and C. D. Cullen, *Global Graphics: Color: A Guide to Design with Color for an International Market*. Gloucester, MA: Rockport Publishers, 2000.

[6] L. Holtzschue, *Understanding Color: An Introduction for Designers*. Hoboken, NJ: Wiley, 2006.

[7] R. W. Harris, *The Elements of Visual Style: The Basics of Print Design for Every PC and Mac User*. Boston, MA: Houghton Mifflin, 2007.

[8] T. Fraser and A. Banks, *Designer's Color Manual: The Complete Guide to Color Theory and Application*. San Francisco, CA: Chronicle Books, 2004.

[9] D. Whitbread, *The Design Manual*. Sydney, Australia: Univ. New South Wales Press, 2001.

[10] E. R. Tufte, *The Visual Display of Quantitative Information*. Cheshire, CT: Graphics Press, 1983.

[11] E. R. Tufte, *Visual Explanations: Images and Quantities, Evidence and Narrative*. Cheshire, CT: Graphics Press, 1997.

[12] E. Finkelstein, *How to Do Everything with Microsoft Office PowerPoint 2003*. Emoryville, CA: McGraw-Hill, 2003.

[13] G. A. Agoston, *Color Theory and Its Application in Art and Design*. New York: Springer-Verlag, 1987.

[14] G. A. Miller, "The magical number seven, plus or minus two: Some limits on our capacity for processing information," *Psychologic. Rev.*, vol. 63, no. 2, pp. 81–97, 1956.

[15] J. L. Doumont, "Magical numbers: The seven-plus-or-minus-two myth," *IEEE Trans. Prof. Commun.*, vol. 45, no. 2, pp. 123–127. Jun. 2002.

[16] Buffalo Wings & Rings. (2010). *Menu* [Online]. Available: www.buffalowingsandrings.com/#/menu Apr. 29, 2010 (last date accessed).

[17] Nuritdinov's. (2009). *Moonshine Bar Menu* [Online]. Available: nuritdinov.co.cc/wp-content/uploads/2009/06/moonshine_menu01.jpg Apr. 29, 2010 (last date accessed).

[18] USA Today. (2010). *Banner* [Online]. Available: www.usatoday.com Apr. 29, 2010 (last date accessed).

[19] Wesleyan University. (2009). *Web Updates Blog* [Online]. Available: http://webredesign.blogs.wesleyan.edu/2009/04/ Apr. 29, 2010 (last date accessed).

[20] J. R. Hays, *Restaurant and Food Graphics*. New York: PBC International, 1994.

[21] A. D. Zwicky and A. M. Zwicky, "America's national dish: The style of restaurant menus," *Am. Speech*, vol. 55, no. 2, pp. 83–92, 1980.

[22] D. K. Farkas and J. B. Farkas, *Principles of Web Design*. New York: Longman, 2002.

[23] B. Karg and R. Sutherland, *Graphic Designer's Color Handbook: Choosing and Using Color from Concept to Final Output*. Gloucester, MA: Rockport Publishers, 2003.

[24] G. Kress and T.Van Leeuwen, "Colour as a semiotic mode: Notes for a grammar of colour," *Visual Commun.*, vol. 1, no. 3, pp. 343–368, 2002.

[25] P. J. Lynch and S. Horton, *Web Style Guide: Basic Design Principles for Creating Web Sites*. New Haven, CT: Yale Univ. Press, 2002.

[26] D. Lawrence and S. Tavakol, *Balanced Website Design: Optimizing Aesthetics, Usability and Purpose*. London, U.K.: Springer, 2007.

[27] M. Sammons, *Document Design for Writers*. Chicago, IL: Parlay Press, 2007.

[28] N. Alton and A. Manning, "Refining specifications of decorative/indicative balance in menu design," in *Proc. Intern. Professional Commun. Conf.*, Honolulu, HI, Jul. 2009, pp. 1–6.

[29] Transportation Safety Administration. (2008). *20 Layers of Security* [Online]. Available: www.tsa.gov/approach/layered_strategy.shtm Aug. 27, 2011 (last date accessed).

[30] W. White, *The Elements of Graphic Design: Space, Unity, Page Architecture, and Type*. New York: Allworth Press, 2002.

PRACTICAL INSIGHTS FROM APPLIED
GLOBAL TECHNOLOGIES

As a growing provider of communication technologies, Applied Global Technologies (AGT) has not only effectively dealt with its own information overload issues but its products also help alleviate these issues for its customers.

AGT was founded in 1992 with a strong corporate culture of providing outstanding customer service. Its team of 75 employees, based in offices in Kennesaw, Georgia; and Rockledge, Florida, work to custom fit solutions to clients' communication needs. Specifically, AGT provides hardware and software technologies for high-quality video-conferencing, allowing interaction between groups that use different conferencing software and providing advanced options such as recording, streaming, scheduling, and automated dialing. But AGT's commitment does not end when a client has purchased the product. Rather, AGT also offers different levels of support, ranging from a help desk to the ability to provide an on-site employee to help clients make the most of the technology. In addition, AGT also offers a Network Operations Center (NOC). Clients can purchase their own servers or use AGT's hosted service, where clients can remotely use AGT's servers to host their conferences.

Information Overload: An International Challenge for Professional Engineers and Technical Communicators, First Edition. Edited by Judith B. Strother, Jan Ulijn, Zohra Fazal.
© 2012 Institute of Electrical and Electronics Engineers. Published 2012 by John Wiley & Sons, Inc.

As do most companies, AGT faces multiple sources of information overload. Interacting with clients through phone, e-mail, and instant messaging gets very busy, especially with new clients who are unfamiliar with the products. To address these challenges, AGT has implemented a first-order response, where trained personnel answer any general questions that clients have about the products and filter out those complaints and bugs that need to be addressed by AGT's R&D team. This shields AGT's core team of engineers and programmers from routine questions and allows them to focus on development. Furthermore, AGT's bug tracking system allows everybody involved to view the history of a bug—from when it was initially reported to when it was resolved.

However, interaction between clients (end users) and the R&D team (developers) is important. As people who have worked on the products, members of the R&D team are intimately familiar with the products' unique features and benefits, allowing them to be able to answer potential client concerns in a way the sales team may not. At the same time, client feedback and concerns allow members of the R&D team to enhance their products to better suit client needs. Effective communication in this case prevents future potential information overload. AGT's R&D team members, therefore, participate in trade shows alongside sales team members, and sales team members receive specific training so they can hone in on client feedback.

Considering AGT's primary offering, it is no surprise that distance is not a challenge for AGT when it comes to communication. In fact, employees from the two main offices are able to communicate with each other and with remote clients using the very technology AGT offers. But AGT also employs its own technologies to record meetings so that team members who are not able to attend or who are in an "overload" situation can view the meeting at a better time.

In an information overload situation, prioritizing becomes necessary for productivity. At AGT meetings, team members prioritize tasks so everybody is on the same page about what needs to get addressed first. These meetings also address client issues or new product features, thus preventing unnecessary chains of individual communication.

While different levels of employees may perceive the information overload challenge differently, it seems that overall AGT makes successful use of its own technologies to manage information effectively.

Contributor to Practical Insights from Applied Global Technologies

• Josh Lott, Software Engineer

7

COST OF INFORMATION OVERLOAD IN END-USER DOCUMENTATION

Prasanna Bidkar

It may be an information age, but . . . it takes more work to earn more money to be overwhelmed by more information that does not equal knowledge or wisdom.

Stephanie Mills

ABSTRACT

This essay puts forward a framework for calculating the cost of information overload in end-user documentation based on user behavior. According to the various definitions of information overload and cognitive load, information overload primarily occurs because of limited abilities of humans to process information. This essay looks at the causes of cognitive load in end-user documentation, which creates the perception of information overload. The effects of cognitive load, which are similar to the effects of information overload, drive a typical user's behavior. Wilson's information behavior model and data from the current survey help establish end-user behavior and the information choices users make when they fail to find or understand information. Referring to the research in understanding user behavior under information overload in other areas, the survey, and the concept of quality costs, I put forward scenarios showing how information overload can increase the cost of documentation failure for both the end user and the organization.

Information Overload: An International Challenge for Professional Engineers and Technical Communicators, First Edition. Edited by Judith B. Strother, Jan Ulijn, Zohra Fazal.

7.1 Introduction

This chapter describes a framework to calculate the cost of information overload in user documentation. I present the various definitions of information overload, the causes of information overload, and the effects as researched by scholars in different areas such as marketing and organizational management.

Information overload is generally defined as too much information. It is variously defined based on the volume of information, the rate at which information is provided, and the capacity of humans to process information. When that capacity is reached, users stop integrating information in decision making [1, 2]. Although there are three main definitions of information overload, in the context of user documentation, this chapter examines information overload from the view of limited processing capacity of users.

As explained in Section 7.2, information overload and cognitive load are essentially the same. Cognitive load in user documentation creates an information-overload-like situation and affects a specific user's behavior. Therefore, all elements of the user manual that load this capacity introduce information overload.

This research studies documentation user behavior, where the user must analyze the available information, decide if the information is relevant, and then use the information to take action. Research in other fields highlights the effects of information overload on the consumer of information. Based on this research, I compared similar situations and behavior that users may experience at their workplace because of perceived information overload. Based on the user's behavior and the basic concept of quality costs, which include the cost to create good quality products and the cost of failure, I put forward a scenario that shows the cost both the consumers and the product organization must incur if the documentation fails because of information overload.

The focus of this chapter is to identify the most important elements in user documents that can introduce noise and create an information overload situation.

7.2 Information Overload

Information is a link in the chain of transformation of data to wisdom. *Data* is defined as something given, meaning facts. *Wisdom*, on the other end, is the ability to discern inner qualities and relationships. In the transformation of data to wisdom, we attach a meaning and context to the data to create information. Information, therefore, is an act or a process of informing the data. That is, information gives a form to data that makes the data identifiable and comprehensible in the mind [3].

Information is all around us. Everything that we sense has the potential to give us some kind of information. The stop sign at the curb gives us information that we must stop before proceeding. Similarly, various billboards and banners that advertise products or sales provide us with some kind of information that we assimilate subconsciously as we move around.

The inability to define information comprehensively has been accepted by many researchers over the years [4, 5]. For simplicity, information is more often defined in the context of a domain such as communication, physics, computing, medicine, and media management. A general definition of information that may be applied across all domains is as characteristics of the output of a process [6]. To expand this definition, the characteristics that make the output information are as follows: the output should provide new information, should be true, and should be about something. Shannon and Weaver [7] defined information in bits in the context of communication engineering; however, this definition lacked any reference to the meaning of the message.

Information overload is a result of different factors and is difficult to define in a succinct manner. A general definition of information overload used loosely is the availability of too much information. Various researchers define information overload in the context of volume, time, and information processing ability of humans. After evaluating all these definitions of information overload, we can see that they are defined in the context of either limited time or limited information processing capacity of humans.

In terms of volume, information overload occurs when the amount of information available to the user is beyond the user's ability to process the information [2]. For example, a simple Google search on "origins of language" returns a whopping 61,900,000 results. Unless users can allocate unlimited time to the task, they will not be able to read all the information presented by the search, nor will all the information be relevant.

Another way of defining information overload is in the context of time. In this context, information overload occurs when the rate of information supply exceeds the time available to process the information [1,3]. All regular television viewers are familiar with the news channels that show a running news ticker. If the script moves a bit faster than the comfortable reading speed of the viewers, even when they can manage to read the script, after some time, the information gets difficult to comprehend because of the forced, consistently high speed.

Yet another definition of information overload is in terms of using the information. In this context, when information overload occurs, the user stops using the information to solve a problem or make a decision. This is often the case when the information is too difficult to understand and demands higher cognitive resources.

Neisser [8] defined cognition as all the processes that we employ to make sense of what we see, smell, and hear. These processes were first put forward in Brodbent's filter model [9]. For example, consider the steps that we go through when we hear sound. We first determine if the sound is worth the attention. This is filtering of information. Next, we try to classify the sound as maybe favorable or unfavorable. After the classification, we try to assign a specific meaning to the sound based on our prior knowledge, and then we act on this information. The information that we gather dictates our action or reaction to the sound. While the sound of soothing music will have a relaxing effect, a loud bang may make us stand up and look around. Therefore, we employ a number of processes to assign meaning and act on the information.

Any interference in these processes loads the processing ability and reduces the efficiency. Cognitive load occurs when the tasks for learning or information comprehension demand more cognitive resources than we have. The multiple processes that we go through to assign meaning to words use our cognitive resources. All the factors that hinder the processes of reading, filtering, and forming a meaning put additional strain on our processing abilities. Therefore, we can infer from all the definitions of information overload that information overload is, in essence, cognitive overload.

Almost all information overload research focuses on consequences of information overload and suggests solutions that the users or receivers of information need to implement. As engineers and technical communicators, we need to find the factors that lead to information overload. Our responsibility as creators of information is to ensure that we do not overload the cognitive resources of the users.

7.3 Causes of Information Overload

The communication process and information are entwined. Successful communication requires that the recipient understand the information and be able to reconstruct the meaning as intended by the communicator. Some definitions restrict the meaning of communication to transmission of the message. More inclusive definitions address communication as an exchange of idea or thought. Other definitions address communication in terms of whether or not the receiver acts on the message that the communicator transmits.

Littlejohn [10] classified various theories of communication in different traditions. In the context of user documentation, the cybernetic tradition and the sociopsychological tradition encompass the relevant communication variables. Cybernetic theories focus on information processing and therefore involve evaluating vocabulary, information characteristics, feedback, and so on. Sociopsychological theories look at the issues of cognition, perception, interaction, and so on. To understand this relationship between information and communication, I studied the systemic model of communication developed by Shannon [4].

According to this communication model, the message is created at the source, transmitted over a channel, and received at the destination. Shannon, in this model, introduces the concept of "noise" in the channel. Although Shannon developed his theorems for telecommunications systems, his concept of noise can be applied to any form of communication. According to Shannon, any interference in the channel that dilutes the information or hinders transmission of information is noise. Applying this model to written communication, we can look for sources of noise in communication or information systems.

User documentation tries to communicate relevant information to the user; in a way, the writer "talks" to the user in specific context in an asynchronous manner. Consistent with Shannon's communication model, all the information is transmitted through a channel to the user. In the case of print documentation, the book and the page are the channels of communication. In the case of online information, the Web pages and other audiovisual aids that carry the information are the channels of communication. Any problem with these channels creates noise.

However, the source of noise is not limited only to the channel. As Chandler [11] pointed out, human communication involves more than just the transmission of information over a wire. The content, context, and user's background are also an integral part of successful communication. As other humanistic communication models outline, a message is created and encoded by the sender, travels over a channel, and then gets decoded before it can be understood by the receiver. Rothwell [12] classified noise in communication on four levels: physical, physiological, semantic, and psychological.

Physical noise in documentation or information systems can stem from visual obstruction and interference in the media that reduce the legibility of the message, while physiological noise can stem from biological influences.

Semantic noise occurs when the receiver is not able to decode the message and fails to understand the intended meaning. In user documentation, semantic noise can occur if the writer uses ambiguous instructions or titles. For example, using "Sheet Metal Drawing" as a title in a document from the manufacturing domain can refer to both the process of manufacturing sheet metal or the process of creating a drawing for a sheet metal part.

Psychological noise in user documentation is noise that occurs due to cultural differences between the author and the reader. Other than the basic reason of cultural difference, many organizations use localization tools to reduce the cost and effort required in the translation of the user documentation and therefore increase the chances of making the instructions ambiguous.

Therefore, noise in technical communication and user documentation can occur at each stage of communication. This noise, created at the source, the channel, or the receiving end, can create cognitive load and hinder communication. Noisy information provided to the user demands more cognitive resources and creates an information-overload-like situation. Section 7.4 identifies the sources of noise that act as major contributors to information overload in each stage of communication.

7.4 Sources of Noise in User Documentation

Based on Shannon's communication model and other humanistic models of communication, the sources of noise in user documentation can be identified. Table 7.1 summarizes the sources of noise in user documentation with the corresponding research on each source.

Any disturbance or content that pollutes the information or distracts the reader from gaining information is noise and that noise contributes to cognitive load. The following section describes various sources of noise that load the readers' processing abilities and may lead to communication failure in user documentation.

7.4.1 Information Content

Information content is the most important part of any communication process. The content that the writer creates becomes the first element of communication that can bring noise into the communication process. The main objective of the user at a workplace is to solve a problem, complete a task, or make a decision [37]. The inherent characteristics of information and strategies of information design used by the writer can interfere with the user's process of understanding the content.

Users of documentation expect to receive clear and concise instructions that will help them resolve the problem and complete the task. The writer can add noise by using inappropriate style, misfit words, and incomprehensible metaphors. The English language follows a specific structure and syntax. The user constructs a meaning based on the way the group of words is structured. Spelling mistakes, grammatical mistakes, and inconsistent vocabulary can overload the user's processing ability. Noise, in the form of information

Table 7.1 Summary of Information Overload Sources in User Documentation

Stage	Source	References
Information source	Language	
	Grammar, vocabulary, style	[13, 14]
	Information characteristics	
	Number of sources, complexity, relevance, quality	[15–20]
	Visuals	
	Resolution, split attention, size and shape, cues	[21–23]
	Design	
	Typography, color, layout, organization, Gestalt principles	[24]
Channel	Navigation, hyperlinks, index, modality, push–pull systems	[25–31]
Receiver	Prior knowledge, schemas, mental model	[17, 32–36]

overload, can also occur if the user has to switch between multiple sources of information, for example, when the information required for completing a task is fragmented across different help files or types of help.

The inherent characteristics of the information can also contribute to information overload. Complexity of information affects the ability of the user to integrate the information in solving a problem, completing a task, or making a decision. If the information is complex, and the time and cost of using the information source outweigh the benefit, the user will stop integrating the information and most likely move on to the next source [20]. However, the complexity of information is relative and depends on available schemas [17].

Schema theory suggests that each individual has a specific set of schemas—concepts or frameworks—gained through experiences or learned from other sources such as reading. People use these schemas or existing mental networks of information units to understand new information. For example, a user with extensive domain knowledge has more schemas available than a novice. Therefore, the writer's understanding of the audience is crucial to providing just the right amount of information.

However, user documentation is always created for a range of audience members rather than tailored for different types of users. This increases the integral complexity of the information and can contribute to information overload. Furthermore, the writer must provide only the information that is relevant to the task and essential to complete the task. Providing extraneous information not only increases the volume of information but also puts additional load on the user to separate relevant information from the extraneous information. Therefore, the quality of content and the inherent information characteristics can overload the user's processing abilities and hinder assimilation of information.

Visuals and illustrations are an important part of user documentation. The writer uses these aids to reinforce as well as clarify the concepts and actions discussed in the text. However, for illustrations to be useful, the visuals must be placed alongside the text that refers to the visuals. If the visuals are placed on a page other than the one that contains the related text, the user has to flip the pages, causing split attention.

Similarly, when a user must refer to a print manual to complete a computer task, attention is divided between the computer screen, the input device, and the manual [38]. Although the user can handle each of these independently, the joint handling of all the three burdens the user's processing abilities [39].

Even when the user successfully navigates online documentation, the positioning, size, and viewing influence the benefits that the user gets from the illustrations in the documentation [23]. Small image sizes or poor image resolution introduce noise in the image. Similarly, lack of cueing elements such as arrows or circles used to highlight specific areas of an image put additional burden on the user to locate the interface elements discussed in the text [22].

Apart from images, other visual elements such as the typography, color, and display text are design aids that the writer uses to guide the user through the document. However, inconsistent use of the display text and colors can confuse users and hinder their ability to process the information. (See Chapter 6 for additional information about the effect of visuals on information overload.)

7.4.2 Channel

In Shannon's communication model, the channel is the conduit or the medium that carries information. In the context of user documentation, the printed document or Web page is the channel of communication because this interface carries the content to the receiver. While using the interface to communicate, the author uses various available aids to deliver the content to the user. We know that noise is anything that distracts the user from gaining information. The cognitive load theory suggests that the focus of instruction should be the instruction itself. Furthermore, load introduced because of the way instruction is organized and presented is extraneous cognitive load [28]. Therefore, anything that the authors use to present the information must not hamper the process of acquiring the skills or understanding the information.

Two extraneous factors are involved in the channel: the ease of finding information and the number of modalities used in explaining the topic. Users refer to the documentation to solve a specific problem or to complete a task. Therefore, they look for specific information whose location within the document may or may not be known to them. To locate this information, users can use either navigational aids, such as the table of contents or the index, or they can use the search function. In an online information system, we can assume that users will go through more clicks to reach the information if the location is not known. Navigation systems that are too deep or too broad reduce the user's understanding of the navigation system and reduce the user's recollection and scanning abilities [29]. An inconsistent organization scheme and classification can also disorient the user. Therefore, the author must organize the information according to the user's expectations and in a way that reduces cognitive load.

Providing audiovisual content to explain the procedures is often more beneficial than providing only the text component. However, the number of modalities used in explaining the topic can also introduce cognitive load. Use of more than one modality to explain the content may introduce cognitive load if the medium is not designed appropriately. For example, consider a screencast demonstrating a software interface. A screencast using audio commentary to explain the actions on the screen will be much easier to follow than actions explained using text balloons on the screen. Since the same sensory mechanism is used to receive information from visuals and text, presenting motion and text at the same time creates split attention, and users may fail to focus on either of the two forms of instruction [40].

Delivering the information to users is also a function of the communication channel. The two main types of information delivery systems are pull systems and push systems. An example of a pull system is online help where the action to find the information is initiated by the user. In a push system, after users initially give consent, information is provided to them at certain intervals of time. A general example of this type of system is a newsletter or an e-mail list subscription that delivers information at specific intervals. In user documentation, an example of this type of system can be the Microsoft Office Assistant used in the Microsoft Office Help.

Although these systems are designed to be intuitive, they introduce the danger of providing information that is not relevant at that particular instance. For example, the action of selecting text in Microsoft Word can lead to various actions from changing fonts to looking up synonyms. However, the Office Assistant may pop up and suggest only a few possible options. Users must first deal with the unwanted suggestions and then seek the information they need or dismiss the Office Assistant. This proactive behavior of the system can interrupt users' work and distract users from the task [26].

7.4.3 Receiver

The last part of the communication model is the receiver. Sociologists have developed communication theories that have added meaning and context to Shannon's mathematics-based communication theory. These theories help explain how the sources of information can cause information overload in user documentation. According to these theories, meaning is constructed based on prior knowledge of the subject or with the help of a metaphor that equates new information with a model that the user knows and understands. For example, an engineer who has never studied philosophy will have to expend more cognitive resources to understand the information than will a philosophy scholar.

Apart from domain knowledge, users' mental models of an information system influence their ability to use the system. When the information architecture matches the users' mental models of the system, this organization facilitates the use of the information system. However, lack of structure or use of a classification system that is radically different from a user's understanding hinders the ability of the user to find information quickly.

What users expect from the manuals may also differ from culture to culture. (See Chapter 5 for more information on cultural influences.) These differences in organization, along with errors introduced by the translation process, can hinder the users' efforts to locate and understand the information as well as load their processing capacity [41].

One more factor contributing to information overload is our limited ability to process information at any given time. Our span of immediate memory is limited, and topics with long instructions and visual clutter on a single online page severely impede the processing ability of users [32]. Similarly, the number of concurrent interfaces and windows on the screen further inhibit the processing abilities of users [35].

7.5 Effects of Information Overload on Users

Information overload affects users in different ways, and as a result, users react in different ways to cope with those effects. No matter what the effect of information overload, it always reduces the users' efficiency and restricts the achievement of the desired objective. This section looks at the effects of information overload experienced by different types of users such as executives, accountants, and consumers. The user behavior and response to information overload by these user groups provide some insight into how the reader of user documentation may behave in a similar situation.

The key to finding the right information is the ability of users to define their information need. Once users define the question, the interaction between the users and the information progresses through three stages. The first stage is the information finding stage. At this stage, the users may use navigational aids or the search function to locate the required information. After the users find the information, the second stage of interaction is the information analysis or filtering stage. The final stage of interaction is the use phase, where the user makes use of the information to solve a problem or complete a task.

For the purpose of our discussion, let us assume that the user is able to define the information requirement and formulate a correct question. Then, during the first stage, the user must find the correct information. However, the information sources or search results presented by the information system may affect the user's ability to find the relevant information. A higher number of search results or available information sources loads the cognitive abilities and reduces the user's ability to conduct a systematic search [42].

For example, a specific search on "change font" in Microsoft Word yields a hundred topic links. Overwhelmed by the available sources of information, the user experiences a declined ability to scan the information [17]. The user may not invest more time in examining the results of the search if the result title is not intuitive and instead try other approaches to find the required information. This situation further affects the user and leads to an increase in the number of alternatives searched by the user to get the required information.

During the second stage, once the user has identified the information that is relevant, information overload affects the ability to critically evaluate the information, which can result in an undesirable impact on the ability to use the information [1].

During the third stage, the user applies the information to solve a problem, complete a task, or make a decision. Failing to find the relevant information quickly leads to stress and lowers the available cognitive resources. This circle of stress and cognitive load takes most users away from the documentation and makes them look for alternative sources of information.

To find out how the readers of end-user documentation would react in a similar situation, I conducted an online survey asking respondents about their information choices in a hypothetical situation where they fail to find or understand the information they need from the user documentation.

7.6 The Current Study

Using the information about the effects of information overload on users reported by other scholars and Wilson's information behavior model [43], the current study sought to discover the information sources that people use at their workplace. While Ellis' information behavior model [44] stated that users often start with evaluating different sources of information, Kuhlthau [45] pointed out that users are often uncertain during the initial search. This uncertainty is reinforced because of information overload, and users look for alternate sources. Wilson's information behavior model is more expansive and includes how users integrate and use the information. Wilson claims that users demand information from various sources depending on the success or failure to find the relevant information.

The survey was designed to reinforce the claims by Wilson and Ellis [43,44] that users will look for alternate sources of information if they cannot find or understand the information in the user documentation because of an information-overload-like situation. The survey was also designed to determine which sources of information the users prefer to solve a problem or complete a task.

7.6.1 The Survey

The survey was conducted online with 41 respondents residing in the United States (45%), Europe (13%), and India (42%). The respondents were typically between 25 and 35 years old, with more than 60% using a computer to perform tasks on the job and 30% using a computer for software development. Ninety percent of the respondents used Microsoft Windows, spending more than 6 hours using a computer each day. The respondents were provided with four sources of information: online help, a colleague, the World Wide Web, and product support. Instructions at the beginning of the survey explained to the respondents that they were in a situation where they needed information to complete a task or make a decision. The survey then asked respondents what their first and subsequent (if the earlier ones failed) choices of information source would be to find the needed information.

7.6.2 Results and Observations

The survey results depicted in Figure 7.1 show that nearly an equal number of respondents preferred user documentation (46%) and the World Wide Web (43%) as their first choice of information. Similarly, nearly an equal number of respondents preferred consulting with a colleague (34%) and using the World Wide Web (32%) to get the required information if

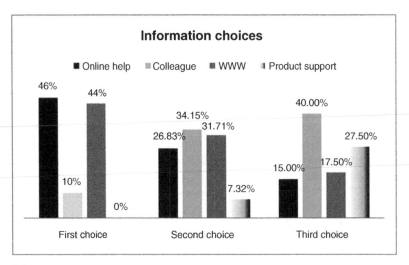

Figure 7.1 Preferred information choices. The survey results show that nearly an equal number of respondents prefer user documentation (46%) and the World Wide Web (44%) as their first choice for information. If that fails, nearly an equal number of respondents prefer consulting with a colleague (34%) and using the World Wide Web (32%) as their second choice. Calling product support was the least popular choice, with only 27% selecting this option, and only if all other options failed.

the first choice failed to satisfy their information need. However, calling product support was the least popular choice, with only 27% selecting this option as the preferred source of information, and only if all other options failed.

The trend in Figure 7.2 shows that respondents who did not find the information in user documentation moved on to the World Wide Web or a colleague as an information source.

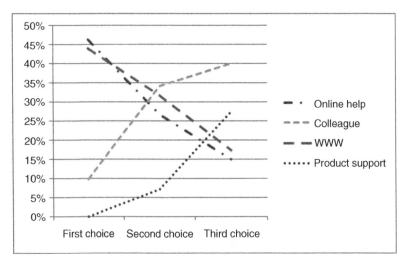

Figure 7.2 Information choice trend. Respondents who did not find the information in user documentation moved on to the World Wide Web or a colleague as an information source. If they were still unsuccessful, they resorted to calling the product support.

If they were still unsuccessful, they finally resorted to calling the product support. The survey results indicate that users first referred to the documentation, then searched the Web, then sought information from a colleague, and finally, called product support.

7.7 Cost of Information Overload

I have attempted to use the cost of quality failure, which includes internal failure costs and external failure costs, to calculate the cost of information overload. The calculation is based on the user behavior that results from information overload. Although I discuss the effects of information overload at various stages of interaction between the user and information, the cost calculation is based on the source that the user first chooses to find information. User documentation can fail because of multiple reasons: the user cannot find the information, the information is difficult to understand, or the information is not relevant. As discussed earlier, deficiencies in user documentation because of noise increase the cognitive load on the user and create a perception of information overload. This induces a specific behavior because of the stress that the user feels.

7.7.1 Cost Framework

Based on previous research and the results of my survey, I put forward a scenario where users fail to get the desired information and thus explore alternative sources until they find the information they need. First, this framework assumes an ideal scenario, where the user gets the information in the first attempt from the user documentation, and calculates the cost based on that result. Other scenarios, based on the survey findings and the information behavior model, show how failure to get the information from user documentation can cumulatively increase the costs for both the user's organization and the product manufacturer. The variables shown in Table 7.2 are used in the calculation.

Table 7.2   Variables Used to Calculate the Cost of Documentation Failure

User Organization	Product Manufacturer (The Organization Manufacturing the Product and Its Documentation)
t: Time the user spends on the documentation in hours	W: Time the writer spends creating the documentation in hours
t_1: Time the user spends on the Web in hours	R: Time the writer spends revising the documentation in hours
U: Time the user spends with a peer in hours	
S: Time the user spends with the support staff in hours	S: Time the support staff spends with the user in hours
C_1: Per-hour wage of the user	C_2: Per-hour wage of the writer
	p: Per-hour wage of the support staff
O: Cost to the user organization	C: Cost to the product manufacturer

7.7.2 Scenario 1: Ideal Scenario

In the ideal scenario, used as a benchmark, the user's cost of finding the information and the writer's cost of making good quality information available to the user are

$$O = t \times C_1 \tag{7.1}$$

$$C = W \times C_2 \tag{7.2}$$

7.7.3 Scenario 2

In this scenario, I have assumed that the user is not able to solve the problem using the documentation. Then, according to the survey, the user will use the Web to find the relevant information. In this case, the time invested by the user is the time spent on the documentation and the time spent on the Web. I included both in the calculation because the documentation has failed, so this situation will require additional time.

$$O = (C_1 \times t) + (C_1 \times t_1) \tag{7.3}$$

Most published user documentation includes a provision to give feedback to the product organization. Assuming the user provides negative feedback, the product organization must now update the documentation for either missing information or redesign based on the feedback. Therefore, the cost to the product manufacturer increases by

$$C = R \times C_2 \tag{7.4}$$

7.7.4 Scenario 3

Continuing the efforts, if Scenario 2 fails (neither user documentation nor the Web are useful), then the user will turn to a colleague to get the information. Here the assumption is that the colleague whom the user consults works in the same area and the per-hour wage for the colleague is the same. The cost, if the user talks to a colleague, now increases to

$$O = (C_1 \times t) + (C_1 \times t_1) + 2(C_1 \times u) \tag{7.5}$$

Therefore, the failure of each source of information to solve the problem cumulatively increases the cost of finding the information for the user and the user's organization. This scenario does not directly affect the cost for the product manufacturer, but ill will and customer dissatisfaction could lead indirectly to future losses.

7.7.5 Scenario 4

The final choice for the user to get information is product support. The costs incurred by the user's organization if the earlier attempts fail increases to

$$O = (C_1 \times t) + (C_1 \times t_1) + 2(C_1 \times u) + (C_1 \times S) \tag{7.6}$$

To simplify further, I assumed that the amount of time spent by the user on the documentation, the Web, the colleague, and the support staff would be equal. As a result

$$O = 5(C_1 \times t) \tag{7.7}$$

Similarly, for the product manufacturer, the cost of providing bad quality documentation increases by

$$C = (S \times p) + (R \times C_2) \tag{7.8}$$

7.7.6   An Example from the User's Perspective: Denim Corp

In this example, Jane at Denim Corp. needs to figure out how to process an order form using a product provided by Logistics Corp. but is unable to figure this out using the user documentation. The following figures were used to compute the cost of the documentation failure for the user.

Time Jane spends on the documentation $(t) = 0.5\,h$
Time Jane spends on the Web $(t_1) = 0.5\,h$
Time John (Jane's colleague at Denim Corp.) spends with Jane to explain $(U) = 0.5\,h$
Time Jane spends talking to Mary, a product support representative $(S) = 0.5\,h$
Wages paid to Jane per hour $(C_1) = US\$16.00$

Now Jane is very happy that Mary, a product support representative at Logistics Corp., has answered all her questions and she can finally process the order and move on with her work. In an ideal scenario, Jane would have found this information in the available documentation and the cost to her organization for this little hiccup would have been

$$O = t \times C_1 = US\$8.00 \tag{7.9}$$

However, because Jane could not find the information and went through all other options to get the needed information, the cost to her organization is now

$$O = (C_1 \times t) + (C_1 \times t_1) + 2(C_1 \times u) + (C_1 \times S) = US\$40.00 \tag{7.10}$$

7.7.7   An Example from the Producer's Perspective: Logistics Corp

Similarly, because the Logistics Corp. documentation failed to answer Jane's information need, the cost of supporting the product has now increased. The cost impact for Logistics Corp. in this case is calculated as follows.

Time spent by Mary on the call $(S) = 0.5\,h$
Time spent by Jason on rework $(R) = 1\,h$
Wages paid to Mary per hour $(p) = US\$20.00$
Wages paid to Jason per hour $(C_2) = US\$35.00$

If the documentation had provided the information needed to Jane, there would have been no additional cost to Logistics Corp. However, because the documentation failed, Mary must answer Jane's questions, and Jason, the software documentation specialist, must now investigate and update the user documentation. Logistics Corp. therefore incurs additional cost to respond to Jane's questions as well as revise the documentation. If Mary spends half an hour to answer Jane's question, and Jason spends 1 hour to revise the documentation, then the additional cost to Logistics Corp. would be

$$C = (S \times p) + (R \times C_2) = \text{US\$}45.00 \tag{7.11}$$

Assuming that a writer spends 3 hours to produce a unit of documentation and that the average time for rework—including understanding the defect, finding a resolution, and correcting the document—is at least 1 hour, the cost of producing the unit is US\$105 and that the combined cost of rework and support is US\$45.

Therefore, in this example, the users' organization pays a heavier cost of less than ideal documentation; however, the producer's organization must also spend approximately 40% more on supporting the product as a result of inadequate documentation. If the quality of documentation does not improve, customer dissatisfaction may lead to losing the customer for current and future products. For the product manufacturer, apart from the cost of rework and the support cost, the quality cost also includes the time invested in planning and checking the quality, which is the prevention and the appraisal cost, and the potential of losing customers.

7.8 Conclusion

Most of the current research on information overload provides solutions that include only the user or the consumer of information. As communicators and information developers, we must investigate the information source and identify potential causes that can lead to information overload. Defects such as grammatical errors, inappropriate vocabulary, and inconsistent/ineffective design in user documentation can lead to cognitive overload. Researchers have identified similar factors in other systems that load the cognitive processes as a reason for information overload.

Engineers, computer scientists, and technical communicators in industry and in academia have invested huge amounts of time and effort in making their documentation user friendly and in increasing users' efficiency. However, a number of issues remain.

This chapter is only a preliminary effort to look at causes and effects in user documentation and the resulting costs. The survey findings indicate that an equal number of users prefer user documentation and the Web as a first choice of information. This preference poses an important question of how user documentation can be augmented to redirect the user from the Web to user documentation and reduce the information overload that the users experience when they look for information on the Web.

Further research in determining when users experience information overload and exactly how much time they spend in exploring each of the alternative sources of information can be more accurately conducted using time study and other ethnographic methods, and the results can be used to calculate the cost. All of these efforts should result in decreased frustration for the user and reduced costs for both the user's organization and the product organization.

REFERENCES

[1] A. G. Schick, *et al.*, "Information overload: A temporal approach," *Account. Organiz. Soc.*, vol. 15, no. 3, pp. 199–220, 1990.

[2] S. R. Hiltz and M. Turoff, "Structuring computer-mediated communication systems to avoid information overload," *Commun. ACM*, vol. 28, no.7, pp. 680–689, 1985.

[3] "What are facts (data) and what is information?" *Essays Inform. Sci.*, vol. 2, pp. 47–48, Mar. 1974.

[4] C. E. Shannon, *Collected Papers*, A. D. Wyner and N. J. A. Sloan, Eds. New York: IEEE Press, 1993.

[5] J. Rowley, "What is information?" *Inform. Services Use*, vol. 18, no. 3, pp. 243–254, 1998.

[6] R. M. Losee, "A discipline independent definition of information," *J. Am. Soc. Inform. Sci.*, vol. 48, no. 3, pp. 254–269, 1997.

[7] C. E. Shannon and W. Weaver, *The Mathematical Theory of Communication*. Urbana, IL: Univ. Illinois Press, 1949.

[8] U. Neisser, *Cognitive Psychology*. New York: Appleton-Century-Crofts, 1967.

[9] D. E. Broadbent, *Perception and Communication*. New York: Pergamon Press, 1958.

[10] S. W. Littlejohn, "The nature of inquiry and theory," in *Theories of Human Communication*. Belmont, CA: Wadsworth Publishing Company, 1999.

[11] D. Chandler. (1994). *Transmission Model of Communication* [Online]. Available: http://www. aber.ac.uk/media/Documents/short/trans.html Aug. 26, 2011 (last date accessed).

[12] D. J. Rothwell, *In the Company of Others: An Introduction to Communication*. New York: McGraw-Hill, 2004.

[13] N. Chomsky, *Syntactic Structures*. The Hague, The Netherlands: Mouton, 1968.

[14] J. Campbell, *Grammatical Man: Information, Entropy, Language, and Life*. New York: Simon and Schuster, 1982.

[15] R. Janssen and H. de Poot, "Information overload: Why some people seem to suffer more than others," in Proc. 4th Nordic Conf. Human-Computer Interact.: Changing Roles, 2006, pp. 397–400.

[16] N. K. Malhotra, *et al.*, "The information overload controversy: An alternative viewpoint," *J. Market.*, vol. 46, no. 2, pp. 27–37, 1982.

[17] S. C. Schneider, "Information overload: Causes and consequences," *Human Syst. Manag.*, vol. 7, no. 2, pp. 143–153, 1987.

[18] K. L. Keller and R. Staelin, "Effects of quality and quantity of information on decision effectiveness," *J. Consumer Res.*, vol. 14, no. 2, pp. 200–213, 1987.

[19] R. L. Ackoff, "Management misinformation systems," *Manag. Sci.*, vol. 14, no. 4, pp. 147–156, 1967.

[20] M. A. Plumlee, "The effect of information complexity on analysts' use of that information," *Account. Rev.*, vol. 78, no. 1, pp. 275–296, 2003.

[21] P. Chandler and J. Sweller, "Cognitive load theory and the format of instruction," *Cognition and Instruction*, vol. 8, no. 4, pp. 293–332, 1991.

[22] H. Van der Meij, "The role and design of screen images in software documentation," *J. Comput. Assisted Learn.*, vol. 16, no. 4, pp. 294–306, 2000.

[23] H. Van der Meij, "Designing the display of the computer screen and the input device(s) in software documentation," in EARLI Conf., University of Twente, Enschede, The Netherlands, 1999, pp.1–15.

[24] L. H. Berry, "Cognitive effects of web page design," in *Instructional and Cognitive Impacts of Web-Based Education*, B. Abbey, Ed. Hershey, PA: Idea Group, 2000.

[25] Y. L. Theng, *et al.*, "Reducing information overload: A comparative study of hypertext systems," in PROC IEEE Colloq. Inform. Overload, Nov. 1995, pp. 6/1–6/4.

[26] M. Franklin and S. Zdonik, "'Data in your face': Push technology in perspective," *ACM SIGMOD Record*, vol. 27, no. 2, pp. 516–519, 1998.

[27] C. T. Chen and W. S. Tai, "An information push-delivery system design for personal information service on the Internet," *Inform. Process. Manag.*, vol. 39, no. 6, pp. 873–888, Nov. 2003.

[28] S. Feinberg and M. Murphy, "Applying cognitive load theory to the design of web-based instruction," in Proc. IEEE Int. Prof. Commun. Conf. Proc. 18th Annu. ACM Int. Conf. Comput. Document.: Technol. Teamwork, Sep. 2000, pp. 353–360.

[29] K. Larson and M. Czerwinski, "Web page design: Implications of memory, structure and scent for information retrieval," in Proc. SIGCHI Conf. Human Factors Comput. Syst., 1998, pp. 25–32.

[30] D. S. Niederhauser, "The influence of cognitive load on learning from hypertext," *J. Edu. Comput. Res.*, vol. 23, no. 3, pp. 237–255, 2000.

[31] H. K. Tabbers, *et al.*, "Multimedia instructions and cognitive load theory: Effects of modality and cueing," *British J. Edu. Psychol.*, vol. 74, no. 1, pp. 71–81, 2004.

[32] G. A. Miller, "The magical number seven, plus or minus two: Some limits on our capacity for processing information," *Psychol. Rev.*, vol. 63, no. 2, pp. 81–97, 1956.

[33] R. S. Owen, "Clarifying the simple assumption of the information load paradigm," *Adv. Consumer Res.*, vol. 19, pp. 770–776, 1992.

[34] C. A. O'Reilly, III, "Individuals and information overload in organizations: Is more necessarily better?" *Acad. Manag. J.*, vol. 23, no. 4, pp. 684–696, 1980.

[35] M. J. Albers, "Cognitive strain as a factor in effective document design," in Proc. 15th Annu. Int. Conf. Comput. Document.,1997, pp. 1–6.

[36] C. Darlene, "Empirical guidelines for writing computer documentation," in Proc. 3rd Annu. Int. Conf. Syst. Document., Mexico City, Mexico, 1984, pp. 44–48.

[37] J. C. Redish, "Reading to learn to do," *IEEE Trans. Prof. Commun.*, vol. 32, no. 4, pp. 289–293, Dec. 1989.

[38] S. C. Kalyuga, *et al.*, "Managing split-attention and redundancy in multimedia instruction," *Appl. Cognit. Psychol.*, vol. 13, pp. 351–372, 1999.

[39] J. C. Sweller, "Why some material is difficult to learn," *Cognit. Instruct.*, vol. 12, no. 3, pp. 185–233, 1994.

[40] R. E. M. Mayer, "Nine ways to reduce cognitive load in multimedia learning," *Education. Psychol.*, vol. 38, no. 1, pp. 43–52, 2003.

[41] T. L. Warren, "Cultural influences on technical manuals," *J. Tech. Writing Commun.*, vol. 32, no. 2, pp. 111–123, 2002.

[42] M. R. Swain and S. F. Haka, "Effects of information load on capital budgeting decisions," *Behavioral Res. Account.*, vol. 12, pp. 171–199, 2000.

[43] T. D. Wilson, "Models in information behaviour research," *J. Document.*, vol. 55, no. 3, pp. 249–270, Jun. 1999.

[44] D. Ellis and M. Haugan, "Modelling the information seeking patterns of engineers and research scientists in an industrial environment," *J. Document.*, vol. 53, no. 4, pp. 384–403, 1997.

[45] C. C. Kuhlthau, "Inside the search process: Information seeking from the user's perspective," *J. Am. Soc. Inform. Sci.*, vol. 42, no. 5, pp. 361–371, 1991.

PRACTICAL INSIGHTS FROM HARRIS CORPORATION

Harris, a publicly traded U.S. corporation with an international presence, is in the business of delivering "assured communications." It employs between 16,000 and 17,000 people worldwide and has several high-profile clients. Harris has a number of business units, many of which operate within Government Communication Systems, with 8000 employees. The organization operates in a matrix environment where programs are the profit and loss centers, and functional management groups are resource centers. As a Program Manager within Government Communications Systems, Amar Patel, who oversees the business aspects for the F-22 Raptor Program and interacts with customers, is no stranger to the challenge of information overload.

Sources of Information Overload

Patel faces three main sources of information overload, each one unique. He oversees a group of 20–25 team members, and because his program is considered an "incubator of talent," where fresh recruits are often placed to gain experience within the company before being moved to other programs, he sees a high turnover rate within his team. Patel therefore receives a lot of requests for direction from team members as well as questions about procedures.

Clients also require Patel's attention. One case that often leads to information overload is when clients change product requirements. As an example, a request from a client to change the sequence in which particular products are delivered usually results in a flurry of

Information Overload: An International Challenge for Professional Engineers and Technical Communicators,
First Edition. Edited by Judith B. Strother, Jan Ulijn, Zohra Fazal.
© 2012 Institute of Electrical and Electronics Engineers. Published 2012 by John Wiley & Sons, Inc.

communication between all parties involved, including team members, management, and the client. Such a situation also needs to be handled delicately so that nobody feels pressured—neither the client nor the team members involved in the production process.

The client's internal communication can also affect Patel's interactions with the client. The client may have buyers, engineers, and price analysts who perform different functions. When these groups within the client organization do not communicate effectively, Patel has to communicate with each group. In this case, he acts like a funnel for the information.

Patel also receives constant communication from management. Leadership's request for data, e.g., can be urgent and must be met. Getting the appropriate level of management to accept a proposal also requires extensive communication with multiple people. Sometimes, customers can communicate requests directly to management, and such requests then reach Patel's team through management.

Strategies for Dealing with Information Overload

Despite the many potential sources of information overload, Patel thrives at Harris. His own strategies for dealing with information overload, along with the tools that Harris provides, help keep the problem in check.

Some strategies are simple yet effective. Blocking time to deal with e-mail, e.g., works well for Patel, although he prefers to meet with people instead of exchanging a long string of e-mails. Patel's Instant Messenger (IM) is also linked with his Outlook Calendar so that it automatically shows him as "busy" when he is scheduled for a meeting.

Patel has also deliberately opted out of having a data plan on his company-issued cell phone as one way of separating his personal life from his professional one. Patel is grateful he has the luxury of doing so because of his administrative assistant, who can get in touch with him if something urgent arises. She also filters his work calls and reduces the amount of information Patel would otherwise have to deal with.

Organizational strategies also help with effective information dissemination at Harris. Harris' own "H-Tube" (think internal YouTube) was developed, e.g., to establish a common ground of communication, so that all members could understand what different aspects of the community were involved in. Quarterly meetings led by division presidents also help members keep in touch with the pulse of the company. Proposal wins, e.g., are announced here. This filters the downward flow of information, reducing the potential of information overload at many levels of the organization.

At Harris, the IT Department lets users know how much data is being transmitted and provides information about potential malware or spyware. Teams also schedule weekly telecoms with customers to avoid daily barrages of e-mails. Shared directory access, MS Sharepoint, and other Supply Chain Management software are used to keep the number of e-mails in check, and full webaccess management allows teams to do away with paper and file cabinets. These are just some of the many strategies that Harris uses to keep information overload in check. As Patel summarizes,

> Harris works hard to limit information overload for its employees and teams by creating a well structured collaborative work environment. Harris incorporates integrated formal data management systems with less formal team work sites, and peer-to-peer communication interfaces to foster this environment. Even with these comprehensive tools, information overload can still have an impact on employees. However, this is typically a symptom of the specific work environment and is often overcome through guidance provided by mentors and peers.

Altogether, while Patel, like other Harris employees, is surrounded by the threat of information overload, it seems that the strategies that he uses have successfully kept it at bay.

Contributor to Practical Insights from Harris Corporation

- Amar Patel, Program Manager, F-22 Raptor Program

Section II

CONTROL AND REDUCTION OF INFORMATION OVERLOAD: EMPIRICAL EVIDENCE

8

TAMING THE TERABYTES: A HUMAN-CENTERED APPROACH TO SURVIVING THE INFORMATION DELUGE

Eduard Hoenkamp

> *The difference between the right word and the almost right word is the difference between lightning and a lightning bug.*
>
> Mark Twain
> Author and Humorist

ABSTRACT

A fear of imminent information overload predates the World Wide Web by decades. Yet, that fear has never abated. Worse, as the World Wide Web today takes the lion's share of the information we deal with, both in amount and in time spent gathering it, the situation has only become more precarious.

This chapter analyzes new issues in information overload that have emerged with the advent of the Web, which emphasizes written communication, defined in this context as the exchange of ideas expressed informally, often casually, as in spoken language. The chapter focuses on three ways to mitigate these issues. First, it helps us, the users, to be more specific in what we ask for. Second, it helps us amend our request when we do not get what we think we asked for. And third, it shows how retrieval techniques can be made more effective by basing them on how humans structure information, since only we, the human users, can judge whether the information received is what we want.

This chapter reports on extensive experiments conducted in all three areas. First, to let users be more specific in describing an information need, they were allowed to express themselves in an unrestricted conversational style. This way, they could convey their information need as if they were

Information Overload: An International Challenge for Professional Engineers and Technical Communicators, First Edition. Edited by Judith B. Strother, Jan Ulijn, Zohra Fazal.

talking to a fellow human instead of using the two or three words typically supplied to a search engine. Second, users were provided with effective ways to zoom in on the desired information once potentially relevant information became available. Third, a variety of experiments focused on the search engine itself as the mediator between request and delivery of information.

All examples that are explained in detail have actually been implemented. The results of the experiments demonstrate how a human-centered approach can reduce information overload in an area that grows in importance with each day that passes. By actually having built these applications, I present an operational, not just aspirational, approach.

8.1 Introduction

In a book on information overload, the subject matter necessarily appears in many forms and guises. Yet, to make headway in practical applications, it is better to choose one form judiciously and focus on it, rather than try and take on all forms and guises at once. What I did not want is to invest in forms that only a handful of people would be interested in or one where information overload is rarely an issue. Nor did I want to conduct experiments with ambiguous results. In the spirit of exploring a topic deeply and well, rather than shallowly and superficially, this chapter looks at domains where a solution to overload is urgent, with an application that can actually be deployed, and accompanied by proof that the application makes a difference.

For a computer application to be considered a useful tool to reduce information overload, it must meet the following requirements.

- Building the software application is an investment, so the form of information under consideration must be of growing importance compared to other forms.
- The application must be useful for an extended period of time. This means that the amount of information in the domain must be steadily growing, yet not too quickly.
- Since applications are useful in the current context only if they lead to a reduction of information overload, an objective measure must be available to assess their effectiveness.

To proceed with the first point, the recent study "How Much Information?" [1] and its predecessor [2] form a good point of departure. Bohn and Short [1] dissected information production on many dimensions and quantified these on the number of gigabytes of information that someone in the United States is exposed to in a day (which is estimated at about 34 gigabytes). On that metric, computer games come out on top, because, measured in bytes, they cover more than half the information content people encounter on average. Yet people do not complain about information overload while playing computer games; on the contrary, the overload is part of the attraction. So computer games would not be a good domain to study information overload.

After eliminating other domains in a similar way, the domain that remains and, at the same time, is the most pressing is information conveyed during verbal, or word-based, communication. Bohn and Short estimated that on an average day, the average person in the United States is exposed to some 100,000 words. This takes into account the many words in the stream of Web pages, newspapers, or text messages that a person is not even aware of. Given that only a fraction of those 100,000 words is actually processed, it seems that people are already good at ignoring such information without technological support. So an application in this area needs only to weed the smaller remaining fraction actually attended to. Written communication, for example, Web pages, chat sessions, e-mail, or any other electronic equivalent of oral communication, has grown in importance through the spread of computers, most importantly through the growth of the Web. In the past, computers were mainly used by programmers and business people; today, however, lay people vastly outnumber those specialists. For example, Web designers have replaced programmers in great numbers. Today, much work that was previously the realm of Database Management Systems (DBMS) can often be done more quickly, more easily, and more effectively with regular search engines [3]. As an example in line with this, all the managers I asked preferred to google for mundane information about their own company (such as a colleague's phone number) rather than to use the available DBMS.

The second characteristic, that information delivery should grow but only moderately, is also true for written communication, which, as the report estimates, has grown only by a factor of 3.5 since 1980 [1]. The so-called information explosion notwithstanding, the amount of information that people consume is not increasing at the pace at which it is produced. Oral information allegedly also grows exponentially, but its share is diminishing according to the report: for example, even though the number of TV channels has grown to about 120 for the average household, people do not watch substantially more TV than several years ago—about 10 channels, according to Bohn and Short [1]. (Perhaps, the reader will empathize with my own experience: the more channels I can watch, the longer it takes me to realize there is nothing on.) So, not only does verbal communication form a substantial part of the everyday stream of information but also it is getting a growing share of communication in general. Also note that contrary to conventional wisdom, people have been reading more rather than less since the advent of the Web; remember that most of the Web's content is rendered in writing, just as every search request starts in writing. Especially visible is the growth of written communication as a result of social networks such as Twitter and Facebook or chat rooms such as Skype and MSN Messenger®. Given that written communication is gradually getting such a large portion of the information pie, it is there that a solution to information overload is the most pressing and hence most worthwhile.

Finally, the third quality in the list above calls for a way to evaluate how effective an application is in reducing information overload. Metrics for this have been available since the 1960s [4], and they are in use today in the realm of information retrieval. The metrics are called *precision* and *recall*. When a search engine returns documents, *precision* measures the proportion of returned documents that is actually relevant. *Recall* measures the proportion of relevant documents that is actually returned. I come back to these metrics later when I propose different search techniques, benchmarking each using precision and recall to quantify precisely how much each of these reduces information overload.

From the above overview, I conclude that Web-based written communication is a valid, viable, and promising domain of inquiry, so let me summarize how I approach this form of information overload in the remainder of this chapter.

Before the advent of the Web, the advice to stem the information tide could be informally stated as follows: "Cancel your newspaper subscription, switch off your TV, and put your cell phone in the dishwasher." The unease about such advice (or more contemporary versions of it) is obvious as well: we do not want to miss out on important information, even at the cost of getting too much. What the Web has changed is not just that we receive so much more unwanted information, but that we continually ask for information and then receive more information than we asked for or thought we asked for.

Based on this observation, this chapter dedicates a section to each of the three parts of my approach. First, it shows us how to be more specific in what we ask for. Second, it helps us amend a request when the resulting information is different from what we think we asked for. And third, it shows us how retrieval techniques can be made more effectual by founding them on how humans structure information.

8.2   Reducing Information Overload by Being Precise About What We Ask for

To understand the approach in this section, consider first how we typically look for information on the Web: we type in two or three key words and wait for the search engine to show a list of potentially relevant documents. We read the snippets of some documents and possibly open some of them. Then we are either satisfied with the result or amend the query, making it more restrictive by adding words or choosing other key words altogether. We seem to have long forgotten that the process requires us, the users, to (1) adapt to a model we have of the search engine and (2) define precisely the information we do not have in the first place. Both factors restrict our ability to be precise about what information we need.

So why do users seem generally so satisfied with search engines? There are several reasons for this. If another user has typed in the same query before, then the search engine can look up at what document the other user stopped searching. In the case of Google, there is an implicit (and patented) voting mechanism: a user who is satisfied with certain documents often links to that document. Google then automatically raises the relevance of such a document. This is the essence of Google's page ranking algorithm [5]. So when a user searches for information, potentially relevant documents are ranked according to how satisfied other users were with them before. No wonder users are generally satisfied with such a search engine; it returns documents that have generally satisfied other users.

In brief, there are several causes for potential information overload when using search engines. First, we have learned to be so thrifty with words that we have to sift through many documents before we can collect those we need. Second, documents we did not ask for distract us away from our original information need. Third, because of an implicit voting system, there are many relevant documents corresponding to the query that are not returned because so few people have seen them. And last, but not least, the search engine may decide to increase the relevance of pages that link to better paying advertisements. And as the users are satisfied anyway, they make do with suboptimal search results.

Compare this to asking another person for information: it would be odd to just utter a few key words; instead, we elaborate our information need as much as is needed to give the other person a good understanding of what we need. We pay attention to the listener's feedback to know when we can stop or when we need to elaborate even more. It is in this way that verbosity helps to reduce information overload. Since many people assume that verbosity increases information overload instead of reducing it, we carefully conducted large-scale experiments, with corpora in the order of millions of documents. What we found is that the more verbose the query, the greater the precision, i.e., the larger the proportion of relevant documents returned [6], thereby reducing information overload.

Hence, in collaboration with some of my Masters degree students at Maastricht University, I built programs to do exactly that: enable users to elaborate their information need as much as they want and in an everyday conversational style. It is more instructive to show the approach by way of examples than to describe the underlying mechanism in detail (which I defer to the last section).

8.2.1   Conversational Query Elaboration to Discover Support Groups

The first example is part of a larger effort to provide medical patients and their relatives with information about their specific disease, based on narratives supplied by other patients [7, 8]. An application was built that aimed in particular at helping women diagnosed with breast cancer to form compatible support groups [9]. With so many potentially relevant Internet sites, mere luck of the draw determines whether patients find others for mutual support. It is simply so hard to define in a few words what would lead to compatible others, that a patient will soon be overwhelmed by irrelevant information.

In our experiment, we collected diaries of patients, in part by searching on the Web but mostly by asking patients participating in the project to make their diaries available. We built a browser interface in which patients could start writing their diary while the underlying system would suggest diaries of patients they were compatible with, continually updating the order of relevance as the narrative proceeded (see Figure 8.1). Each patient, meanwhile, was free to look at other diaries, possibly amending her own story according to what she found there.

Note that instead of typing just a few words into a search engine, the patient elaborates her query on the basis of live feedback. Because the engine presents only small sets of results, the user is less distracted by potentially irrelevant material, thus reducing information overload. In our example of patients writing diaries, the process increases the odds of a patient's finding compatible support, as opposed to being overwhelmed by all sorts of irrelevant information not pertaining to her individual case. For these women, it can mean building virtual support groups to complement friends and family. And in the absence of those, it can mean the difference between a network of friends and potential isolation.

8.2.2 Constructing Verbose Queries Automatically During a Presentation

The application in the previous section shows that users are able to formulate their queries in a normal everyday conversational style. The whole apparatus of the underlying search techniques is hidden from the users, and at no time during the process do they get a feeling of information overload. The conversational style is effective and feels very natural, but obviously, the users have to be prepared to type in their input, which may be tedious. The verbosity stands in sharp contrast to the mode in which we normally interact with search engines. That mode is a tradeoff between being succinct, requiring mental preprocessing, and reducing the amount of typing, hence saving time and reducing the number of typing errors. So I thought of building an application that obviates typing altogether by allowing spoken input. What follows describes a prototype of such a system.

We all know how even the best-prepared speaker may encounter unforeseen questions during or after a presentation. Such is the case, for example, when someone in the audience has more recent data or more detailed knowledge than was presented. It would help if someone could assist the speaker in such a case.

In the application shown in Figure 8.2, a speech recognizer monitors and processes a speaker fielding a question. The explanation is analyzed by the speech recognizer and sent to a search engine, which, in turn, searches for documents related to what the user is saying. The returned information continually updates the speaker's display while the answer unfolds. The speaker can read information (about game theory in this example) that would help illustrate the answer or perhaps select part of the display to show to the audience. As part of their work for a course on information retrieval, my graduate students collaborated with me to build the prototype. The project focused on the viability of spoken input, with only minor attention to interface aspects.

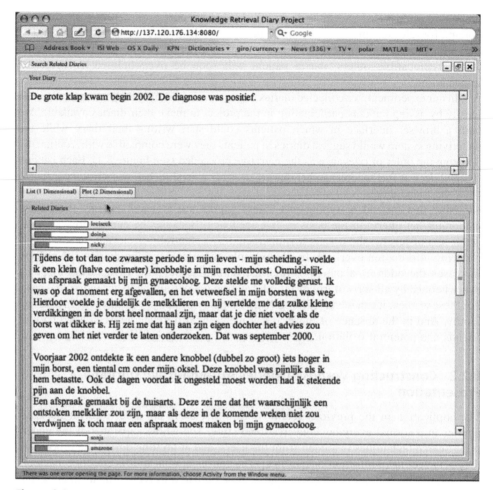

Figure 8.1 Query elaboration as an ongoing conversation.Patients in The Netherlands diagnosed with breast cancer are able to write a diary directly into the upper pane of the Web interface. During the elaboration, the system continually updates a list of existing diaries in order of relevance to the diary written so far. The patient is free to look into other diaries and amend her story or discover other compatible patients with whom to form a support group. This is a screen shot of an ongoing interaction. The upper pane is where the user inputs her story. The lower pane shows (fictitious) names of patients who have contributed diaries. The names are ranked in order of relevance to the input. When the user selects a name, the diary belonging to that person is displayed. (The example diary starts with the patient writing that during the hardest time of her life, a divorce, she discovered a lump in her breast. Her doctor told her not to worry. When two years later, she discovered another lump in her breast, her doctor referred her to a gynecologist.)

8.3 Steering Clear of Information Glut Through Live Visual Feedback

The previous section showed two examples where the conciseness required by search engines was replaced by means that help users elaborate a query in the language of everyday communication with a fellow human. This, in turn, helped users be more precise

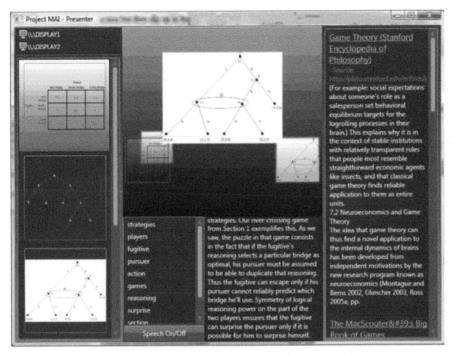

Figure 8.2 A second example of query elaboration.The picture shows the speaker's display while fielding a question about one of the slides after a talk on game theory. The audience sees only the original slide, here appearing center top. The other information results from a speech recognizer that sends what is spoken to a search engine. The left pane is reserved for diagrams and graphs found by the search engine, while verbal explanations go to the right. The speaker can select images to display or check the textual boxes for assistance.

about what they asked for and thus helped them avoid information overload. Interestingly enough, the Web seems to be the one place where people usually want what they get rather than get what they want—so much so that Google has an "I'm Feeling Lucky" button under the search window. Is the engine's performance superb? Or do people easily and willingly adapt to whatever new technology comes along?

These two explanations for why people are so happy with their search results (superb search engines and users' adaptability) leave open a third explanation that I explore next, namely that people are ignorant about recall. They appreciate the relevant documents they get and obviously cannot fret about the possibly more relevant ones they do not even know exist. So they easily accept the result as the best they can get. On the other hand, users can see snippets of returned documents and see how many of them look promising. But once a certain precision is achieved, users are in limbo about whether searching further would uncover documents that are more important. Yet, continuing to search will return other potentially irrelevant information as well. Would it not be best if users could assess the value of continuing to search?

To allow users to make such an assessment, I developed a system that provides a visual representation of documents such that they can literally focus on promising documents and steer clear of irrelevant ones. The system is based on the most prevalent representation of documents used by search engines, the so-called *document space* model. This has led to the

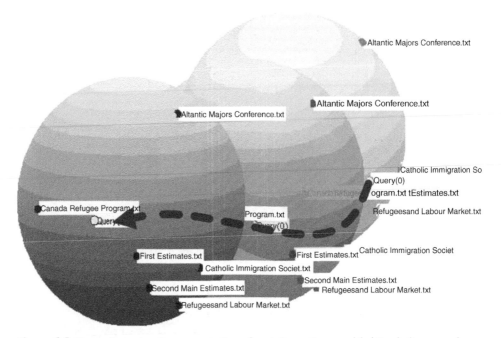

Figure 8.3 Three-dimensional representation of an information need (white dot) among documents pertaining to it (black dots). The documents that may be of interest to the user are mapped from the high-dimensional document space onto the surface of a sphere in 3-D. The user can move and rotate the sphere at will to bring the query in sight if it is not clearly visible. The three spheres represent a single sphere that has been rotated by the user to bring the query to the front to observe its location with respect to related documents. The underlying algorithm preserves the distances among documents in the high-dimensional space. The dashed line indicates the trajectory of the query during rotation. The sphere can be freely rotated to allow the user to visually zoom in on desired information (even without reading the documents) and gradually explore its neighborhood. The user can thus avoid an overload of irrelevant documents.

definition of the *document space*. This space is a vector space that has the words (more generally the "terms") as coordinates and the documents as points in that space. Using words as coordinates means there are as many coordinates as there are words, so the document space is usually of very high dimension.

Conceptually, a search engine views a query as a very small document of just a few words located somewhere in the document space. The search engine then is a means to return documents that are nearest the query and to rank them according to some ordering algorithm. The list of documents that Google returns gives the user a 1-D view of the document space, without relative distance between two documents. The actual spatial distribution, while known to the search engine, is lost this way. I developed an interface that, instead, places the documents on a sphere in three dimensions [10]. The user can freely rotate the sphere to form an impression of how similar the query is to the returned documents, as depicted in Figure 8.3.

There are more ways to explore the document space near the returned documents [11]. For example, the user can ask the system to indicate clusters of documents by a single point (the centroid), which is helpful especially when there are so many documents that the user cannot see the forest for the trees anymore. Alternatively, contours can be drawn around

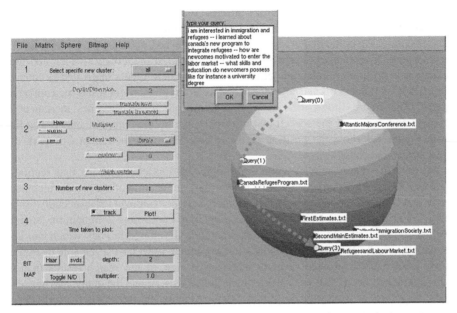

Figure 8.4 The interface for query tracking. In the panel on the left, a user can set parameters such as the degree of dimension reduction or the number of clusters to be computed and displayed. The right side displays the documents under study, centroids for clusters if desired, and the position of the query after each elaboration.

similar documents. The sphere in the interface can show a *still* picture as in Figure 8.3, to visually explore the neighborhood of the query, and a *live* picture as in Figure 8.4, to get an update of how elaborating a query influences its location in the document space. The distance between two points on the sphere corresponds to the similarity between documents as computed by the search engine. In summary, the interface allows a user to do the following:

- Select a point on the sphere. The program then opens a window showing the text of the associated document.
- Rotate the sphere to get an impression of the location of documents or to discover document clusters near the query.
- Set the number of clusters to partition the space into. This divides the documents into groups that are probably mutually related. In addition, the user can zoom in on an area and get a new sphere with a more refined view of the query's vicinity. (This looks like the zoom in Google maps, except that, for a higher resolution, the system recomputes a dimension reduction in the vicinity of the query, a technique akin to multidimensional scaling.)

In one experiment, my students and I collected online political pamphlets and official party programs in the days prior to the election of a new Dutch parliament (where many parties present their political positions). It was striking how well the political similarities were reflected by the points on the sphere that represented the party programs. In another experiment, we showed how the same interface could be used to zoom in on collections of e-mails. This procedure is shown, but for a different topic, in Figure 8.4. The documents are taken from the site of a Canadian member of parliament who made her stump speeches available online. We computed the sphere for these speeches, and then put in the following query, sentence by sentence: "I am interested in immigration and refugees—I learned about Canada's new program to integrate refugees—How are newcomers motivated to enter the labor market?—What skills and education do newcomers possess, for instance, a university degree?" The figure shows the query meandering along the speeches increment by increment and ending at the speech that most pertained to the complete query. The light dots in Figure 8.4 show the position of the query after each increment (the dark dots represent the speeches).

We used the representation while developing the application of the first section in Figure 8.1. There is a technical issue that plagues any application of the present technique, namely that continual updating is computationally very demanding [12], especially for live feedback. Fortunately, we could replace the most time-consuming computation (singular value decomposition) by the very efficient technique developed in [10], which can keep pace with the input. With the live feedback just presented, users have a versatile instrument to their avail that (1) provides a more informative view of the search results than the traditional ranked list, (2) allows them to more judiciously amend the query, and hence (3) avoids the return of irrelevant documents, thus reducing the information overload in a variety of search situations.

Mengis and Eppler can shed additional light on the use of our interface in reducing information overload (see Chapter 10). The interface just discussed addresses the aspects of what they call "quantity" and "uncertainty" (see Table 10.1). The major impediments to using visualization they mention are "a perceived lack of visualization skills, a need for simple tools to use visualizations effectively, and an insufficient amount of time to develop visuals." These impediments barely apply in our case since users found the interface very intuitive, could use it effectively through direct manipulation, and did not need time for development other than connecting the interface to a search engine. So, even if we described the interface as just one example of reducing information overload, the studies by Mengis and Eppler bode well for its adoption by users if it were deployed in practical situations.

8.4 Improving Search Engines by Making Them Human Centered

This chapter began by introducing the premise that in order to reduce information overload, it is important to be precise about what we ask for. I described several programs that implement that idea by allowing a conversational style. Yet, often, we have to describe the very information we do not possess and then there is little feedback on how close we are to describing what we need. The second section, therefore, uses the devices and tools of real-time visual feedback, a facility to gradually get nearer to the information we seek.

In this final section, I focus on the process mediating these two: the search engine that takes a query and produces search results. This section does not propose a particular technique, but rather a different approach to existing search engine technology.

The approach is based on the observation that only we, the human users, can judge whether a search engine returns the information that we are looking for. Hence, retrieval techniques could be more effective if they incorporate how humans structure information. Few, however, share this viewpoint. Browsing through proceedings of the major conferences on information retrieval, one is confronted with an approach to search that I call the hard approach: ever faster machinery, increasingly refined statistics, algorithms tweaked to specific domains, applied linear algebra, and more and more powerful classification methods. Even if research in information retrieval were to suddenly disappear, the hard approach would improve for the sole reason that computers would get faster and better.

Although great strides are still being made, one cannot escape the impression that more and more effort is needed to yield relatively small results. In economic terms, the hard approach seems to have reached a point of diminishing returns. This could mean that the information retrieval problem is practically solved or, perhaps, that the field is ready for a paradigm shift away from the dominant hard approach. The former will raise the eyebrows of anyone who looks up information on the Internet. The latter conclusion is perhaps premature. Whichever is the case, an alternative route is gaining attention as apparent in several introductions to information retrieval such as "Finding out about" [13] and "The Turn" [14]. These publications show that while in traditional structured database search there are no cognitive issues to consider, the advent of the World Wide Web is bringing these issues to the fore. The publications barely make a beginning in solving the cognitive hurdles by cognitive means. Yet, I have ardently defended this approach for many years now. Over those years, I showed again and again that cognitive principles can lead to more effective information retrieval. Compared to the hard approach just mentioned, such a softer approach reduces information overload by returning a much higher proportion of documents relevant to the query. Moreover, I found that no new cognitive theories need to be developed because there are enough cognitive principles readily available in the literature, as shown in the following discussion.

Several areas of cognitive science have obvious potential for applications in information retrieval. Linguistics comes to mind, since most searching is still text based. Indeed, compared to search in other media, text search has the longest tradition, is the most widely used, and is by far the most successful. Another area would be psychology, as it is people who search for and who evaluate the results. So applying what we know about how people structure information seems crucial for the future of information retrieval and hence for any attempt to stem the information tide. A third potential area in cognitive science is metaphors. The best example in this case is Google's document ranking heuristics [15], based on the metaphor of reputation (a source recommended by reputable sources is itself reputable).

The three examples that follow illustrate how combining the potential of linguistics and psychology can reduce information overload. The first shows how the connection between language and imagery at the *basic level* can be used for multimedia retrieval on the World Wide Web. The second applies the cognitive status of *complex nominals* to improve search results by automatically constructing specialized queries. The final example employs the notion of "semantic space" to make retrieval more effective, especially for large-scale corpora.

I am not the first to contend that cognitive science is important to reduce information overload. But what we dearly need are examples to actually support this position. The following case studies demonstrate that a comparison of the soft and hard approach can be quantified. To do so, the following steps are crucial:

- using an established principle from cognitive science;
- applying that principle to search in large corpora and the World Wide Web;
- evaluating the results on the accepted information retrieval metrics.

(The last point emphasizes the ambition to beat the hard approach on its own turf.)

8.4.1 Case 1: The Basic Level Category

An important impetus for the rapid growth of the Web has been that it made publishing practically free. At the same time, the required search technology has blossomed because it could build on several decades of research in the library sciences. As we well know, the price of storage devices has been dropping so quickly that storing large amounts of multimedia has come within almost everyone's reach, which, in turn, has motivated a growing interest in multimedia retrieval. Different from text retrieval, however, multimedia retrieval has much less proven technology to build on. So, new retrieval techniques are under development, geared toward images, video, and sound. (For an accessible overview of multimedia retrieval, see [16] or [17] in particular for multimedia evaluation metrics.)

The success of my approach (as quantified on the accepted metrics for search engines) relies on the fact that people searching for information generally look for content, not text. Hence the approach started with a representation of content and the remaining machinery was used to locate text representing similar content. As search engines still basically search in text, multimedia search on the Web uses closed image libraries that are annotated with text describing the images (e.g., Google Images™). This substitutes multimedia search with text search. Similarly, people who want their photos on Flickr® or videos on YouTube to be found need to carefully annotate the products of their artistic expression with text. Of course, that leaves out the vast number of illustrations available in texts, which remain unannotated and hence cannot be found so easily. To find such often diligently selected material, several approaches have been advanced, for example, learning algorithms [18]. Another is to recruit people to annotate manually: at one time, Google designed a game for people to "label random images to help improve the quality of Google's image search results" [19]. I follow a different avenue here: the cognitive associations between language and images. This might help in annotating automatically or obviating the annotation problem altogether.

Let us demonstrate the steps outlined above to show how cognitive science is important in reducing information overload. First, I need to base my search for multimedia on a well-established principle from cognitive science.

Principled, Cognitive Approach The principle used is known as the *basic level category*. I focus on images rather than the full spectrum of multimedia (explicitly omitting an admittedly voluminous supply of entertainment). Let me introduce the principle with an example.

When you hear the word "chair," you may imagine a chair. You can mention its parts, indicate its height, or perhaps mime how you would sit down in it. If you hear the word "kitchen chair" instead, not much will change. But for the word "furniture," the situation changes radically, no simple image pops up, and there is little to mention in terms of parts or specific motor activity. The level in a hierarchy of concepts at which such sudden proliferation of attributes occurs was investigated by Rosch [20], who called this the *basic level*. Several of the initial claims about the basic level have been subject to debate [21, 22]. Uncontested, however, are Rosch's observations about the striking connections at the basic level between imagery and language, to wit.

A. *Findings from Imagery.*
- The basic level appears to be the most abstract level for which an image can represent a class as a whole [23].
- When just the name of an object is mentioned to a subject, about the same attributes are listed as when that object is visually present [20].

B. *Findings from Language Studies.*

- Many more attributes are listed for words at the basic level than for the superordinate, and only a few are added for the subordinate [20].
- For physical objects and organisms, *parts* notably proliferate at the basic level [22] and far fewer are added at the subordinate level.
- When people have to name a picture of an object at the subordinate level, they choose the word for the basic level [20].

For example, it is difficult to imagine a picture of "furniture," let alone mention its parts. But hearing only the word "chair" (the basic level) evokes a picture, and parts can readily be named, such as legs, a seat, and perhaps armrests. But for the subordinate "kitchen chair," it is hard to name new parts not yet available for "chair."

As key words have become a serious limitation when searching nontextual material on the Web, my students and I hypothesized that the basic level might be a way to derive appropriate key words for image retrieval. This hypothesis takes this cognitive principle out of the amber in which it has been preserved for quite some time.

Automated Detection of "Information Underload" Users search the Web because they lack information and assume that it is available on the Web. While browsing, they have to deal with all the information they did not ask for, either because it is irrelevant or because it leads them astray to places they did not initially intend to visit. But most people can recall an occasion when they happened upon the information they could not find the previous time or that they suddenly realized was welcome information, without explicitly looking for it. The situation is similar to meeting someone at a conference who tells you about some work that you did not know existed, but which is very relevant to your own research. Unfortunately, such occurrences are usually rare. It is almost the opposite of information overload; instead of drowning in too much and often irrelevant information, one recognizes right away that this is information one would not want to have missed. Perhaps, the word "information underload" would be an apt description.

The good news of the next example is that such occurrences do not have to depend solely on luck. The property used here is the connection between words and images by combining (1) the evidence that objects are identified by recognizing their parts [24] and (2) the evidence that parts are important in distinguishing the basic level per se [22].

The task for the system was to find parts of an airplane (the concept) on the Web via the basic level word "airplane." I asked a student to express what he knew about parts of an airplane, and the relevant part of this knowledge is represented in Figure 8.5.

A search engine (Excite, in this case) was queried under program control until about 10,000 documents were retrieved that gave a hit on "airplane" (the basic level word). These documents were indexed and then underwent singular value decomposition for dimension reduction [25]. This technique helps reveal words that are related in meaning. The algorithm set a threshold such that it retained the words expressing parts (as defined in Figure 8.5). Words indicating parts of an airplane were prominently present. Especially noteworthy is that several parts (italicized in Figure 8.5) were found absent in the original diagram. This shows how a knowledge gap could be filled by a fully automated process [26].

This sheds interesting light on fighting information overload. Note that part of the reason all the information around us is so unsettling is that we are afraid to miss important

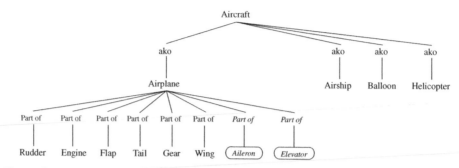

Figure 8.5 Part of a user's knowledge for the aircraft domain. The words at the basic level are airplane, airship, balloon, and helicopter. The parts in italics were not known to the user, but they were automatically derived by searching the Web at the basic level. (Note that "ako" stands for "a kind of.")

information. If we could be confident to be automatically served the information we need, there would be no reason for such fear. So, mechanisms such as the one just explained could diminish the feeling of information overload by ensuring us that we will not miss any important information.

Retrieving Images The diagram shown in Figure 8.5 is only a tiny fraction of what in the area of knowledge engineering is called an "ontology," which represents a domain of interest in a hierarchy of concepts together with their properties. In our research, we made extensive use of the Stanford ontology server to encode the knowledge we needed in our experiments. Other online ontologies we used are Cyc [27] and WordNet [28]. The latter encodes the Oxford Dictionary but with its implicit relationships made explicit. In the next example, I use the observation by [20] that the outlines of objects within the same basic level look alike. In our prototype, to each basic level concept, two pointers are added. One pointer associates the concept with a stereotype image and the other pointer associates the concept with a so-called "word sense." (Word sense is the term used in WordNet to denote the meaning of a word as distinguished from its rendering in a particular language.) Images are stereotypical in the way they are used to illustrate texts, namely, such that the objects are easy to identify [29]. (For example, a picture of a plane will typically show wings and body, instead of being viewed head on.)

The image retrieval prototype we designed [26] proceeded in two stages: a retrieval stage and a selection stage. In the *retrieval stage* (Figure 8.6), a user uses a tablet to sketch the outline of some aircraft (simplified as the oval in the lower rectangle). The outline is compared to stereotype images [29] stored in the ontology. The concept corresponding to the best matching image is looked up (the oval in the upper rectangle) and its index in WordNet is followed. The word(s) expressing the word sense are then sent to several search engines.

In the *selection stage* (Figure 8.7), the hits are retrieved and the images are isolated by looking for file extensions such as .jpg and .gif. These images are compared to the outline the user had originally drawn, and the pictures that match are shown to the user for verification.

On the surface, the procedure is similar to what is known as Query By Example (QBE) in image retrieval. In QBE, an example sketch or an image is entered and the program ranks other images based on color, shape, or texture, using some similarity measure [17]. Doing

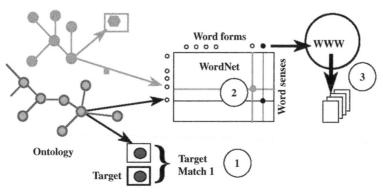

Figure 8.6 The retrieval stage for images. The ball and stick figures to the left represent a tiny part of the ontology. In the example, we assume two basic level concepts (distinguished as dark gray and light gray) with their connections to related concepts. A basic level concept has a pointer to a stereotype picture and to a word sense in WordNet. The picture is depicted as a box with a figure, for example, the hexagon in the upper left. When a user draws a target picture, here represented by the lower oval in the box labeled "Target": (1) that picture is matched to images in the ontology and the best match (here represented by the oval in the box above the target) is followed back to the concept at the basic level (the tail of the arrow), (2) a linguistic rendering of that basic level concept is found by following the pointer to a word sense in WordNet, and (3) that action associates word senses that can be used as key words to retrieve documents from the Web. All the user sees and interacts with is the pad to draw on.

this for the whole Web would mean matching potentially billions of images to the target sketch. While text search is relatively successful, at this time, for images, it is inconceivable for lack of effective indexing techniques. The power of our basic level approach is that an example sketch needs only be matched to a large but manageable number of basic level images in the ontology. The user can then select relevant pictures and let the system present only the documents containing the relevant pictures, thus preventing having attention diverted to irrelevant documents.

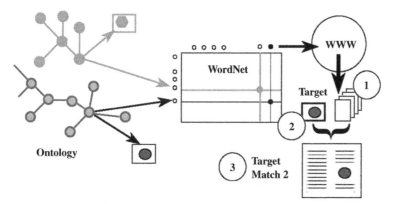

Figure 8.7 The selection stage for images. (1) Documents that contain images are collected, (2) the images are matched to the original target (not shown again here), and (3) matching images are collected and shown to the user for evaluation.

This concludes my first case of applying a cognitive principle to the core of information retrieval. In the next section, I turn to text search again, partly because that is still the dominant area of application and partly to show how a reduction in information can be objectively measured. Without such measurement, any proposal to reduce information overload would be as good as the next, or, worse, it may be an interesting proposal that does not result in reducing information overload at all.

8.4.2 Case 2: The Complex Nominal

Before going on, let me give some advice to help the reader avoid information overload from this particular chapter. Perhaps, by now, I have already convinced some readers that a more human-centered approach is viable in reducing information overload. In that case, I would advise skimming what follows for section headings that seem interesting given the reader's background. What I do in the next sections is repeat the methodology and apply it to other areas of cognition. My reason is the paucity of the approach in the literature and my desire to show that the previous case is not just one lucky strike. Also, the next sections go into more detail to actually measure the reduction of information overload and are recommended for readers interested in the cognitive approach to information retrieval.

Our next case involves the use of noun phrases, a humble grammatical category but with interesting properties. Noun phrases are a richer and more precise representation of meaning than separate key words, for example, "horse race" and "race horse" carry more circumscribed meanings than the words "horse" and "race" in isolation. For this reason, the intuition of several researchers has been that using noun phrases (instead of separate key words) should make retrieval more effective. A second reason for using noun phrases as queries is that people use them all the time to identify referents in the information they want to convey. So, retrieval systems sometimes employ a syntactic analyzer to match noun phrase queries to the documents. This could become even more powerful if such a noun phrase is extended with meaning preserving syntactic transformations (e.g., from the noun phrase "horse race" to "race for horses"). Unfortunately, empirical studies using noun phrases as queries, or using the more relaxed definition of phrases (e.g., that key words be in the proximity of one another in the text), did not confirm this intuition [16, 30].

When the performance of an algorithm clashes with intuition, one could suspect the intuition or blame the implementation (which people tend to prefer). An alternative is to trust neither and start the study from well-documented empirical data. This is the road I have taken in my own work and in the sections that follow. The research I show here isolates a general, ubiquitous, and productive subcategory of noun phrases for which the intuition does turn out to be accurate. To introduce the principle, let us first look at an instance of a noun compound: *The Hillary Clinton health care bill proposal.*

Note that the meaning of the expression is immediately clear. Yet, considering the dozens of ways the nouns might qualify one another in principle, it is truly amazing that the intended reading is picked out so quickly and easily. Such is a sign to expect difficulty in trying to equal this performance by machine. Indeed, deriving meaning from noun compounds was found to be a recalcitrant problem, from early on in artificial intelligence [31] to later attempts in linguistics [32]. But assume for a moment that search engines were clever enough to know the preferred meaning—would that not make noun compounds ideal as queries to unequivocally define referents? I show that this is indeed the case, even for the more general category of complex nominals, of which noun compounds are a special case.

The complex nominal subsumes three kinds of expressions, which have mostly been studied as separate phenomena: noun compounds such as *cellar door*, *mailbox*, and *tattoo regret,* nominalizations such as *film producer* and *convention planner,* and nonpredicate noun phrases such as *electrical shock* and *musical comedy.* These expressions share so many aspects that it is worthwhile to study them as a single category. First, note that complex nominals usually have a preferred reading where there could be many. Take "horse doctor," obviously a doctor for horses. But by analogy [33], it could be a doctor who is a horse (e.g., "witch doctor"), who uses horses (e.g., "voodoo doctor"), or who has horses (e.g., "peg leg doctor"). So for just two nouns, there are already several potential readings. But even the six nouns of the "Hillary Clinton" example above seem to evoke just one reading.

Much is known about complex nominals, and it was one of the devices my students and I used in a project that built an application for proactive information filtering [34]. Everyday, we look up information on the Web, thinking we know what information we need. However, not only do we have to describe information we do not possess, but that information may not even exist. This almost inevitably results in an overload of information that we have to weed through in search of relevant parts. The situation for proactive information filtering is almost the opposite.

There is much information available that may be useful to us, without our realizing that we need it or that it even exists (e.g., a job opening that fits our current situation exactly or a computer program that would allow us to set up a new experiment). Instead of a search engine that bombards us with irrelevant information, a proactive filter would alert us when information became available that would fill our information need. Such a filter is left on its own to construct the query. Hence, it may have to express concepts for which it does not have a vocabulary. This, however, is exactly the situation in which people use complex nominals all the time. They construct new complex nominals on their feet when needed, and may never use them again (a reviewer proposed "solid rocket fuel regression rates").

Much is known about the way people process novel complex nominals [35, 36]. Underlying our mechanism for generating complex nominals from a conceptual representation is Judith Levi's theory [33]. She proposed her theory in the framework of transformational generative grammar, in which a sentence is generated from a deep structure and a set of meaning preserving transformations.

We used a separate conceptual representation to define or derive a user's information need, and this was mapped into the representation used by Levi. From there, all grammatical renderings of the concept were generated. To find relevant passages in texts, we used an option in search engines that we called *proximity matching.* Search engines often provide this option to confine the matching of key words to passages where the key words are within proximity, for example, AltaVista's NEAR operator. (AltaVista was the search engine of choice at the time of these experiments.) We introduced what we called *coherence matching* to express that the words must not only be proximal but also influence each other's interpretation. To take examples from actual retrieval results, in proximity matching, "picture NEAR book" matches not only "She bought a well-known picture book of Italian Art" but also "He had his picture taken carrying a book."

Coherence matching allows the first phrase (". . . picture book of Italian Art") but not the second ("He had his picture taken . . ."). It also matches queries that proximity matching would not. More precisely, it matches syntactic variations, for example, "picture book" matches "book with pictures" as in "She had chosen the book with the rare pictures." It also matches semantic variation, for example, "picture book" matches "volume of portraits" as in "He published a volume of excellent portraits of contemporary scientists."

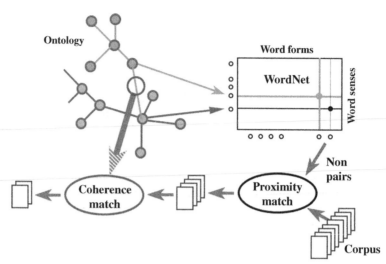

Figure 8.8 Coherence matching for a compound concept containing two subconcepts and a relation between them. Subconcepts are translated via WordNet to pairs of key words, thus losing their original relationship. Proximity matching locates potentially relevant passages in the corpus. Coherence matching subsequently reinstates the original relationships (dashed arrow), removing passages where the relationship does not hold. What remains are only the documents for which the words stand in the relationship that was true in the conceptual representation of the information need. It thus removes the information glut that would otherwise reach the user.

Different from the way search engines match queries, for coherence matching: (1) the query has to be generated first, starting from a conceptual representation, and (2) the phrase that it matches has to reflect the original relationships of that conceptual representation. Search engines normally have no access to the possible relationship between the key words because they are not mentioned by the user. As coherence matching conserves this relationship, it is more effective in locating concepts rather than words in proximity. This causes a substantial reduction of the information received by the user, as illustrated in Figure 8.8.

In the example previously mentioned, proximity matching with "picture book" located "He had his picture taken carrying a book." Coherence matching removed the irrelevant result for the reasons I discussed. The conjecture, then, is that *coherence matching reduces information overload by improving precision*. The example illustrates the method to actually prove such a conjecture. An ideal collection for our experiment would be a corpus that (1) is grammatically tagged, making the experiment independent of the power and correctness of a parser; (2) contains a substantial portion of spoken language, so that novel complex nominals are likely to occur; and (3) contains a variety of subject areas, to ensure generality of the results. Such a collection is the British National Corpus (BNC) [37], a 100-million word collection of samples of written and spoken English from a wide range of sources. Next, complex nominals had to be chosen, so they would represent a wide range. For this, we took Adams' extensive overview of word formation as reference point to collect a representative set of complex nominals [36]. Using a proximity matcher, every complex nominal exemplar we had constructed was matched against each document in the "BNC sampler," a subset of the BNC with manually checked grammatical tagging. All

T a b l e 8.1   Comparison of Proximity and Coherence Matching for the "BNC Sampler," a Corpus with Manually Checked Grammatical Tagging

	Relevant Passages	Irrelevant Passages	Average Precision
Proximity matching	36	152	0.43
Coherence matching	27	28	0.72

Precision was averaged over the 13 types of complex nominals. Matching was performed with exhaustive semantic expansion and limited stemming (-ing, -er, etc.). Coherence matching shows a precision gain from 0.43 to 0.72, which is significant (Wilcoxon, $N = 13$, one-tailed, $p < 0.01$).

matching passages were manually scored by two raters and marked as relevant if the raters agreed. Next, precision was calculated. This was repeated for all sets of NCs. The results are in the top row of Table 8.1.

In a second pass, coherence matching was performed over the documents already found during proximity matching. The reason was twofold. First, coherence matching is a stricter match than proximity matching, so there were no other documents it could match. Second, we wanted to eventually exploit the powerful search engines on the Internet for harvesting documents from the Web and see if the method is effective. The matching passages were scored by two raters, as before, and the results are summarized in the bottom row of Table 8.1. For example, the cell "proximity matching" versus "relevant passages" gives the total number of relevant passages found for the 13 types of complex nominals, matched over the total "BNC sampler." This positive result encouraged us to step up the size of the corpus, and we repeated the procedure for the complete BNC. The improvement was again substantial as the precision increased from 0.53 to 0.77. This is, in itself, not so surprising as precision usually increases with the size of the corpus. In terms of information overload, after the first pass, roughly half the information was useless overload, and the second pass reduced this to about one quarter. The human-centered approach in the last example was to search on the basis of the user's conceptual representation rather than blindly matching words.

8.4.3   Case 3: Exploiting Natural Language Properties

For a final example of the human-centered approach, I need to briefly explain the search paradigm that has only recently become a challenger of the traditional document space model. The latter represents documents as points in a high dimensional vector space, and documents that are near each other in that space are assumed to be similar in meaning. It is a so-called "bag of words" approach in that it depends almost entirely on counting words in documents, taking little or no heed of word dependencies. The new paradigm is called probabilistic language modeling or just *language modeling* in short. Documents are represented by a probability distribution over the words they contain [38, 39]. The speaker or writer is seen as source that produces words according to some distribution, and a document is just a sample of that source. Documents that have a similar probability distribution are assumed to be similar in meaning. The task of a search engine in this model is to compute distributions over documents and compare these with the distribution of a query. As opposed to the document space model, the approach makes very explicit assumptions about the mutual dependencies of words [40, 41]. The paradigm does not make any assumptions about the identity of the source. It could be anything from a mechanical machine to an alien from outer space.

My contribution from the human-centered orientation is to make the extra assumption that the source is a human speaker or writer. Imagine (or perhaps recall) that you just came back from a well-deserved vacation in the South Pacific. When someone asks you about your vacation, you are happy to recount how it was. First, you tell it to the people at home, then to your neighbors, and then to your colleagues at work. At first, there will be much variation in your story, but, by and by, when all has been said, the rendition of your experience becomes stable, mentioning only the essential parts. Or think of an event that lands as late breaking news on your paper's front page. As days go by, the story may reappear a few times, but eventually, all has been said.

Now suppose a search engine would need to return the most relevant (as opposed to the most entertaining) story about your vacation. Should it be one from the earlier stages where it still meandered haphazardly along all that happened? Or should it be one of the later more concise and orderly accounts? Our contribution is to compute that final stage from a sample in the beginning. That this is possible is not at all self-evident. But it becomes evident by posing very simple restrictions on the language that the source produces. It describes the source as a Markov chain subject to two intuitive observations about natural language [6, 42].

- Words in a natural language corpus can be separated by any number of intermediate words. (Think of adding an extra adjective before a noun. This is a property of the language and not meant to define what a reader could grasp in its entirety.)
- You can always get from one word to another by continuing to produce text (words can never be used up).

Without going into details (which can be found in [42]), these properties guarantee that the final distribution can be computed given the initial distribution as found in the document. Still, the initial distribution has to be computed; here again we use the human-centered approach by borrowing from models of human memory. Again without going into detail, we used the so-called hyperspace analog to language [43, 44], which assumes that words are closer in meaning if they are found closer together in texts. We called our model *epi-HAL* (for "ergodic process interpretation of HAL").

What is left to be proven is that the human-centered approach is more effective than one that does not incorporate it. So we compared the best performing general language model with our human-centered variation and found ours to provide superior precision on large corpora. For evaluation, we used the relevance scores that are available from the Text Retrieval Conference (TREC) Program [45], which is the official source from the National Institute of Standards and Technology (NIST) for experiments in information retrieval. It contains corpora, for example, containing the collections from the Associated Press and *Wall Street Journal*, and it also provides standard queries. These consist of a title, a short description, and a more elaborate narrative to circumscribe what documents to search for. Finally, it provides relevance scores so that precision can be computed for each query.

Table 8.2 gives results for a very large corpus (ROBUST4) consisting over 500,000 documents. The "baseline" gives precision values for a simple search engine. The columns give results when the query is the title only, the description only, or the two together. The table shows that after averaging over hundreds of different queries, the longer, more verbose query gives the most precise results. We used this finding in the beginning of this

Table 8.2 ROBUST04 Results, Comparing Mean Average Precision (MAP) for Title, Description, and Their Combination, for Baseline and epi-HAL

ROBUST04	\<title\>	\<desc\>	\<title, desc\>
Baseline	25.7	24.8	28.7
epi-HAL	31.1	31.0	33.1
Bendersky and Croft [46]	25.28	26.2	—

Number of documents: 528,155.

chapter where we replaced the standard short queries that we are accustomed to for Googling, with a more verbose, conversational style. With "conversational style" we mean mimicking the way we talk to other people, not Googling with more key words.

Although I left out the technical details of the model, the example should demonstrate the value of the human-centered approach. In this case, this approach means (1) incorporating the constraints of natural language into the search engine and (2) using a conversational style for querying the corpus. When we performed this experiment in 2008, the method proved to be superior to the best performing probabilistic language model [46] at that time. (The term *language model* is confusing. In information retrieval, the term is used to distinguish it from the traditional document space model that we introduced earlier.) In the language modeling approach, a document is represented as a probability distribution over words. That approach makes use of different ways to measure similarity between documents. Hence, in terms of reducing information overload, our model is superior as a search engine as well.

8.5 Conclusion

This chapter approaches the problem of information overload in the context of verbal communication on the World Wide Web. This is likely the fastest and most important growing source of information, and hence of information overload, for years to come. We showed how applications built on the basis of a human-centered approach could mitigate information overload by assisting the user in formulating an effective query, by incorporating the human-centered elements in the search engine, and by providing the user with live feedback on query results for amending and improving the query. This way, the human-centered approach showed its merits in the complete cycle of search on the World Wide Web.

Acknowledgments

The experiment about patients' narratives was conducted in collaboration with Regina Overberg, University Hospital of Leiden, Leiden, The Netherlands. All other experiments were conducted with students I advised for their Master's degree. They were Rob de Groot, Gijs van Dinther, and Danny Bloemendaal, University of Nijmegen, Nijmegen, The Netherlands, and Roman Atachiants, University of Maastricht, Maastricht, The Netherlands.

REFERENCES

[1] R. E. Bohn and J. E. Short,"How much information? 2009 Report on American consumers," Global Inform. Indus. Center, Univ. California, San Diego, 2009.

[2] P. Lyman, *et al.*, "How much information?" School Inform. Manag. Syst., Univ. California, Berkeley, 2003.

[3] M. Stonebraker and J. Hong, "Saying good-bye to DBMSs: Designing effective interfaces," *Commun. ACM*, vol. 52, no. 9, pp. 12–13, 2009.

[4] C. Cleverdon, *et al.*, "Factors determining the performance of indexing systems," ASLIB Cranfield Res. Project, Tech. Rep., 1966.

[5] A. N. Langville and C. D. Meyer, *Google's PageRank and Beyond: The Science of Search Engine Ranking*. Princeton, NJ: Princeton Univ. Press, 2006.

[6] E. Hoenkamp, *et al.*, "An effective approach to verbose queries using a limited dependencies language model," in *Proc. 2nd Int. Conf. Theory Inform. Retrieval (ICTIR)*, Cambridge, U.K., 2009, pp. 116–127.

[7] L. Wolf, *et al.*, "Design of the narrator system: Processing, storing, and retrieving medical narrative data," *J. Integr. Des. Process Sci.*, vol. 10, no. 4, pp. 13–30, 2006.

[8] R. Overberg, *et al.*, "Illness stories on the Internet: Features of websites disclosing breast cancer patients' illness stories in the Dutch language," *Patient Educ. Counsel.*, vol. 61, no. 3, pp. 435–442, 2006.

[9] E. Hoenkamp and R. Overberg "Computing latent taxonomies from patients' spontaneous self-disclosure to form compatible support groups," in Medical Inform. Europe, Maastricht, The Netherlands: 2006, pp. 969–976.

[10] E. Hoenkamp, "Unitary operators on the document space," *J. Am. Soc. Inform. Sci. Tech.*, vol. 54, no. 4, pp. 314–320, 2003.

[11] E. Hoenkamp and G. van Dinther, "Live visual relevance feedback for query formulation," in *Proc. 28th Annu. Int. ACM SIGIR Conf. Res. Develop. Inform. Retrieval*, 2005, pp. 611–612.

[12] M. W. Berry and R. D. Fierro, "Low-rank orthogonal decompositions for information retrieval applications," *Numeric. Linear Algebra Applicat.*, vol. 3, no. 4, pp. 301–327, 1996.

[13] R. K. Belew, *Finding Out About: A Cognitive Perspective on Search Engine Technology and the WWW*. Cambridge, U.K.: Cambridge Univ. Press, 2000.

[14] P. Ingwersen and K. Järvelin, *The Turn: Integration of Information Seeking and Retrieval in Context (The Information Retrieval Series)*. New York: Springer-Verlag, 2005.

[15] L. Page, *et al.*, "The PageRank citation ranking: Bringing order to the Web," Stanford InfoLab, Stanford, CA, Tech. Rep., 1999 [Online]. Available: http://ilpubs.stanford.edu:8090/422/ Sep. 4, 2011 (last date accessed).

[16] A. Smeulders, *et al.*, "Content-based image retrieval at the end of the early years," *IEEE Trans. Patt. Anal.Mach. Intell.*, vol. 22, no. 12, pp. 1349–1380, Dec. 2000.

[17] N. Vasconcelos, "Minimum probability of error image retrieval," *IEEE Trans. Signal Process.*, vol. 52, no. 8, pp. 2322–2336, Aug. 2004.

[18] K. Barnard, *et al.*, "Matching words and pictures," *J. Mach. Learn. Res.*, vol. 3, pp. 1107–1135, 2003.

[19] Google. (2011). *Google Image Labeler* [Online]. Available: http://images.google.com/image-labeler Aug. 27, 2011 (last date accessed).

[20] E. Rosch, *et al.*, "Basic objects in natural categories," *Cognit. Psychol.*, vol. 8, no. 3, pp. 382–439, 1976.

[21] G. Murphy and E. Smith, "Basic-level superiority in picture categorization," *J. Verbal Learn. Verbal Behavior*, vol. 21, no. 1, pp. 1–20, Feb. 1982.

[22] B. Tversky and K. Hemenway, "Objects, parts and categories," *J. Experiment. Psychol.*, vol. 113, no. 2, pp. 169–193, 1984.

[23] M. Peterson and S. Graham, "Visual detection and visual imagery," *J. Experiment. Psychol.*, vol. 103, no. 3, pp. 509–514, 1974.

[24] J. Biederman, "Recognition-by-components. A theory of image understanding," *Psychologic. Rev.*, vol. 94, no. 2, pp. 115–147, 1987.

[25] G. Eckart and G. Young, "The approximation of one matrix by another of lower rank," *Psychometrika*, vol. 1, no. 3, pp. 211–218, 1936.

[26] E. Hoenkamp, "Spotting ontological lacunae through spectrum analysis of retrieved documents," presented at the 15th European Conference on Artificial Intelligence (ECAI), Workshop on Applications of Ontologies and PSMs, Brighton, U.K., 1998, pp. 139–148.

[27] D. Lenat, "CYC: A large-scale investment in knowledge infrastructure," *Commun. ACM*, vol. 38, no. 11, pp. 33–38, 1995.

[28] G. Miller, "WordNet: A lexical database for English," *Commun. ACM*, vol. 38, no. 11, pp. 39–41, 1995.

[29] L. Schomaker, *et al.*, "Using pen-based outlines for object-based annotation and image-based queries," presented at the Visual Information and Information Systems 3rd International Conference (VISUAL), Amsterdam, The Netherlands, 1999, pp. 585–592.

[30] W. Croft, "Effective text retrieval based on combining evidence from the corpus and users," *IEEE Expert*, vol. 10, no. 6, pp. 59–63, Dec. 1995.

[31] A. Gershman, "Conceptual analysis of noun groups in English," in *Proc. 5th Int. Joint Conf. Artif. Intell.*, vol. 1. 1977, pp. 132–138.

[32] M. Johnston and F. Busa, "Qualia structure and the compositional interpretation of compounds," presented at the ACL SIGLEX Workshop on Breadth and Depth of Semantic Lexicons, Santa Cruz, CA, 1996.

[33] J. N. Levi, *The Syntax and Semantics of Complex Nominals*. New York: Academic Press, 1978.

[34] E. Hoenkamp, *et al.*, *"Profile: A proactive information filter,"* Comput. Sci. Inst., Univ. Nijmegen, The Netherlands, 1996.

[35] M. E. Ryder, "Ordered chaos: A cognitive model for the interpretation of English noun-noun compounds," Ph.D. dissertation, Univ. California, San Diego, 1994.

[36] V. Adams, *An Introduction to Modern English Word Formation*. London, U.K.: Longman, 1973.

[37] BNC. (2011). *British National Corpus* [Online]. Available: http://www.natcorp.ox.ac.uk/Aug. 27, 2011 (last date accessed).

[38] J. M. Ponte and W. B. Croft, "A language modeling approach to information retrieval," presented at the 21st Conference on Research and Development in Information Retrieval, Melbourne, Australia, 1998, pp. 275–281.

[39] F. Song and W. B. Croft, "A general language model for information retrieval," in *Proc. 22nd Conf. Res. Develop. Inform. Retrieval*, Berlekey, CA, 1999, pp. 279–280.

[40] V. Lavrenko and W. B. Croft, "Relevance-based language models," in *Proc. 24th Annu. Int. ACM SIGIR Conf. Res. Develop. Inform. Retrieval*, 2001, pp. 120–127.

[41] D. Metzler and W. B. Croft, "A Markov random field model for term dependencies," in *Proc. 28th Annu. Int. ACM SIGIR Conf. Res. Develop. Inform. Retrieval*, Salvador, Brazil, 2005, pp. 472–479.

[42] E. Hoenkamp and D. Song, "The document as an ergodic Markov chain," in *Proc. 27th Conf. Res. Develop. Inform. Retrieval*, Sheffield, U.K., 2004, pp. 496–497.

[43] K. Lund and C. Burgess, "Producing high-dimensional semantic spaces from lexical co-occurrence," *Behavior Res. Methods, Instruments, Comput.*, vol. 28, no. 2, pp. 203–208, 1996.

[44] C. Burgess, *et al.*, "Explorations in context space: Words, sentences, discourse," *Discourse Processes*, vol. 25, nos. 2–3, pp. 211–257, 1998.

[45] TREC. (2011). *Text Retrieval Conference* [Online]. Available: http://trec.nist.gov/ Aug. 27, 2011 (last date accessed).

[46] M. Bendersky and W. B. Croft, "Discovering key concepts in verbose queries," in *Proc. 31st Annu. Int. ACM SIGIR Conf. Res. Develop. Inform. Retrieval*, New York, NY, 2008, pp. 491–498.

LaQuSo
Laboratory for Quality Software

PRACTICAL INSIGHTS FROM THE LABORATORY FOR QUALITY SOFTWARE

ON THE DEVELOPMENT OF THE WEB AND THE UNCONTROLLABLE DELUGE OF UNREFINED DATA

The growth of the Internet is a steadily ongoing process; only 25 years ago, it connected about a thousand hosts, but it has grown ever since to link billions people through computers and mobile devices. Although the penetration of the Internet in North America and Europe is already high with 78.3% and 58.3%, respectively, over the last decade, it has been steadily growing at rates of 151.7% and 353.1%, respectively. For other parts of the world, this growth has even been larger, for instance, 706.9% in Asia and 1987.3% in the Middle East. In 2011, more than 2 billion people were using the Internet [1].

Most people using the Internet are actually using the World Wide Web, or the Web, which started out at the end of the 1980s as a system of interlinked hypertext documents accessed via the Internet. It evolved in the 1990s into a popular public system for data interchange in which with a Web browser, one can view Web pages that may contain text, images, videos, and other multimedia and navigate between them via hyperlinks.

The Internet and the Web in particular have been developing from a predominantly read-only source of information where individual users can only passively interact by reading the content (often referred to as Web 1.0) to a virtual community where users have the opportunities to interact and collaborate with each other in a social media dialogue as creators of user-generated content (often referred to as Web 2.0). Next may be the transition to a more Semantic Web, integrating technologies such as TV quality video, immersive

Information Overload: An International Challenge for Professional Engineers and Technical Communicators, First Edition. Edited by Judith B. Strother, Jan Ulijn, Zohra Fazal.

3-D, and augmented reality using the so-called widgets (small components that constitute a graphical user interface) and mashups (applications that take data from multiple sources and represent the combination of these), often referred to as Web 3.0.

The Web has been growing almost exponentially, and, in 2008, it already measured over a trillion unique URLs [2]. Although the exact size of the Web is considered extremely hard to measure, at this point in time, it contains more than 19.6 billion pages [3] and the amount of user-generated information is increasing with enormous growth figures. These growth figures will further increase dramatically due to people's mobile use of the Web, which has been made possible due to recent developments in smart phones and tablets and by common people, not professional publishers (often referred to as *prosumers*). However, more importantly, perhaps, is the fact that with this growth of user-generated content, we need to question the trustworthiness of the source, the quality of the data, the credibility of the original authors, and so on, as well as the sanity with which people keep modifying and forwarding it. The syndication and distribution of content has basically become uncontrollable.

Given the huge amounts of data available on the Web, people need to use search technology to retrieve and access the valuable data they are looking for [4]. (See Chapters 8 and 9.) Such technologies, however, may in the long term be contested as recent studies also indicate that these may have an adverse effect on people's capability to recall the information itself; instead, they recall *where* they can access that information. As such, the Internet may actually be becoming "a primary form of external or transactive memory, where information is stored collectively outside ourselves" [5].

One major next step in the development is to progressively evolve from a network of interconnected computers used by people to a network of interconnected objects, from books to cars, from electrical appliances to food, and thus create an *"Internet of Things,"* enabling any noninteractive objects or systems around us to be replaced by almost invisible, intelligent interactive systems, an ambient intelligence forming a natural part of our everyday lives [6]. The Internet of Things, once considered a disruptive technology for 2025, is coming of age with the potential of connecting literally trillions of devices. In fact, in 2008, there were already more devices connected to the Internet than people [7]. The Web of Things is a vision inspired from the Internet of Things where everyday devices and objects that contain an embedded device or computer are connected by fully integrating them to the Web by using Web technologies. Examples are wireless sensor networks, ambient devices, and household appliances.

The convergence of the Internet of People and the Internet of Things will allow messages from devices to be interwoven with messages from people. Until recently, such convergence was still part of active research, but now, various initiatives are underway to enable such innovative convergence within a few years. A large number of people, for instance, use personal health and well-being monitoring devices to measure (albeit sometimes with limited quality) their vital signs and to (automatically) store these on (social networks on) the Web and share the information with their family and friends, often not for medical reasons, but just for leisure and fun. Similarly, using location-based services, people are actually autonomously broadcasting their whereabouts to the Web using their smartphone, tablet, or personal navigation device, which can then be followed by other people.

Some people are also using Internet-based surveillance systems in combination with *smart homes* to notify them via the Internet or the Web when something unexpected happens in parts of their house or in the house of a family member. Also, more and more vehicles, trucks as well as cars, are monitoring their vital signs (and occasionally also those of the driver) and sending relevant data to the Internet, allowing others to observe the state of the vehicle for various purposes. We are actually indiscriminately flooding the Internet with large amounts of data, some of which is very valuable for businesses [8] and some of which has only extremely limited value (or even a negative value). Regrettably, once it is on the Web, it may take an incredible or even prohibitive amount of effort to get it removed.

However, it will become crucial to provide means of limiting the amount of information actually provided to people in a context-sensitive manner, possibly seriously limiting any noncrucial information during stressful circumstances. For instance, when a car notices that the traffic or road situation is deteriorating, it should consider no longer displaying any irrelevant messages on the windshield display. Many modern day cars already contain a large array of sensors and processors and are able to determine many of the necessary characteristics related to their speed, acceleration, direction, position, tire slippage, and so on, and various displays. In the very near future, cars will be interacting with each other as well as with intelligent traffic systems, thus providing each car access to an integral perspective on the traffic and road situation and even allowing cooperative driving. The necessary technologies are already available and are being field tested.

The proliferation of multimodal interfaces has often occurred without a solid consideration of the fundamental limitations, as well as capabilities, of human multi-sensory information processing. There remains an increasing need to refine multimodal design guidelines and principles so that they meet the needs and actual capabilities of human users. Aside from this, there is also a need to make sure that, although people are presented with information in a context-sensitive manner, eventually, they should receive all relevant information they are subscribed to. For instance, when the aforementioned car senses that the traffic or road condition has improved, it should still make sure that the driver gets any other signals that are deemed crucial, such as the trigger of a burglar alarm in his house.

Eventually, the Web (and the underlying Internet) will potentially contain trillions of actors (people as well as devices) who will be interacting with each other via all kinds of (social networking) services, and people will retrieve and access information (albeit part of it may be just data) whenever and wherever they want with a wide range of multimodal interfaces, possibly embedded in their homes, cars, offices, trains, and airplanes. When people retrieve the information, they should be offered the information in a context-sensitive way, possibly filtering and reordering the information (messages) to allow for prioritized and preferential processing. All this will have serious ramifications for how we create and distribute information as well as how we can control the level of information overload resulting from the sheer volume of unrefined data available to us.

Contributor to Practical Insights from Laboratory for Quality Software

- H. T. G. Weffers, Director, Laboratory for Quality Software, Department of Mathematics and Computer Science, Eindhoven University of Technology

REFERENCES

[1] World Internet User and Population Statistics. (2011, Mar. 31). *Internet Word Stats: Usage and Population Statistics* [Online] Available: http://www.internetworldstats.com/stats.htm Sep. 12, 2011 (last date accessed).

[2] Google. (2008, Jul. 25). *We Knew the Web Was Big* [Online] Available: http://googleblog. blogspot.com/2008/07/we-knew-web-was-big.html Sep. 11, 2011 (last date accessed).

[3] M. de Kunder. *The Size of the World Wide Web (The Internet)*, Tilburg University, Tilburg, The Netherlands [Online]. Available: http://www.worldwidewebsize.com Sep. 11, 2011 (last date accessed).

[4] J. Bughin, *et al.* (2011). *The Impact of Internet Technologies: Search*, High Tech Practice, McKinsey & Company [Online]. Available: http://www.mckinsey.com/~/media/mckinsey/dotcom/client_service/High%20Tech/PDFs/Impact_of_Internet_technologies_search_final.aspx Sep. 2011 (last date accessed).

[5] B. Sparrow, *et al.* (2011, Jun.). Google effects on memory: Cognitive consequences of having information at our fingertips, *Sci. Mag.* [Online]. Available: http://www.sciencemag.org/content/early/2011/07/13/science.1207745 Sep. 10, 2011 (last date accessed).

[6] E. Aarts and S. Marzano, Eds. *The New Everyday: Views on Ambient Intelligence*. Rotterdam, The Netherlands: 010 Publishers, 2003.

[7] Cisco. (2011, Jul. 15). The Internet of things infographic, *Cisco Blog*. D. Evans, Ed. [Online]. Available: http://blogs.cisco.com/news/the-internet-of-things-infographic Aug. 12, 2011 (last date accessed).

[8] J. Manyika, *et al.*, (2011, Jun.). *Big Data: The Next Frontier for Innovation, Competition, and Productivity,* McKinsey Global Institute [Online]. Available: http://www.slideshare.net/victori98pt/big-data-the-next-frontier-by-mckinsey Sep. 11, 2011 (last date accessed).

9

TECHNOLOGIES FOR DEALING WITH INFORMATION OVERLOAD: AN ENGINEER'S POINT OF VIEW

Toon Calders, George H. L. Fletcher, Faisal Kamiran, and Mykola Pechenizkiy

> *We're flooding people with information. We need to feed it through a processor. A human must turn information into intelligence or knowledge.*
>
> Grace Hopper
> Mathematician and
> U.S. Navy Rear Admiral

ABSTRACT

In this chapter, we provide an overview of the technological side of the information overload problem. We discuss the challenges and opportunities offered by the ever-growing and emerging stream of information from an engineering point of view. More concretely, we survey storage and querying techniques for semistructured data, data mining, and information retrieval for analyzing large data collections, and then survey stream processing techniques for online handling of continuously flowing data. In this way, we cover a whole spectrum of different levels of "structuredness" of data. At one end of the spectrum, there is data that, although only loosely structured, can still be organized in some way. At the other end, there are data streams that come in at such a fast pace that even storing them is no longer a valid option. Rather, we need to rely on immediate processing and approximate methods in order to be able to distill information from them. In between these two extremes, we have information retrieval and data mining. For all four domains, we survey the main challenges and give some insights into recent developments and techniques. We also show that from an engineering point of view, information overload is not always considered to be a problem, but that it can also offer many opportunities.

Information Overload: An International Challenge for Professional Engineers and Technical Communicators,
First Edition. Edited by Judith B. Strother, Jan Ulijn, Zohra Fazal.
© 2012 Institute of Electrical and Electronics Engineers. Published 2012 by John Wiley & Sons, Inc.

9.1 Introduction

We are confronted with an ever-growing amount of available data that can no longer be handled without technological tools. A 2010 study by the International Data Corporation (IDC), which was sponsored by the data storage company EMC, estimated that the world generated 800,000 petabytes of digital information in 2009, and that we were on track to generate 1.2 zettabytes in 2010 (1 zettabyte equals approximately 1 million petabytes, and 1 petabyte equals approximately 1 million gigabytes, so 1.2 zettabytes approximately equal 1,200,000,000,000 gigabytes!) [1].

These numbers, however, do not necessarily mean that the amount of available information has increased at the same rate; in other words, much of the data referred to above contain replicated data with huge overlaps, data with high redundancy, or simply meaningless data. For example, imagine a network of temperature sensors. Such a network may generate huge amounts of data in the form of detailed temperature readings, but be sparse in information content if temperature is stable or if the readings of the different sensors are highly correlated.

Much of the data is available in forms such as networks monitoring traffic, machines logging alarm events, and customer transactions taking place in a supermarket. Even though the information density may have decreased relatively, finding information has not become easier; more data is available, but finding the piece of information one is looking for has become the main bottleneck. From an engineering point of view, this complex information overload problem has been approached from different angles, each dealing with a different facet.

In this chapter, we give an overview of these facets and the approaches toward them. In summary, we identify the following technological challenges and opportunities connected to the information overload problem.

- Data is not only abundant but often also heterogeneous. The structure of data elements is unpredictable or data is so abundant that it is not convenient to fully study its structure and foresee all potential exceptions. Semistructured database formats [2] such as extensible markup language (XML) [3, 4] and resource description framework (RDF) [5] make it easier to describe and query such data without imposing strong restrictions on the data format. For example, the complete DBLP Computer Science Bibliography, containing more than 1.2 million bibliographic records, has been made available in the XML data format [6]. This dataset is clearly semistructured; not all publications listed in the bibliography have editor information, the number of authors differs among publications, and depending on the venue, different information may be available. XML, however, makes it possible to record some of the structure of the document. In this way, using XML query languages, such as XQuery and XPath, questions such as "Give all authors who published more papers in journals than in conferences" or "Give all pairs of authors who published more than half of their papers together" can easily be answered.
- Finding the right piece of information in large data repositories, such as the Internet, often resembles looking for a needle in a haystack. Information retrieval (IR) is the study of tools to assist information seekers. The most well-known examples of information retrieval systems are the search engines we use daily. How to connect a string of key words to the relevant documents on the Internet is one of the main research topics in this field.

- An abundance of data is not only a challenge but also an opportunity. *Data mining* [7] is the study of tools to extract models and patterns from large databases. For example, there exist tools that learn, based upon large repositories of mail classified into spam and non-spam, classification rules to sort new incoming mail into genuine mail versus spam. These classification rules are currently being used in many spam filtering systems. The process of extracting the classification rules directly from data is what constitutes data mining—the automatic extraction of information from large amounts of data with minimal human intervention.
- Some devices, such as high-frequency sensors or network traffic monitors, generate data at such a high pace that it is no longer possible to store and analyze the complete data set. Such continuous flows of data are often referred to as *data streams* [8]. Analyzing data streams requires a significant paradigm shift as traditional methods cannot work in such a context. One example of stream processing is automatically tracking when the IP traffic over a network link suddenly increases. Such an increase could indicate, for example, that a network attack is taking place. As such, it is important to process the data immediately, without first storing it.

The structure of the chapter is as follows: in Section 9.2, we survey the challenges and opportunities of information overload more in depth and illustrate them with examples. In Section 9.3, we discuss the research efforts in semistructured data management. By considering the decrease in the amount of structure in the data, we arrive at information retrieval in Section 9.4. Then, we present the tools to automatically extract useful information and patterns from the data mining field in Section 9.5. We finally arrive at the other end of the spectrum in Section 9.6 with data streams—data sources generating data at such a fast pace that traditional data processing techniques fail and we have to rely on approximate computing. We conclude the chapter with a discussion in Section 9.7 on the future of data processing as it is envisioned by the authors of this chapter.

9.2 Information Overload: Challenges and Opportunities

The traditional approach toward data storage assumes that data is homogeneous and stable over time and that its structure is predictable. Due to the limited amounts of data, these features were true for a very long time. As data was generated at a modest pace, it paid off to carefully design data gathering, thoroughly study the structure of the generated data, and design special-purpose storage systems. In many fields, however, we are now confronted with massive and diverse flows of information. We can compare the situation with that of a physical library versus the Internet; physical libraries have a steady, yet modest, input of new books. These books are categorized by the librarian and indexed on title, author, and genre. Keeping this index up to date is a time-consuming and laborious job. For the Internet, such a structured index is not an option. The information sources are simply too diverse to capture in one index, the data is not stable, the content can constantly be changed, and the amount of data is unprecedented. From this perspective, the distinction between data versus information is often made; having access to data does not necessarily imply that one has information. The best place to hide a leaf is in the forest; the best place to hide information is in data.

The first problem we have to cope with is the heterogeneity of data. Traditional database technology requires an a priori knowledge of what data can be expected in which form. For example, if a company wants to use a database to store its customer data, the first step is determining in detail which information about the customers will be stored and in what form. Here, it is important that we assume that the data organization is fixed before the data is gathered. This assumption, however, is no longer valid. All too often, data is available already, but it is spread over different sources, is in many different formats, and is often incomplete.

This reality required a paradigm shift in database researchers' minds and opened up the whole new research area of *semistructured data*. The mission of this research area can be summarized as follows. Accepting the reality that data often lacks a solid and fixed structure and is heterogeneous and often incomplete, what can we do as a database community to support storage and querying? Often geared by industry, new, less rigid data storage structures, such as XML [3, 4] and RDF [5], and flexible querying paradigms, such as XPath [9] and XQuery [10], have been developed. In Section 9.3, we provide more detail on the issue of semistructured data management and give some examples of data representation and querying when the data structure is unknown.

Particularly illustrative for information overload is the plethora of news services. While in the old days there were tens of newspapers being printed daily, currently, almost all newspapers constantly update real-time news sites and virtually everyone with Internet access can post blog messages. Indisputably, this has led to a greater freedom of speech and diversification. But the freedom has come at a price. Even though media and news are more diverse and accessible, it has become increasingly more difficult to find the right piece of information, its source, and its trustworthiness. The need for support in finding the right piece of information and assessing its quality are the main driving forces behind the information retrieval [11] research field, which is mainly concerned with textual documents, yet approaches have also been developed to deal with areas such as image retrieval [12] and music retrieval [13]. For assessing the quality of a document, techniques such as PageRank [14] and HITS [15] have been developed. In Section 9.4, we provide more detail about key-word-based searches and the intuition behind the PageRank and HITS algorithms.

Data mining is another example of using technology to deal with information overload. The goal of data mining is literally to find useful and novel patterns and trends in large amounts of data [7, 16]. In data mining, researchers consider the abundance of data rather as an opportunity than a challenge. Some artificial intelligence researchers consider the Internet as the realization of an old artificial intelligence dream: a database storing the collective knowledge base of humankind. Several applications have already shown that publicly available and collaboratively generated information repositories such as Wikipedia can be used to semantically enrich information retrieval queries and construct ontologies, i.e., a formal representation of sets of concepts within a domain and the relations between them [17]. For example, based upon an information repository of paintings, it might be possible to automatically build an ontology expressing that a painting is a piece of art, a painter is an artist generating paintings, an artist is a person generating artworks, and so on. In other words, an ontology for a certain domain represents concepts and relations between them in a form that can be used by a computer. Besides providing the benefit of making large quantities of data available, data mining can also be used to help manage data and focus on interesting parts.

The best illustration of how data mining can assist us in ordering continuous flows of information is spam filtering. Many of the popular spam filtering techniques, including Spam Assassin [18] and Spam Bayes [19], apply data mining and machine learning techniques to screen e-mail messages. Spam Assassin uses genetic algorithms to select the best set of rules, whereas Spam Bayes uses a naïve Bayesian network in which the weights have been trained on a training corpus [19].

The last area we discuss is data stream analysis, which originates from the problem that sometimes data is generated at such a high pace that it is no longer possible to store it all, let alone analyze it. Examples are monitoring IP traffic [20] in computer networks and sensor readings from larger sensor networks [21]. Analyzing the data generated by these networks is, however, very important for detecting anomalies, which could indicate, e.g., a network attack. Data available under such a context is often called a data stream. Under this context, seemingly simple tasks, such as identifying the IP address with the most network traffic, are already quite challenging, and often, only approximate methods can be used. In Section 9.6, we use an example to illustrate the difficulty of some tasks that are trivial in a static setting and provide some intuition on how data stream research helps alleviate those problems.

9.3 Storing and Querying Semistructured Data

Traditionally, data management technologies have focused on data produced and consumed in relatively static and isolated environments. A prototypical use case is the data management component of the information system internal to some business activity. Such data management solutions have evolved from the ad hoc specially tailored technologies in the 1960s and 1970s to general-purpose solutions, which are based on the relational database model and which have been dominant since the 1980s. The relational database model is targeted for formally representing and reasoning over rigidly structured data, such as that found in monolithic business information systems, for example, a system storing orders of clients and the delivery status of these orders. As data sharing and integration across previously isolated information systems became increasingly common in the 1980s and 1990s, technologies for evolving and transforming structured data within the relational model emerged [22, 23]. The development of such technologies remains an active area of research and development. A more detailed discussion of the interesting back history of data management technologies can be found in [24].

As discussed in Section 9.1, while such monolithic data management solutions continue to play a vital role in contemporary information ecosystems, they are inherently ill-suited to deal with many aspects of the deluge of data experienced by each of us, leading to information overload. Typically, this deluge consists not only of highly structured data but also of semistructured data often encountered on the Web, such as Web pages, e-mails, blogs, chat messages, pictures, and office documents. In this context, structure is no longer predictable. Although traditional technologies for managing structured data are quite mature, fully supporting efficient querying over extremely large data sets, such scalability and efficiency are even more crucial in management of Web-scale semistructured data collections.

In this section, we discuss two contemporary approaches to modeling and working with semistructured data, namely, XML and RDF. XML is the foundation of a wide range of industrial strength information management tools and has been adopted in a broad variety of domains. RDF, which is currently in an early stage of development, serves, among other numerous applications, as the basis for the linked open data vision for publishing and sharing data on the Web [25]. Both data models originate and are used in the context of Web information management. Driven by the need for flexible query techniques, we also discuss typical languages for these data models, namely, XPath for XML data and SPARQL for RDF data.

9.3.1 XML as a Data Format for Semistructured Data

XML is a syntax, governed by a W3C international standard, for representing hierarchical data [4]. XML has proven highly effective in a wide variety of environments for the creation and sharing of loosely structured data. We illustrate XML via the sample document fragment presented in Figure 9.1. Here we have information encoded about students. Note that XML cleanly supports semistructured features such as missing information (e.g., neither Hayes nor Kawasaki have known mailing addresses) and variation in structure (e.g., unlike the other students, Kawasaki's contact information is nested in a contact field). Clearly, data like this is flexible and not as easily representable in structured data models.

```
<students>
      <student id="00001"> <name>A. Smyth</name>
            <address>Rue del'Hopital 511, Paris, France</address>
            <courses><course name="Advanced databases"/></courses>
      </student>

      <student id="00002">
            <name>B. Hayes</name>
            <courses>
                  <course name="Advanced databases"/>
                  <course name="Introduction to art history"/>
            </courses>
      </student>

      <student id="00003">
            <contact>
                  <name>C. Kawasaki</name>
                  <phone>9998887777</phone>
                  <email>kawasaki@example.org</email>
            </contact>
            <courses><course name="Philosophy of science"/> </courses>
      </student>
</students>
```

Figure 9.1 Fragment of a student XML data set. XML is effective for the creation and sharing of loosely structured data and supports semistructured features such as missing information (e.g., neither Hayes nor Kawasaki have known mailing addresses) and variation in structure (e.g., Kawasaki's contact information is nested in a contact field).

In addition to capturing hierarchical semistructured data, related XML technologies support flexible querying over XML documents. XPath [9], also a W3C standard, is a core language for retrieving relevant pieces of an XML data set. XPath is the basic building block of other, more powerful languages, such as XQuery. XPath supports querying data with an a priori unknown structure. As an illustration, suppose we want to retrieve student names from the data of Figure 9.1 but do not know beforehand exactly where they are located in the data. In XPath, we can describe such pieces of data in the data set navigationally as

```
//student//name
```

Here, the double slash "//" represents navigation down the document hierarchy to some arbitrary depth. We use this because we do not know exactly where names are located beneath students in the data. The output of this query on the data in Figure 9.1 is A. Smyth, B. Hayes, and C. Kawasaki.

To further illustrate the power of XPath, consider the following query: of these student names, suppose we want only those for which we also have a mailing address. In XPath, we could write this query as

```
//student[.//address]//name
```

Here, the brackets [] indicate a predicate (i.e., a filter) on students, requiring that somewhere below the student, we must have an address. The output of this query on the data in Figure 9.1 is just A. Smyth.

From XML and its most important query language, XPath, we move to another semistructured data format, i.e., RDF.

9.3.2 RDF as a Data Format for Semistructured Data

Often, semistructured data does not abide by a natural hierarchical structure. Indeed, a wide variety of naturally occurring data exhibits a looser graph structure, e.g., social networks; biological, physical, and chemical networks; and technological networks. Although one can encode such data in XML, its hierarchical syntax is often unnatural and unwieldy for these applications. Therefore, we discuss a relatively new data model for such semi-structured data, called the RDF, which further relaxes the structure found in XML.

The RDF data model is also governed by an international standard of the W3C [5]. Indeed, RDF data is often serialized in an XML syntax for data sharing over the Web. The basic idea of RDF is to represent all data as statements that consist of triples of data objects often denoted by URIs. A triple <subject predicate object> is typically interpreted as "subject has relationship predicate to object." We illustrate RDF with a fragment of a student data set in Figure 9.2. Here, for example, we see that Smyth and Hayes are students, and Smyth knows Hayes.

```
<00001 typeOf    student>

<00002 typeOf    student>

<00003 typeOf    student>

<00004 typeOf    student>

<00001 knows     00002>

<00001 friendOf  00003>

<00002 knows     00003>

<00002 enemyOf   00004>

<00004 friendOf  00001>
```

Figure 9.2 Fragment of a student RDF data set. RDF data is often serialized in an XML syntax for data sharing over the Web. The basic idea of RDF is to represent all data as statements, which consist of triples of data objects often denoted by URIs. A triple <subject predicate object> is typically interpreted as "subject has relationship predicate to object." Here, e.g., we see that Smyth and Hayes are students, and Smyth knows Hayes.

```
<knows typeOf    socialRelationship>

<friendOf typeOf    socialRelationship>

<enemyOf typeOf    socialRelationship>

<00001 name      "A.  Smyth">

<00001 address   "Rue  de  l'Hopital  511,  Paris,  France">

<00002 name      "B.  Hayes">

<00003 name      "C.  Kawasaki">

<00004 name      "D.  Gonzales">

<name       domain    student>

<name       range     UnicodeString>
```

Figure 9.2 (*Continued*)

RDF is at the heart of a suite of standards for modeling and using semistructured data. W3C's standard query language for RDF is SPARQL (SPARQL protocol and RDF query language) [26, 27]. SPARQL provides a simple and intuitive syntax for extracting relevant information from an RDF data set. The construct of SPARQL is a basic graph pattern, which is essentially a small RDF data set wherein some data values are replaced with variables, represented as ?x for some string x. The semantics of a basic graph pattern, evaluated on an RDF data set, is a set of bindings for the variables, which makes the pattern hold in the data. For example, if we wish to locate student names in the data set of Figure 9.2, we can express this query in SPARQL as in Figure 9.3.

```
SELECT   ?name

WHERE  { ?id typeOf  student.

         ?id name    ?name }
```

Figure 9.3 SPARQL query to locate student names. SPARQL (SPARQL protocol and RDF query language) provides a syntax for extracting relevant information from an RDF data set. For example, if we wish to locate student names in the data set of Figure 9.2, we can express the query as shown here. The output of this query would be A. Smyth, B. Hayes, C. Kawasaki, and D. Gonzales.

The output of this query would be A. Smyth, B. Hayes, C. Kawasaki, and D. Gonzales. If we wish to retrieve again only the names of those students with known addresses, we can express this as in Figure 9.4.

```
SELECT   ?name

WHERE  {  ?id     typeOf    student  .

          ?id     name      ?name .

          ?id     address   ?address   }
```

Figure 9.4 SPARQL query to find the names of students with known addresses. If we wish to retrieve the names of only those students with known addresses from the data set in Figure 9.2, we can express the query as shown here. The output of this query would be only A. Smyth.

The output of this query would be A. Smyth, i.e., there is now only one valid binding for the variable ?name in the data set. Readers familiar with SQL, the standard language for querying relational databases, will immediately recognize the similarities in syntax between SQL and SPARQL.

A crucial aspect of RDF is that, unlike in the relational and XML models, there is no strict distinction between data and metadata. For example, while name is a tag value in our XML student data, in our RDF data, we can see that name itself is at times metadata (e.g., describing a relationship between 00001 and A. Smyth) and at other times a piece of data itself (e.g., participating in a relationship with student). To illustrate the power of this flexibility, consider the query "Who are the people socially related to those people known by A. Smyth?" We can express this in SPARQL as in Figure 9.5.

```
SELECT     ?person2

WHERE  { ?id        name        "A.  Smyth".

         ?id        knows       ?person1   .

         ?person1   ?relation   ?person2   .

         ?relation  typeOf      socialRelationship}
```

Figure 9.5 SPARQL query to find people socially related to people known by A. Smyth. Unlike the relational and XML models, RDF does not make a strict distinction between data and metadata. For example, we can see that name itself is at times metadata (e.g., describing a relationship between 00001 and A. Smyth) and at other times a piece of data itself (e.g., participating in a relationship with student). Consider the SPARQL query shown here: "Who are the people socially related to those people known by A. Smyth?" We are querying over the "meta"-data itself, as data, via the variable ? relation. The output of this query would be the students 00003 and 00004.

Here, we are querying over the "meta"-data itself, as data, via the variable ?relation. The output of this query would be the students 00003 and 00004.

9.3.3 Remarks on the Use of XML and RDF

XML technologies for storage and querying are reaching a mature level and are incorporated into many major commercial data management products. These technologies are clearly major competitors of traditional structured data management solutions for XML data sets. RDF technologies, on the other hand, are still in the early stages of research and development. There are relatively few RDF data management tools available commercially. Given the high level of activity of research, development, and use of RDF data and tools, both in academic and industrial communities, we can reasonably expect to see RDF technologies achieving a similar level of maturity in the next decade.

9.4 Techniques for Retrieving Information

IR was defined in the 1960s as "a field concerned with the structure, analysis, organization, storage, searching, and retrieval of information" [11]. Boolean and parametric search were the main IR paradigms of that time. This type of IR is very efficient and precise; users get exactly what they are asking for. For professional searchers who use Boolean IR in specialized applications that have relatively small collections of textual documents, constructing queries such as "Find me all the documents containing *democracy* and *freedom* but not *Mubarak* as key words" has been acceptable practice.

During the 1970s and 1980s, the focus was still on document (textual) retrieval. However, as the repositories were becoming larger, it was becoming more and more problematic to analyze the complete set of results. Since each document available in the repository was considered to be either relevant or not relevant, ranking of the results was hardly possible, and, consequently, a searcher found it difficult know which of the relevant documents are more or less relevant. Key-word-based retrieval relying on vector space, on probabilistic, and on (statistical) language models came to the rescue. Now, instead of considering the presence or absence of particular query terms in a document, the importance of each key word in describing the document and distinguishing it from the other documents in the collection is taken into account. At the same time, a query is matched to each of the documents.

Term frequency–inverse document frequency (tf–idf) weighing of terms has been one of the most popular weighing schemes in the vector space retrieval model. This model assumes that each document and each query is presented in a multidimensional vector space and the relevance of each document to the query is computed as a cosine similarity between the query and the document vectors. The higher the term frequency (how many times the word appears in a document) and the inverse document frequency (how many documents contain the word) are, the more important the term is for characterizing the document in the collection. As an example, if *democracy* appears often in Document A but does not appear in many other documents, this term is good for characterizing A. In probabilistic models and language models, similar ideas are explored, but they come from different foundations. An important aspect here is that documents could be ranked with respect to their similarity to a query, sim (document$_i$, query), in vector space retrieval, to the probability of relevance of the document to a query, P(relevant | document$_i$, query), in probabilistic retrieval, or to the probability of generating the query from the document language model, P(query | Model$_{document\ i}$), in statistical language retrieval. Consequently, a user searching for some important documents on a subject would be able to start reading from the most relevant documents and stop where needed. In addition, with the vector space retrieval model, querying by example ("find me other documents similar to this one") becomes possible.

The aim of IR systems, in general, is to retrieve only relevant documents. The fewer irrelevant documents an IR system retrieves, the higher precision it has. The more relevant documents it can retrieve, the higher recall it has. To complicate things even more, what is relevant or irrelevant is highly subjective and depends on the user's preference. Since matching of a query to documents is no longer exact and the strength of a document's relevance is typically expressed by a score between 0 and 1, an IR system can be optimized to have a higher precision (by retrieving only those documents that have scores closer to 1) or higher recall (by selecting more documents with lower relevance scores and thus trading off some of the precision).

Depending on the user's needs from an IR system, either precision or recall can be of higher importance. In legal document searches, lawyers may want to be sure that they go through all of the relevant cases and can therefore tolerate the appearance of irrelevant documents in the result output. In a more general case, when we search for documents on a particular topic, we want to be sure that all or most of the top results are the most relevant to our information need, and we can live with the fact that we do not examine some of the other hundreds of relevant documents. However, when we give a query to an IR system that is too specific or when we do not know the magic combination of the right key words associated with the documents we are interested in, it is desirable that the system try to optimize for higher recall. Query expansion and query reformulations are useful mechanisms for achieving this. When an IR system recognizes that the query is too narrow or likely does not use the right key word, it may decide to substitute some of the query terms with more appropriate synonyms or more general terms, for example, a "penicillin resistance" query can be substituted with "antibiotic resistance."

Existing domain and general language knowledge captured in domain taxonomies, WordNet, or thesauri can be used automatically by the algorithms for query expansion or reformulations. Latent semantic indexing (LSI) is an automated statistical analysis of relationships between the occupancies of the words within and across the documents. When particular actions of a user (e.g., selection of a document) can be interpreted as a positive or negative reaction to the output of the system, this information can be utilized in a relevance feedback mechanism for re-ranking relevant documents. For instance, the Rocchio algorithm can be used in the vector space retrieval model to increase the weights of those terms that appeared in the documents that the user already considered useful, i.e., while the user is browsing the returned set of ranked documents, the weights of the terms and the ranking itself are being updated. For example, suppose the user asks for all documents with the key word "Bush;" the system may return documents on both the former president as well as of the type of vegetation. Subsequently, the user starts retrieving and browsing through only those documents about the former topic. In that case, the algorithm will decide that the documents of the former president are of more interest to the user and the ranking of the documents will be changed, moving the documents about the vegetation down the list. This reweighing affects relevance scores and, consequently, the ranking of the documents. In general, IR (or search) is being considered more and more as an iterative and interactive process during which both a user and an IR system understand better and better what the desired output is.

Data mining approaches for pattern mining, classification, and clustering discussed in Section 9.5 have been actively used to enhance the performance of IR systems, for example, by classifying documents into topics, by clustering the output results on the fly for better navigation, or by modeling user interests for improving precision and recall of IR. Take again the example of the user querying for "Bush." Based on the word occurrences, the system can recognize that there are different types of "Bushes" being discussed in the documents, create groups, and present the different groups as such to the user, for example, one about the former president and one about the vegetation. The user can then quickly identify which group is of interest to him or her and concentrate the search on the group of interest only.

During the last two decades, with the appearance and steadily increasing popularity of the Internet, the number and variety of information objects and the variety of users and uses of IR systems have exploded. Therefore, it is not surprising that during this period, IR as a field has expanded into many different topics, including Question Answering, Cross-Lingual IR, Topic Detection and Tracking, Summarization, Multimedia IR, Text Structuring, Text Mining, Geographical IR, and Personalization/Adaptation/Recommendation. Although being considered more widely as a field, IR is still primarily considered to be a set of processes aimed at satisfying vague information needs imprecisely specified in ambiguous natural language by users, by matching them (approximately) against a large collection of information objects created and specified (in the same ambiguous natural language) by authors.

The explosion of information in the Web era took the problem of information overload to a new level. IR technology responded with new approaches for tailored and personalized filtering and ranking of content. Personalized IR, recommender systems, adaptive content filtering, and personalized search are just some of the key words describing a set of techniques aimed at better understanding various and possibly changing information needs and preferences of users and better matching them to the available information resources. What we get in the results list may depend on what the modern search engine we are using thinks our information need is—is it informational (e.g., "Give the definition of the term information retrieval"), navigational (e.g., "Show me the URL of the TU/e website that I want to reach"), or transactional (e.g., "Show me websites where I can buy an iPhone or download a ringtone for it")? For most of us, it is no longer surprising that the search engine uses information about our country of residence current location, language, or some geographic names we typed in the query to refine its search results.

Modern IR systems attempt to satisfy user information needs by

- helping to find (or refind) information objects on the Web;
- facilitating a vertical search (given a query, we find and select some relevant document, and then we can identify paragraphs or sentences that are directly related to our query);
- extracting pieces of information (rather than complete documents) that are relevant to a user, maintaining links between the extracted information and the original documents for evidence and context.

Besides understanding user information needs and the current context, modern IR systems try to guarantee that the information they provide access to is trustworthy. One of the main problems in the early days of the Internet was the large number of spam pages cluttering the search results. The democratic nature of the Internet and low costs of publishing information online allowed many of these spammers to mislead search engines of that time simply by ingeniously hiding key words inside their pages in order to attract as much traffic as possible from key-word-based searches.

The first step taken initially by Google and later by other search engines was a simple link analysis. Such an indexing of anchor text and the hyperlinks themselves were only successful for a short time until spammers changed their tactics and started to generate millions of links pointing to their Web pages. This failure suggested that besides a relevance score assigned to a query–document pair, there was a need to find a way of computing some quality score for each page on the Web.

Hubs and authorities (HITS) [15] and PageRank [14] algorithms are the most well-known approaches to achieve this. PageRank is based upon the following recursive definition: a Web page is important if many important Web pages refer to it. More precisely, its importance is proportional to the cumulative importance of other pages that directly point to it. This seemingly problematic definition works well in practice [14]. PageRank is query independent and its score for a page corresponds to a long-term visit rate for someone randomly surfing the Web.

HITS is built on the idea that there are two kinds of Web pages relevant for broad-topic searches. There are trustworthy or authoritative sources of information on the topic (e.g., the National Institute of Health pages are an authoritative source for queries on antibiotic resistance) or simply authorities. There are many so-called hub pages containing lists of links to authoritative Web pages on a specific topic, which already gives us a hint that hub pages can be used to discover authority pages. This observation leads to a similar recursive definition as used for PageRank: a good hub is one that points to many good authorities and a good authority is one that is pointed to by many good hubs. HITS is query dependent; given a query, every Web page is assigned two scores: a hub score and an authority score. Then, pages with the highest hub scores and pages with the highest authority scores appear at the top of the list.

PageRank and HITS are not fully resistant to spammers who may use a particular type of link spamming—so-called link farms, i.e., Web pages and links among them created for the sole purpose of fooling the search engine—to boost the PageRank score for target pages. Usually, an additional mechanism (such as blacklisting, i.e., keeping explicit lists of recognized spammers) is used to fight such behavior.

It should be understood that the problem of Web spam is difficult to solve with just one shot. There is a continuous battle between content providers (or their agents), who are interested in promoting some information, ideas, products, or services, and so-called content aggregators (general-purpose and specialized search engines, recommender systems, question-answering systems on the Web, sentiment analyzers, or opinion miners). Content providers have to either pay search engines for placement or contextual advertisements or find a way to boost their PageRank or other indicators to appear at the top of the result list.

Despite the technological solutions for information retrieval, recently, the idea that human interaction is needed in the information retrieval loop has gained interest. Often for a human user, it is difficult, if not impossible, to clearly and unambiguously state an information need as a set of key words. Also, from a cognitive perspective, algorithms such as PageRank and frequency measures such as tf–idf represent only a very idealized and simplified vision of the complex process of identifying interesting information. Therefore, an increasing subcommunity of the information retrieval community aims at the development of human-centered approaches in which the technology is no longer seen as the solution to every information need, but rather as a tool that can serve to extend the limited human capacity of dealing with huge information loads. See Chapter 8 for an explanation of a prototype system of a human-centered tool.

9.5 Mining Large Databases for Extracting Information

The step from information retrieval to data mining [7, 28, 29] is a small one. The main difference between the two is usually considered to be the absence of the notion of a user query in the data mining context. Put roughly, in data mining, a user provides a dataset with the implicit request to extract useful information from it without too much user interaction. Usually, data mining is defined as " . . . the use of sophisticated data analysis tools to discover previously unknown, valid patterns and relationships in large data sets" [7].

Most data mining methods can be divided into one of the following three categories.

1. *Classification* [30]. Learning to predict based on examples.
2. *Clustering* [31]. Dividing a given dataset into logical coherent groups.
3. *Pattern Mining* [32]. Discovering regularities, trends, or patterns.

Clustering and pattern mining are often referred to as unsupervised methods as they require only data, whereas supervised methods, such as classification, require annotated or labeled data. Data mining is heavily used in many applications, including marketing [33] and fraud detection [34].

A standard example of the first category, classification, is learning a model for predicting whether an e-mail is a spam mail or not. This task is of great importance, since the huge bulk of e-mails being sent is actually spam. In the most recent edition of "Spam and Phishing: a Monthly Report," the Internet security firm Symantec reports spam percentages that are steadily around 85% of total mails sent [35]. Given an e-mail, we first transform it into a set of scores, according to its characteristics. Such a score could be, for example, the number of times the word "cheap" is used, the frequency of upper case letters, or whether the sender is listed as a known spammer. The number of characteristics is huge in most cases; several thousands of characteristics are not uncommon. Based on these characteristics, it is decided whether the mail is marked as spam or not. For an example, see Figure 9.6. This figure contains the report of a spam filtering tool and lists some of the scores. The true challenge now is to determine exactly how important a specific characteristic is. In classification, we try to learn how important these characteristics are based on some given training examples. This approach has several advantages; it takes away the responsibility of an expert to manually set and learn the best weights by trial-and-error. By eliminating this time-consuming step, we can more adequately and quickly respond to the evolving behavior of spammers who are constantly trying to circumvent the filters, as the role of the expert is taken over by the computer merely using data rather than human intelligence.

The second data mining method we discuss, clustering, on the other hand, is often used in situations in which we have no idea about categories of the data objects. In contrast to classification, where explicit labels are given (spam vs. non-spam), in clustering, we try to automatically assign objects to groups. An example could be a supermarket database containing purchase information per customer. A very first analysis task could be the identification of "types" of customers. Such analysis requires an algorithm that looks at the characteristics of the different customers and subsequently groups them according to similarity. An example output could be a division of the customers into the following

```
Content analysis details: (10.7 points,    7.0 required)

0.0  STOX_REPLY_TYPE        STOX_REPLY_TYPE

4.4  HELO_DYNAMIC_IPADDR2      Relay HELO'd using suspicious hostname(IP addr 2)

3.5  HELO_DYNAMIC_SPLIT_IP     Relay HELO'd using suspicious hostname (Split IP)

2.1  RCVD_NUMERIC_HELO      Received:  contains an IP address used for HELO

-2.6 BAYES_00    BODY:    Bayesian spam probability is 0 to 1% [score: 0.0000]

0.5  RAZOR2_CHECK  Listed in Razor2 (http://razor.sf.net/)

0.1  RDNS_NONE     Delivered to trusted network by a host with no rDNS

2.8  DOS_OE_TO_MX Delivered direct to MX with OE headers
```

Figure 9.6 A spam report for one e-mail. Classification is learning a model for predicting whether an e-mail is a spam mail or not. Given an e-mail, we transform it into a set of scores, according to its characteristics. The scores help determine whether the mail is marked as spam or not. This figure contains the report of a spam filtering tool and lists some of the scores. Every line represents one characteristic measured for the e-mail. The optimal weights assigned to every characteristic and the threshold for the final aggregated value are determined by a data mining algorithm.

categories: those who come everyday and mainly buy food products, those who shop only on the weekend and buy mainly the supermarket's own brand, and those who come once, on Sunday, and buy mainly luxury goods. As such, clustering is often applied as a first analysis step, when we do not have any idea of what our data looks like. Overwhelmed by the sheer volume of the data, clustering the data items automatically into groups and aggregating the dataset at this group level makes the information more manageable. Other examples are automatically grouping a large collection of documents into consistent categories based on the appearances of specific words in them or the grouping of stocks that have a similar price behavior.

Closely related to clustering is outlier detection [36], the identification of objects that do not fit well together with the other objects. Outlier detection has been used successfully in the context of credit card fraud detection.

The third type of data mining technique, pattern mining, is used for detecting patterns and associations. This technique is again unsupervised in the sense that no target label is given. Instead, it aims at the detection of unusual relations between attribute values. An often quoted story in this context is that of a supermarket in the United States that applied data mining to the purchase data of its customers in order to detect groups of items that are often sold together. It was claimed that the analyst found the following surprising relationship in the data: on Friday evening, people who buy beer also tend to buy diapers.

Although generally assumed to be an urban legend [37], this example illustrates very well the aim of pattern mining; even though this finding is easy to verify against data using the traditional statistical hypothesis testing techniques, one would not easily come up with this hypothesis independently. In pattern mining, however, intelligent search techniques are being used in order to either exhaustively or heuristically explore the whole space of all possible hypotheses. In this way, no bias toward the background knowledge of the analyst is introduced and thus potentially surprising patterns may be generated.

Although this principle is very nice and pattern mining best embodies the true mission of data mining, it would not be fair to claim that these techniques are free of problems. Often, the application of pattern mining results in huge collections of patterns being generated, many of which are redundant with respect to each other. Similarly, often the strongest patterns found are also the least interesting ones, as they are already well known by the analyst. Therefore, considerable effort has gone into the development of sophisticated techniques for measuring the interestingness of patterns [38–42]. Quantifying how interesting patterns are, however, turns out to be a highly nontrivial problem that has not been solved satisfactorily yet.

Interestingly, we can extend the observation on the difficulty of defining what is interesting for pattern mining to the whole of the information overload problem. It is fair to say that usually, the absence of data is not the problem. The absence of suitable and efficient analysis techniques is not usually the problem either, nor is the absence of a query expressing which piece of information we need. The true limiting factor is our inability to define in an unambiguous way what is interesting. Often the answer to this question is context dependent. Yet, a large research effort is taking place in the data mining community to find more satisfactory answers to this question.

In summary, data mining can be considered as the next level in the taxonomy we are building in this paper. In contrast to the information retrieval case, in data mining, the notion of a query is missing. The input to the algorithms is merely the data and the implicit question is, "find me something of interest." Overwhelmed by the data, the analyst does not even know anymore what questions are the right ones to pose.

9.6 Processing Data Streams

In data stream processing [43, 44], it is assumed that data is generated at such a fast pace that it is no longer feasible to store the data for off-line processing. Instead, data has to be processed instantly, preferably by devices with modest hardware requirements. A prototypical example of such systems is a set of monitors that registers IP traffic on computer networks. In many networks, for example, the Internet, the amount of information sent over the network is so huge that it is not possible to store it for analysis with the traditional statistical and data mining methods. In addition to that, it is critical to get up-to-date information instantaneously. Imagine, for example, a network monitor developed to detect a sudden increase of traffic from or to a specific IP address. Such an increase could indicate specific types of network attacks, such as denial of service attacks, in which a malicious party tries to overload a service or a website with numerous requests in order to make it unavailable. In such a situation, it is of great importance to find out about such an increase immediately; finding out tomorrow that there was a network attack today is obviously of little interest.

The solution to this particular problem seems straightforward; why don't we just keep a counter that keeps track of all sources and destinations of Internet packages? A substantial increase in one or more of these counters would imply a network attack. This approach, unfortunately, is too simplistic and fails because it is impossible to allocate enough memory to keep counters of every possible IP address. This example illustrates the immense problems we are faced with when developing algorithms in a streaming context.

Based on the network increase detection example above, one may be convinced that under such conditions, it is impossible to perform meaningful computations. This is not that far from reality. Many of the main results in stream processing are actually negative results, stating that certain operations can no longer be performed exactly. Luckily, in many cases, it is still possible to give certain probabilistic guarantees. This means that even though it is not possible to solve a problem exactly with reasonable means, often it can be guaranteed that in the majority of cases, the correct answer can be approximated with a limited error. In the case of the example in the last paragraph, it is not possible to always find all "heavy hitters" exactly, but it is certainly possible to almost always find most of them. For example, we can make algorithms that are able to report instantaneously 95% of the heavy hitters, 99% of the time [45]. For many practical cases, such approximate algorithms with strong guarantees are sufficient.

The streaming paradigm is increasingly gaining importance due to the enormous number of data sources that are generating data at a large pace. Some popular examples include sensor networks, both in scientific and military applications, cell phone networks, and traffic streams. From an engineering point of view, it can be seen as the most extreme way of dealing with information overload; instead of trying to carefully design a storage structure for the data such that it can be enclosed for analysis purposes, it is processed right away at the source.

9.7 Summary

This chapter gives an overview of an engineering point of view of information overload and how to deal with it. Depending on the level of overload, different techniques are appropriate. At one end of the spectrum, there is the semistructured data paradigm. It is no longer assumed that we can organize the data in a structured way; we have to deal with unexpected constructions and heterogeneous data. Still, we want to be able to store and query this data in order to extract the information we need. One step further, we find the information retrieval paradigm, where it is assumed that we no longer have control over the data organization. The best we can do is to build indices over it and develop smart retrieval techniques in order to detect the needle in the haystack.

The next level is data mining. In data mining, the notion of a query is also missing; the input to the algorithms is merely the data and the implicit question is "Find me something of interest." Overwhelmed by the data, the analyst does not even know anymore what questions are the right ones to pose. The final level we identify is that of the data stream paradigm. At this level, we have even lost the capability to store the data, and therefore we have to process the data at its source; immediate processing and information extraction is required.

We expect that the explosion of available information will continue into the future. Getting access to large amounts of data will become increasingly easier, although data quality will remain the bottleneck in the information society. Although recent technologies, such as XML and RDF, offer rich possibilities for adding additional structure to data, we often observe that users tend not to use or to only partially use these possibilities. Also, in many cases, such as online information on the Internet, the extra information can be used by malicious parties to delude search engines for commercial goals. Because of that, we expect that acquiring the appropriate technological tools for dealing with information overload will become a necessity for information-driven activities. Automatic tools, such as information retrieval tools for locating specific pieces of information and data mining tools for transforming large amounts of data into actionable knowledge (e.g., ontologies), will further gain importance and become natural extensions for individuals dealing with information overload.

REFERENCES

[1] J. West. (2010). *Zettabytes, Petabytes, and All That: The Intersection of Supercomputing and all the Data We Create* [Online]. Available: http://insidehpc.com/2010/10/14/zettabytes-petabytes-and-all-that-the-intersection-of-supercomputing-and-all-the-data-we-create/ Aug. 25, 2011 (last date accessed).

[2] S. Abiteboul, "Querying semi-structured data," in Database Theory ICDT, 1997, pp. 1–18.

[3] E. R. Harold and W. S. Means, *XML in a Nutshell*. Sebastopol, CA: O'Reilly & Associates, 2002.

[4] T. Bray, *et al.*, *Extensible Markup Language (XML) 1.0*, W3C Rec. REC-xml-20001006, 2008.

[5] G. Klyne and J. Carroll, *Resource Description Framework (RDF): Concepts and Abstract Syntax*, W3C Rec. REC-rdf-concepts-20040210, 2004.

[6] M. Ley, "DBLP: Some lessons learned," in *Proc. VLDB Endowment*, vol. 2, no. 2. 2009, pp. 1493–1500.

[7] J. Han and M. Kamber, *Data Mining: Concepts and Techniques*, 2nd ed. Burlington, MA: Morgan Kaufmann, 2006.

[8] L. Golab and M. T. Ozsu, "Issues in data stream management," *ACM Sigmod Record*, vol. 32, no. 2, pp. 5–14, 2003.

[9] J. Clark and S. DeRose, *XML Path Language (XPath) 1.0*, W3C Rec. REC-xpath-19991116, 1999.

[10] S. Boag, *et al. XQuery 1.0: An XML Query Language*, W3C Working Draft 15, 2002.

[11] R. Baeza-Yates and B. Ribeiro-Neto, *Modern Information Retrieval*, 2nd ed. New York: ACM Press, 2011.

[12] Y. Rui, *et al.*, "Image retrieval: Current techniques, promising directions, and open issues," *J. Vis. Commun. Image Represent.*, vol. 10, no. 1, pp. 39–62, 1999.

[13] Y. H. Tseng, "Content-based retrieval for music collections," in *Proc. 22nd Annu. Int. ACM SIGIR Conf. Res. Develop. Inform. Retrieval*, 1999, pp. 176–182.

[14] T. H. Haveliwala, "Topic-sensitive PageRank: A context-sensitive ranking algorithm for web search," *IEEE Trans. Knowl. Data Eng.*, vol. 15, no. 4, pp. 784–796, Jul.–Aug. 2003.

[15] S. Chakrabarti, *et al.*, "Experiments in topic distillation," in ACM SIGIR Workshop Hypertext Inform. Retrieval Web, Melbourne, Australia, 1998, pp. 13–21.

[16] U. Fayyad, *et al.*, "The KDD process for extracting useful knowledge from volumes of data," *Commun. ACM*, vol. 39, no. 11, pp. 27–34, 1996.

[17] D. A. Grossman and O. Frieder, *Information Retrieval: Algorithms and Heuristics*. Dordrecht, The Netherlands: Kluwer Academic, 2004.

[18] J. Zdziarski, *Ending Spam: Bayesian Content Filtering and the Art of Statistical Language Classification*. San Francisco, CA: No Starch Press, 2005.

[19] T. A. Meyer and B. Whateley, "SpamBayes: Effective open-source, Bayesian-based, e-mail classification system," in Proc. 1st Conf. E-Mail and Anti-Spam (CEAS), vol. 98. 2004.

[20] S. McCreary and K. Claffy, "Trends in wide area IP traffic patterns: A view from Ames Internet exchange," in Proc. ITC Specialist Seminar IP Traffic Model., Measure. Manag., 2000.

[21] A. Deshpande, *et al.*, "Model-driven data acquisition in sensor networks," in *Proc. 30th Int. Conf. Very Large Data Bases*, vol. 30, 2004, p. 599.

[22] G. H. L. Fletcher and C. M. Wyss, "Towards a general framework for effective solutions to the data mapping problem," *J. Data Semantics*, vol. 14, 2009.

[23] M. Arenas, *et al.*, "Relational and XML data exchange," *Synthesis Lectures Data Manag.*, vol. 2, no. 1, pp. 1–112, 2010.

[24] T. Haigh, "A veritable bucket of facts: Origins of the data base management system," *SIGMOD Record*, vol. 35, no. 2, pp. 33–49, 2006.

[25] T. Heath and C. Bizer, *Linked Data: Evolving the Web into a Global Data Space*. San Rafael, CA: Morgan and Claypool, 2011.

[26] E. Prud'hommeaux and A. Seaborne, *SPARQL Query Language for RDF*, W3C Rec. REC-rdf-sparql-query-20080115, 2008.

[27] J. Perez, *et al.*, "Semantics and complexity of SPARQL," *ACM Trans. Database Syst.*, vol. 34, no. 3, pp. 1–45, 2009.

[28] L. Torgo, *Data Mining with R: Learning with Case Studies*, New York: Chapman and Hall/CRC Press, 2010.

[29] E. Frank, *et al.*, "Weka: A machine learning workbench for data mining," *Data Mining and Knowledge Discovery Handbook*. New York: Springer, 2010, pp. 1269–1277.

[30] R.O. Duda and P. E. Hart, *Pattern Classification and Scene Analysis*. New York: Wiley, 1973.

[31] A. K. Jain, *et al.*, "Data clustering: A review," *ACM Comput. Surveys (CSUR)*, vol. 31, no. 3, pp. 264–323, 1999.

[32] J. Han, *et al.*, "Frequent pattern mining: Current status and future directions," *Data Mining Knowl. Discovery*, vol. 15, no. 1, pp. 55–86, 2007.

[33] M. J. Berry and G. S. Linoff, *Data Mining Techniques: For Marketing, Sales, and Customer Relationship Management*. New York: Wiley Computer Publishing, 2011.

[34] C. Phua, *et al.*, "A comprehensive survey of data mining-based fraud detection research," Arxiv preprint arXiv:1009.6119, 2010.

[35] Symantec. (2010, Dec.). *State of Spam and Phishing Report* [Online]. Available: http://www.symantec.com/connect/blogs/spam-and-phishing-landscape-december-2010 Aug. 26, 2011 (last date accessed).

[36] V. Hodge and J. Austin, "A survey of outlier detection methodologies," *Artif. Intell. Rev.*, vol. 22, no. 2, pp. 85–126, 2004.

[37] K. Hermiz, and S. Manganaris, "Beyond beer and diapers," *DB2 Mag.*, 1999.

[38] A. Gionis, *et al.*, "Assessing data mining results via swap randomization," *Trans. Knowl. Disc. Data*, vol. 1, no. 3, pp. 1–14, 2007.

[39] S. Hanhijärvi, *et al.*, "Tell me something I don't know: Randomization strategies for iterative data mining," in *Proc. ACM SIGKDD*, 2009, pp. 379–388.

[40] N. Tatti, "Maximum entropy based significance of itemsets," *Knowl. Inform. Syst.*, vol. 17, no. 1, pp. 57–77, 2008.

[41] T. De Bie, *et al.*, "A framework for mining interesting pattern sets," *ACM SIGKDD Explorations Newsletter*, vol. 22, no. 2, pp. 92–100, 2011.

[42] M. Mampaey, *et al.*, "Tell me what I need to know: Succinctly summarizing data with itemsets," in *ACM SIGKDD*, 2011, pp. 573–581.

[43] G. Cormode and M. Hadjieleftheriou, "Methods for finding frequent items in data streams," *VLDB J.*, vol. 19, no. 1, pp. 3–20, 2010.

[44] H. Liu, *et al.*, "Methods for mining frequent items in data streams: An overview," *Knowl. Inform. Syst.*, vol. 26, no. 1, pp. 1–30, 2011.

[45] R. Berinde, *et al.*, "Space-optimal heavy hitters with strong error bounds," *ACM Trans. Database Syst.*, vol. 35, no. 4, pp. 26:1–26:28, 2010.

Florida Institute of Technology
High Tech with a Human Touch™

PRACTICAL INSIGHTS FROM THE COLLEGE OF AERONAUTICS, FLORIDA INSTITUTE OF TECHNOLOGY

THE ART OF MANAGING INFORMATION IN THE NATIONAL AIRSPACE ARCHITECTURE

The U.S. National Airspace System (NAS) is a complex structure that accommodates approximately 50,000 flights daily. An array of equipment, facilities, systems, airports, and personnel are dedicated to the safe and efficient transportation of people and goods by air. The NAS includes

- over 750 air traffic control (ATC) facilities;
- over 18,000 airports;
- 4500 air navigation facilities;
- nearly 50,000 FAA employees.

The system handles

- over 7000 takeoffs and landings per hour;
- over 2 million instrument approaches each year [1, 2].

Information Overload: An International Challenge for Professional Engineers and Technical Communicators,
First Edition. Edited by Judith B. Strother, Jan Ulijn, Zohra Fazal.
© 2012 Institute of Electrical and Electronics Engineers. Published 2012 by John Wiley & Sons, Inc.

Within the NAS, over 660 million passengers and 37 billion cargo revenue ton miles (an important transportation industry measurement to describe 1 ton of cargo moved 1 mile) are transported safely each year [1]. In one brief century, the historic 12-s flight of the Wright Flyer has transformed the United States and the world into a highly mobile and interconnected society reliant upon the safe and efficient operation of the U.S. NAS. Technological advancements in jet power-plant design and airfoil advancements have played the most vital role in the development of the aircraft since that historic moment. Yet, in spite of these noteworthy advancements in aircraft design, the NAS does not rely upon the prolific advances of power-plant design or airfoil advancements to keep its complex aeronautical system operating safely and efficiently each day; instead, it is dependent upon information—massive amounts of information.

From Data to Information to Situational Awareness to Decisions

Aviation information is vigorous, dynamic, and complex. Yet, aviation information is short lived—it expires quickly. In the NAS, data are handled and transformed into information, meaningful communicative elements ready for access. Eventually, the access to this information will bring about knowledge—what the aviation community opts to call situational awareness. Situational awareness therefore requires the ability of the aviation professional to transform complex spatial and temporal data into information. It is this information that is used to make critical decisions; both pilots and air traffic control personnel rely upon accurate and timely information.

Historically, airspace has been controlled by tactical control protocols between an air traffic controller and an aircraft pilot. All instructions between the controller and the aircraft are conducted by means of radio voice transmissions. The pilot and controller are bombarded by audio information that needs to be *sorted* for accuracy, relevancy, and timeliness; *encoded* into working memory; and *assimilated* into knowledge to form a mental model of a highly dynamic 4-D environment (situational awareness). If these tasks are completed successfully, an accurate (safe and efficient) *decision* can be made.

Needless to say, cognitive loads upon the controller and pilot are significant in this arrangement. Therefore, the controller is the limiting element that constrains the number of aircraft that can fly in an assigned sector. For the non-aviation professional, a descriptive analogy may represent this informational quandary better—the task of a controller in the NAS is similar to that of a traffic police officer being assigned to direct traffic on a high-speed highway. The "police officer" must be flexible enough to account for an unannounced lane closure (surprisingly, "lane closures" do happen in the sky; dynamic weather hazards force NAS controllers to shut down routinely used aircraft arrival and departure paths). The "police officer" must also be collaborative enough to subjugate individual traffic priorities to the priorities of multitudes of other nearby police officers who have been assigned similar tasks with dissimilar goals and priorities, while each possesses dissimilar pieces of the whole informational picture (the NAS controller, armed with "pieces" of information, must work collaboratively with multitudes of other operational entities to ensure his or her decisions are in accord to achieve local and system-wide air traffic priorities). This picture captures the real cognitive challenges that daily confront and limit a controller in the NAS. Without a doubt, this is a picture of information overload par excellence!

If this process of assimilation of information fails and situational awareness is lost, poor decisions occur. In the dynamic environment of aviation, poor decisions, known as errors, are very costly and sometimes catastrophic. Therefore, it is critical that information that is handled by professionals in the NAS possess attributes of accuracy, relevance, and timeliness. Without these three informational attributes, participants in the NAS will experience information overload. The NAS decision maker becomes overloaded when

- costly time is dedicated to discerning the accuracy of information when inaccurate or dissimilar sources are mixed with trusted information products;
- relevant information becomes irrelevant when burdened by redundancy or when the wrong decision makers are presented with the right information;
- lack of timeliness ages accurate information to staleness and clutter.

Transformative Airspace Architecture

In spite of the informational challenges that face the NAS, travelers in the United States have come to rely upon an airspace system that is efficient, safe, and dependable. Yet inefficiency does exist in the system. The U.S. Government Accountability Office (GAO) in a May 2010 report found that one quarter of all domestic flights in the United States were cancelled or delayed. Furthermore, the GAO reported the sobering consequential statistics from a domestic airspace system under tremendous capacity strain and resulting delays; in 2007, there were a collective 320 million hours of passenger delays and a cost to the U.S. economy of $41 billion. The GAO report goes on to paint a grim outlook for the future: "Over the next decade, the number of flights, and accordingly the number of delays in the U.S. aviation system is predicted to increase" [3].

Present domestic aviation activity has already placed the NAS under extreme stress, and the FAA's forecast sounds a portent prediction of a brisk 3.0% annual growth through the year 2030 [4]. The NAS will be saturated with system usage exceeding present day demands. Although the limiting constraints include physical infrastructure (e.g., runways, terminals, and facilities), they are mainly characterized by a systemic weakness to leverage information effectively to manage the massive and complex NAS.

A systemic weakness requires a systemic solution that transcends physical architecture—namely, distributed and collaborative decision making and shared situational awareness. Neither of these qualities is possible in an information overloaded environment. The obvious reason is that collaboration for decision making and shared situational awareness are only possible in an environment that richly facilitates information sharing. Department of Transportation Secretary Norman Y. Mineta stated, "The changes that are coming are too big for incremental adaptations of the infrastructure . . . we need to modernize and transform our air transportation system" [5]. Deployment of technology and procedures is key to the transformational movement of the NAS beyond evolutionary advancements.

Robust, Agile, and Intelligently Responsive Information-Sharing Architecture

Any architectural feat requires massive amounts of information and collaboration. The longest main span bridge in the world, the cable-stayed Sutong Bridge, spans over a half mile across the Yangtze River in China, connecting Nantong and Changshu [6]. Only the lavish exchange of information between international partners, highway planners, academic researchers, and architectural and computer engineers, not to mention monetary resources, made such a collaborative feat possible. The famed and legendary narrative of the Tower of Babel (that has parallels in many faith traditions) transmits a similar imperative of the inseparable nature of information and collaboration made possible through a communicative conduit—language. Workers and architects attempted an "impossible" architectural feat to build a tower with "its top in the sky"—their attempts were dashed by their inability to share their common vision and information effectively. The architectural feat of constructing a safe and efficient NAS is no less impressive than the endeavor of Babel; both seek to construct a massive architecture in the sky. But the amazing fact here is that transforming the domestic airspace is even more of a challenge to information sharing and collaboration. Varied aircraft flight profiles and ever-changing weather patterns eclipse any conceptual task of erecting a 3-D structure. The daunting task of erecting a safe and efficient 4-D structure, combined with the challenges of the elements of time and movement, dynamic information sharing, and collaboration can easily fall prey to inaccurate, irrelevant, and untimely information (IIUI).

IIUI is invalid information, i.e., it can pile up quickly, asphyxiating decision makers, blocking their sight to a common operational picture, and confusing their ability to possess a shared situational awareness, ultimately robbing their ability to practice collaborative decision making. Conversely, valid information in the NAS is that information that can be systemically acquired, accommodated, and acted upon quickly. Efficient, safe, and secure operations are the ultimate goal of the NAS. Like oxygen to the pulmonary system, valid information feeds the NAS with the necessary element to achieve the system's operational goals. Therefore, the system must become resistant to the stated triaxial predator of valid information, IIUI. For at the heart of the capacity-limited NAS rests the limited capacity of the system to breathe—acquire, accommodate, assimilate, and act upon valid information—the system is overloaded by information rather than aircraft.

The challenge then to the FAA is not to indiscriminately add physical infrastructure to address the issue of system overload or even to add novel technology to address information overload, but to transform to an enterprise solution based upon the restructuring of technology and procedures, i.e., enterprise practices utilizing enterprise technology to deliver upon enterprise goals. FAA Administrator Marion C. Blackey sounded the call for such a transformation claiming, "This plan isn't something that's nice to have. If we don't move forward, we won't be able to catch up. Transformation is a must" [7]. The NAS must transform to a robust, agile, and intelligently responsive information-sharing architecture that gives organizational form and substance to its dynamic operation.

Next Generation Efforts to Manage Information

Next Generation (NextGen), the FAA's enterprise solution, is the transformative system-wide movement of the capacity-limited NAS to an architecture that accommodates air transportation's requirement for increased capacity. NextGen's goals are clearly articulated: increase the safety, security, and capacity of the NAS. To achieve these goals, eight transformational capabilities that fundamentally change the approach to air transportation have been identified. These capabilities include improved aircraft position and tracking known as ADS-B, deployment of data-link capabilities that reduces radio voice transmissions, real-time weather forecast and publishing capabilities that fuse together thousands of observations for accurate shared situational awareness, and an information-sharing platform built upon a service-orientated architecture (SOA) that ties together all the NextGen transformational capabilities into a unified and accessible network. Although these capabilities are varied, they all employ technological innovation and procedural alignment to the enterprise effort of the NAS. By recognizing that stakeholders in the NAS have a vested interest in the success of the NAS, NextGen seeks to form a collaborative approach to the enterprise endeavor. Stakeholders include the FAA, controllers, pilots, air carriers, military, and law enforcement. At the heart of NextGen is a fundamental shift from air traffic control to airspace management.

Distributed Decision Making

Airspace management depends on making good decisions—good decisions cannot be made without good situational awareness, and good situational awareness is not possible without good information. Airspace management is a collaborative endeavor of decision making among parties that are distinguished by varied geographical distances, functional mandates, and organizational identities. Therefore, it is no overstatement to stress that the free flow of good information is critical to facilitate distributed decision making. Although it may be implicit from this statement that an increase in information with a corresponding increase in access can lead to an increase in the NAS capacity, the necessary free flow of information does not imply a disorganized flow of data. While the transformational capabilities of NextGen depend upon the free-flow of information, the organizing principle of "valid information" mitigates the risk of information overload by discouraging the injection of IIUI.

Valid information can be defined by three attributes: the right information, transmitted to and from the right people, and at the right time. Stated another way, information must be tailored to the individual and must be accurate and timely. The lack of tailoring has limited decision makers to the access of information by erecting mounds of "invalid" information as an obstacle to valid information. Such antiquated systems and processes have contributed to an NAS that is capacity limited. This is because antiquated systems have been structured to "broadcast" information that is sometimes redundant, out of date, inaccessible, unsearchable, or disorganized.

Furthermore, legacy information exchange in the NAS has depended upon fixed networks. Point-to-point data interfaces have limited information transfer. In fact, applications and services have often been duplicated because one application is unable to communicate with another or is unaware of the information found in another application. NAS producers and users of information have suffered at the hands of poor information assurance when tasked with sorting through rogue, dissimilar, and untrusted sources for relevant information. Redundancy has encumbered information consumers with duplicate information.

It is no wonder that improved information sharing has been described in the FAA's *Next Gen NAS Concept of Operations* as being crucial to increasing the capacity of the NAS [8]. The flexibility of information exchange unshackled from point-to-point constraints, concomitant with the tailoring of information, are significant characteristics of a transformed NAS because they provide the necessary context for distributed decision making.

System-Wide Information Management (SWIM)

The mechanism that serves as a conduit to ensure distributed decision making is the FAA's system-wide information management (SWIM). SWIM underpins NextGen's transformational initiatives and goals by serving as an information broker, thereby making possible critical operational NAS information exchange. Based upon the network-centric principle, SWIM transcends a data-sharing bus by providing core services: registries and directories, messaging and brokering, information assurance, and systems management. Producers of information as well as users of information become loosely coupled in information exchange relationships brokered by SWIM. Therefore, both providers and users of information can be confident that the information provided is searchable, accessible, timely, and accurate. SWIM also facilitates information enhancement—many consumers of information will become information providers—upon consuming information, additional information may be added and made available to other SWIM users.

SWIM directly addresses information overload by eliminating the opportunity of accidental redundancy or redundancy by design due to interoperability barriers. SWIM's many-to-many information exchange eliminates redundancy, while SWIM's interoperability makes possible seamless, secure, organized, and flexible exchange of information. SWIM eliminates the many point-to-point interfaces and serves as a one-stop shop with the commerce and exchange of critical NAS information. As shown in Figure 1, instead of a confusing web of discreet information exchanges between applications or systems, SWIM serves to functionally consolidate services to a central location. It is important to stress the inherent benefit of this net-centric design: neither consumers nor producers need be hampered by coupling constraints—the tailoring of information occurs in the context of publishing and subscribing of information instead of through a unique, inflexible, and permanent interface. In this case, it is not essential to know the audience. A publisher of information makes information available, while a user subscribes based upon his or her needs. Here, the free flow of information is encouraged in a loosely coupled and flexible environment based upon informational needs so that trusted information can be

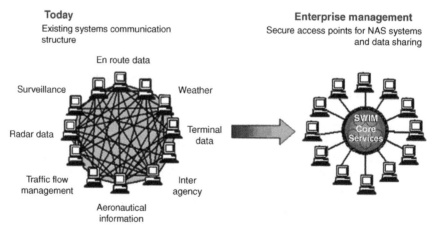

Figure 1 Current NAS communication technology updated for enhanced information sharing [9]. System-wide information management eliminates the many point-to-point interfaces and serves as a one-stop shop with the commerce and exchange of critical NAS information. SWIM directly addresses information overload by eliminating the opportunity of accidental redundancy or redundancy by design due to interoperability barriers.

presented to provide a common picture among other users of services provided within the information network.

Shared Situation Awareness and Collaborative Decision Making

The concept of *shared situational awareness* is uniquely a part of aviation operations. An example in this case would be helpful—weather provides aviation decision makers with a unique information challenge. Assume that one decision maker gathers relevant information to develop situational awareness of a highly dynamic weather event and makes a decision based upon that awareness. The complexity of this process becomes apparent when we realize that a pilot, a controller, or another functional group within the air traffic control organization may each possess dissimilar pieces of information, thereby assigning differing probabilities of the likeliness of the weather event or forming a different mental model of the weather situation. In short, the required ingredient of situational awareness remains unaligned among the participants in critical aviation decisions of safety and efficiency. In the case of weather, gateways in the sky, called fixes, may close or open to air traffic based upon one decision maker's particular meteorological situational awareness and resultant decision only to aggravate the goals of another controller with a different mental model. Worse is the possibility of a controller directing an aircraft to enter a hazardous area because of a dissimilar mental model from the pilot.

To address this challenge, SWIM provides for information exchange from ground-to-ground facilities and air-to-ground facilities. When integrated with a common and trusted weather source, each user of the NAS will have a common awareness. Not only will this aid in operations, but it will also ensure alignment of operational air transportation system goals because all stakeholders will have the same situational awareness. *Collaborative decision making* is now possible with the alignment of information resources.

Automation and Information in the NAS

Automation can be brought to bear upon situational awareness, distributed decision making, and collaboration in the SWIM context. Transformative automation is built into the transformed processes of NextGen. Enhanced trajectory management tools, probabilistic weather forecasts, and data communications make possible air traffic control by exception. Planning and predictive automation are brought to bear upon aircraft spatial orientation and trajectory in order to precisely plan aircraft position, optimize fuel profiles, maximize arrival and departure capacity, and de-conflict en-route travel. The key is that automation is built into the enterprise goal of sharing information. However, the prerequisite for vibrant exchange of information is an information strategy and architecture. Only then can information be accessed and used to increase safety, security, and efficiency in the NAS.

Summary

The NAS is a complex network that facilitates the daily safe and efficient operation of air traffic throughout the United States. It is composed of varied organizations, people, and technologies that are daily fused together to seamlessly operate in a highly dynamic environment and context. Each "player" in this choreographed operation of the NAS is interrelated and essential to the other—collaboration must occur! This collaboration depends upon the flow of large amounts of information to decision makers in the NAS. It is obvious that the need for the flow of massive amounts of information invites the unavoidable challenge of information overload. In the NAS, information can become bottlenecked through irrelevancy and obscurity if transformative procedures and technology are not embraced to facilitate information sharing. Interface flexibility combined with a tailoring strategy that is founded upon a net-centric platform is desirable to assure accurate information is delivered to the right person at the right time. Until recently, the lack of this strategy has limited the capacity of the NAS.

Now, NextGen seeks to ensure information is accessed and used by stakeholders in the NAS to make the operational decisions. This transformed approach to information exchange is designed to support the forecasted demand and complexity that will challenge the NAS in the decade to come. The FAA's NextGen initiative of SWIM will ensure that the information necessary to operate the NAS is accumulated, assessed, and used to create a common mental model by distributed decision makers in a collaborative fashion to make continued air travel safer, more efficient, and more secure for decades to come.

Contributors to Practical Insights from College of Aeronautics, Florida Institute of Technology

- Captain Nicholas Kasdaglis, U.S. Air Force, Retired, Aerospace and Human Factors Researcher
- Dr. John E. Deaton, Chair, Human Factors, and Director of Research, College of Aeronautics, Florida Institute of Technology

REFERENCES

[1] FAA. *Air Traffic Organization* [Online]. Available: http://www.faa.gov/about/office_org/headquarters_offices/ato Jan. 3, 2012.

[2] FAA, *Instrument Procedures Handbook*, FAA-H-8261-1A, Washington, DC, 2007.

[3] GAO, *National Airspace System: Summary of Flight Delay Trends for 34 Airports in the Continental United States, an E-supplement to* GAO-10-542, GAO-10-543SP, Washington, DC, May 2010.

[4] FAA, *FAA Aerospace Forecast Fiscal Years 2010–2030*, Washington, DC, 2009.

[5] N. Y. Mineta, Secretary of Transportation, Speech to the Aero Club, Jan. 2004.

[6] D. Janjic. *Structural Analysis of the Sutong Bridge* [Online]. Available: ftp://ftp2.bentley.com/dist/collateral/Web/Civil/StructuralAnalysis-Sutong_Bridge.pdf Jan. 13, 2012 (last date accessed).

[7] M. C. Blackey. FAA Administrator. (2005, Apr. 21). *Problems and Peanuts*, Speech, Washington, DC [Online]. Available: http://www.faa.gov/news/speeches/news_story.cfm?newsId=6052 Jan. 13, 2012 (last date accessed).

[8] JPDO. (2007, Jun.). *Concept of Operations for the Next Generation Air Transportation System*, v2.0, Washington, DC [Online]. Available: http://www.jpdo.gov/library/ngats_transformation.pdf Jan. 13, 2012 (last date accessed).

[9] FAA. *SWIM Program Overview* [Online]. Available: http://www.faa.gov/about/office_org/headquarters_offices/ato/service_units/techops/atc_comms_services/swim/program_overview Jan. 14, 2012 (last date accessed).

10

VISUALIZING INSTEAD OF OVERLOADING: EXPLORING THE PROMISE AND PROBLEMS OF VISUAL COMMUNICATION TO REDUCE INFORMATION OVERLOAD

Jeanne Mengis and Martin J. Eppler

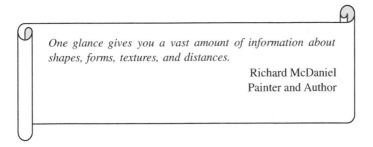

One glance gives you a vast amount of information about shapes, forms, textures, and distances.

Richard McDaniel
Painter and Author

ABSTRACT

Information overload is often simply conceived as receiving too much information—a mere problem of dealing with quantity. In this chapter, we argue that overload is not only the result of the quantity of information but also the result of other information characteristics (such as novelty, ambiguity, or uncertainty). We describe this qualitative dimension of overload and show how information visualization—an alternative way to represent and structure information—can minimize cognitive load. The term information visualization designates the graphic representation of information through spatial mapping in two or three dimensions, using such elements as position, color, shape, texture, and direction to distinguish information elements.

Our survey among 636 communication professionals shows that while they see and value the potential of information visualization to reduce information overload, they do not apply it frequently in their own communication. We explain this knowing–doing gap by drawing on the Technology Acceptance Model (TAM) and developing a revised version for visualization.

Information Overload: An International Challenge for Professional Engineers and Technical Communicators, First Edition. Edited by Judith B. Strother, Jan Ulijn, Zohra Fazal.
© 2012 Institute of Electrical and Electronics Engineers. Published 2012 by John Wiley & Sons, Inc.

The comparison between the model and the survey data suggests that, in addition to usefulness and ease of use, personal visualization aptitude and availability of simple tools that facilitate visualization attempts influence the use of these methods. In conclusion, we highlight that—when introducing countermeasures to information overload in organizations—it is key to build on what is already partially practiced in organizations and start with highly usable and easy to use visualization methods.

10.1 The Qualitative Side of Information Overload

On a normal day, a typical office worker spends 2 hours managing his or her e-mail inbox [1] and another 30 minutes searching within the approximately 346 million active websites of the Internet [2]. With the ongoing exponential growth of information and the ever more difficult task to quickly find relevant pieces of information, the phenomenon of information overload is often understood as a problem of simply "too much" information. Employees feel paralyzed by analysis [3] or at a loss to recognize or understand the big picture of an issue [4] when they are faced with what seems to be an infinite sea of information. In technical communication, lengthy instructions that dive into details too quickly and provide an overabundance of pointers without an overall structure may also leave readers with a paralysis by analysis, unable to achieve what they have set out to do amidst the myriad of tips.

Quantity, however, is only one part of the phenomenon. The growing body of research on information overload indicates that managers and employees do not experience information overload only because of the sheer amount of information with which they are faced [5–9]. We know, for example, that a sense of information fatigue [10] comes about when information reaches employees with a high *intensity* and through a multitude of *channels* [11]. The constant inflow of information (instant messaging, spam e-mail, telephone calls, etc.) leads to frequent and often unnecessary interruptions. According to a study by Basex [12], the latter costs businesses in the United States around US$588 billion each year.

Information overload also comes about when information is diverse, complex, and ambiguous [7]. Additional time is needed to evaluate contrasting messages and/or to make sense of particularly complex information. This shows that the perception of being overloaded depends not only on the quantity of information but also on the *flow* of information and its *quality*. Thus, reducing overload is equally a question of *how much* information is provided (i.e., to contain or limit its quantity), *when* it is delivered (i.e., to reduce its intensity), *what kind* of information is presented (i.e., to increase its quality), and *how* it is communicated (i.e., to improve its delivery). This is why in technical communication, information designers have been preoccupied since the 1990s with *shaping* the content of information, i.e., making its structural features visible through text, graphics, photography, typography, or video. The idea is that these elements enable viewers and readers to discern between the various elements of the piece of information: fact from interpretation, detail from overview, and must-read information from nice-to-have information. This allows for a more personalized use of information and ultimately reduces information overload [13].

In this chapter, we focus on the *qualitative* characteristics of information that contribute to the phenomenon of information overload and propose a solution to act on the quality of information. Within this endeavor, we believe it is necessary to achieve three

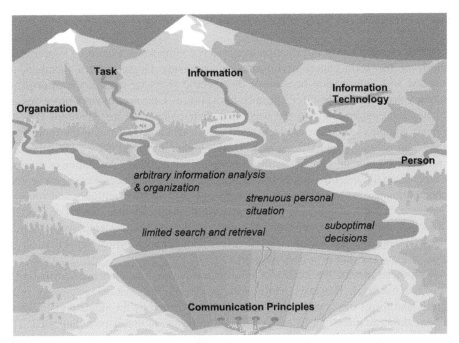

Figure 10.1 The **causes** and *symptoms* of information overload. The visual metaphor of the dam lake exemplifies how information visualization can represent abstract data through concrete means and make it more easily understandable.

things. First, we show—based on the current literature on information overload—which information characteristics increase the perceived load. Second, we discuss one way to address these attributes of information that lead to overload. Specifically, we argue that information visualization (as the domain is generally known in academia) is a particularly effective way to reduce overload. This is also true for the domain of technical communication where visualization has typically been examined only within the context of depictions of apparatuses or to graphically represent manual procedures. (For such a typical treatment of visualization in the technical communication domain, see [13].) By information visualization, we mean the "visual presentation of abstract data," often through computer-supported, interactive means [14]. Figure 10.1 provides an example of information visualization using a visual metaphor to illustrate the main causes and symptoms of information overload.

Finally, the third thing we need to address is how information visualization can become a widely accepted organizational practice to reduce overload in domains such as technical communication, employee or corporate communication, or training. By drawing on a survey that we have conducted among communication professionals worldwide, we explore why information visualization is still not frequently used, even though its potential for reducing information overload is acknowledged by practitioners. Because there is a considerable gap between what communication professionals know and what they do in practice [15], it is not sufficient to propose technical solutions. Rather, we have to propose ways in which the knowing–doing gap can be reduced so that an effective overload remedy can be translated into practice.

10.2 Causes of Information Overload

The picture of what causes information overload is complex (see Figure 10.1 for a visual synthesis of the main causes of information overload). Organizational members are overloaded with information because of the specific characteristics of the information technology that supports them (i.e., push systems) [16, 17], the type of work processes and tasks they are engaged in, and the ways that their companies are organized [18–20]. The more complex and interdependent the tasks to tackle [21, 22] and the more work organized in cross-disciplinary teams [23], the higher the need for coordination and communication—and thus the higher the information intensity. The perception of overload also depends on the specific characteristics of users, such as their degree of expertise, their psychoemotional state, their attention capacity [11, 24], or their attitudes and routines. For example, novices in a domain feel overloaded more quickly than experts do [25, 26]. On the other hand, highly motivated [27] and emotionally positive employees [28] can handle more information without feeling overloaded. (For a critical reflection on the relation of expertise and overload, see [29].)

Given these factors, the quality of information is only one element contributing to information overload. However, while work processes, information technology infrastructure, and forms of organizing are unlikely to be shaped by the single organizational members, they can directly affect the quality of information that is communicated. This is why the *quality of information* becomes a central aspect when we aim to focus on solutions to information overload that are workable in practice.

Table 10.1 outlines the various characteristics of information that affect information overload. On one level, it shows the *quantitative drivers* of information overload, such as the *amount or quantity* of messages reaching organizational members, the *density* of information with which they are faced (the amount of information items within a given message, i.e., text), and the *intensity* with which the messages reach them. On a second level, it shows five *qualitative* information characteristics that contribute to overload. These include the level of *uncertainty* (or reliability) of information, its *novelty*, its *complexity*, the *ambiguity* associated with the interpretation of a piece of information, and the *diversity* among various information items (e.g., with regard to format or style).

The first information attribute that contributes to information overload is information *uncertainty*. Information is uncertain when its validity is not clear. This is the case, for example, when the empirical basis for a specific conclusion cannot be determined [35]. Employees exposed to this type of information need more time to decide whether they can trust it and how they can put it into context.

A similar, yet distinct, attribute is information *ambiguity*. Ambiguity refers to a situation where the basis of a piece of information might be clear; however, various interpretations and conclusions are possible and equally likely [38]. For example, we have a good empirical basis that global warming is taking place; however, what causes the warming and what we can sensibly do about it are still up for discussion. Ambiguity also increases the resources needed to make sense of a piece of information as well as the perceived information load.

Closely related to information ambiguity is information *diversity*. Our understanding of diversity of information differs slightly from that expressed in existing studies inquiring into the relationship between information diversity and load. Iselin [43], for example, distinguishes between repeated dimensions and different dimensions in a set of cues and defines diversity as the number of different dimensions. On a content level, diversity increases when various and sometimes conflicting findings and opinions

T a b l e 10.1   Key Attributes of Information Contributing to Information Overload

	Information Attributes	**References**
How much information? (quantitative drivers of information overload)	*Intensity:* Number of messages per time unit and temporal enfolding (rhythm) of messages.	[4, 8, 30]
	Quantity: Number of messages communicated.	[3, 7, 31–35]
	Density: Number of information items per message, i.e., in a text or visual.	[36, 37]
What kind of information and how is it communicated? (qualitative drivers of information overload)	*Uncertainty* of information: The value of information is unclear as its sources are questionable, evidence is contradicting, and the validity period uncertain.	[7, 22, 35]
	Ambiguity of information: On the basis of a piece of information, multiple interpretations are possible and equally likely.	[7, 35, 38–40]
	Diversity of information: Contradicting information is at hand (e.g., studies with different findings, different sources) and similar information is presented in different styles and presentation formats.	[3, 9, 21, 30, 41–43]
	Novelty of information: Information entails new and unknown insights and is represented in unusual style or format.	[7, 31]
	Complexity of information: Number of information items and their (types of) interrelations are high.	[7, 42]

regarding a topic are at hand—for example, is global warming the result of human actions or natural causes? Diversity also arises when the same or similar pieces of information are represented in various styles and formats, i.e., when there is no standard structure and design for a specific type of report.

Regarding *novelty*, the more the format or content of a message is unfamiliar to a reader, the less quickly he or she can compare these ideas with already acquired knowledge [7]. The new information might not fit existing schemata and beliefs, and time is needed to make sense of the message by aligning the schemata to the new piece of information or choosing a new schemata altogether [44]. In order to reduce information overload, we have to find ways of relating novel information more effectively to already known contents.

The last qualitative attribute of information that increases information load is the *complexity* of a message [7]: the more demanding a text and the more interconnections between the various information items, the quicker a sense of overload emerges.

It is important to note that these qualitative attributes of information relate not only to the content of a piece of information but also to its form, structure, and style. For example, information is not novel only when we are not very familiar with its content. If we take well-established company guidelines and change their structure, users have to spend part of their resources to navigate through the novel structure.

In the following section, we show how information visualization (not to be confused with mere mental visualization or imagination) can shape these attributes in a direction to contain or reduce overload and thus improve information quality.

10.3 How Information Visualization Can Improve the Quality of Information and Reduce Information Overload

Representing information visually can facilitate the retrieval, perception, elaboration, and recall of information in multiple ways [45–48]. Some of the benefits of visualization positively address the causes of information overload. In an early study, Larkin and Simon [49] showed, for example, that diagrams and other visual formats allow people who are less experienced in a domain to process information more like an expert could. In particular, they found that visual representation facilitates the recognition of patterns, the capability to relate various pieces of information and focus on the most relevant ones, and finally, the easier retention of information. In our context, this finding is of particular interest if combined with the studies by Swain and Haka [26] and Agnew and Szykman [25], which show that experts are less prone to be overloaded than novices when exposed to the same amount of information. Taken together, these two strands of research suggest that visualization helps to contain overload.

Studies that empirically investigate the relationship between visualization and information overload are still scarce [50–52], although there are multiple studies (mainly from the computer science or information systems domain) that conceptually claim that visualization can reduce information overload [53–56]. There are, however, interesting indications on how visualization can act on the information quality attributes we have outlined above. With regards to the *quantity* and *complexity* of information, for example, Shneiderman [57], as well as Burkhard and Meier [58], suggested that visuals can be read both in overview and by zooming in on details. Visuals provide a "detailed overview" by interconnecting details to an overall image, and viewers can decide the level of detail and complexity at which they want to understand it. With regards to *novelty*, representing novel information on a known, visual metaphor (e.g., a representation of a bridge) can activate an understanding of the novel information in relation to the already familiar [59, 60]. Similarly, standardized, visual formats (e.g., pictograms for specific types of information) can give orientation and structure [61]. Through relating the novel with the known and making it a bit more familiar, visualization can reduce the novelty of information, thus positively shape its quality. In this way, visualization can reduce information overload.

This is true only if the use of visual representations is judicious and if the visualizations themselves are clear, simple, and informative. Like any piece of information, images too can be ambiguous, misleading, overloaded with unnecessary detail, or simply inappropriate. An image can indeed be worth 10,000 words, as Simon and Larkin [49] pointed out, but sometimes it requires 10,000 words to understand a picture. Inaccessible, messy, or cluttered graphics can thus aggravate the problem of overload—if certain design guidelines are not respected. (For a concise overview of such guidelines, see [62] and Chapter 6.) To illustrate this point, we invite our readers to have a look at some of the confusing, but authentic, diagram examples presented in [63–65]. Inaccessible graphics that contribute to overload, rather than reduce it, can also be found using a simple Internet image search with the key words "bad diagram." The root cause of such a counterproductive use of visualizations is often a lack of design knowledge. We explore this and other aspects in the next section.

10.4 Using Visualization in Practice: Understanding the Knowing–Doing Gap

Knowing the benefits of information visualization for reducing overload has not yet made it a workable solution in practice. In fact, according to Pfeffer and Sutton [66], "knowing 'what' to do is not enough." They show that there is a consistent gap between what we know would be best to do and what we actually do in practice. This suggests that we have to find ways to understand and address the knowing–doing gap regarding overload reduction mechanisms in general, and information visualization in specific.

One stream of research that has systematically examined the knowing–doing gap can be found in the literature on the Technology Acceptance Model (TAM). TAM suggests that the intention to use a new technology or technique is determined mainly by two beliefs: *perceived usefulness* of a technology (i.e., belief of the degree to which the use of a technology will enhance one's own job performance) and *perceived ease of use* (i.e., belief that the technology will not require much effort to be used) [67]. TAM further states that these two elements are related to each other: the perceived usefulness is influenced by the ease of use, which means that the easier the system, the more useful it is perceived to be [67]. Much of the work on TAM aims to understand the external variables that impact both perceived usefulness and ease of use [68]. Such variables are, to name a few, technical and managerial support provided [69], prior use and experience [70, 71], subjective norm (perception that the people important to a person think it is appropriate to use a technology) [67], or expected quality of the output. (For comprehensive reviews, see [67, 68].)

Although TAM has been developed and tested almost exclusively in the field of information systems, we suggest that we can use the model to understand potential resistances to the use of information visualization in organizational communication. In particular, we draw on TAM to better understand why solutions to information overload, such as information visualization, might not be used by practitioners even if they deem them as useful. In Figure 10.2, we translate some of the core ideas of TAM to the context of information visualization.

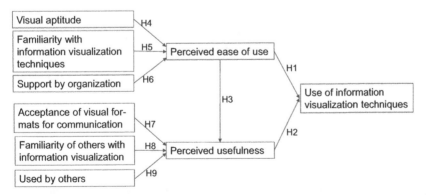

Figure 10.2 A model for understanding the knowing-doing gap for using information visualization to reduce information overload (adapted from the technology Acceptance Model, TAM, for the context of information visualization [67]). It suggests that there are two main variables that explain whether information visualization techniques are used: How easy is it to use the technique and how useful is it perceived to be?

Figure 10.2 shows an adaptation of TAM for the context of information visualization. Like in the classic version of TAM, we propose that there are two main explaining variables for the use of information visualization techniques: on the one hand, the perceived ease of use, and on the other, the perceived usefulness. The higher the perceived ease of use, the more information visualization techniques are used (H1); the higher the perceived usefulness, the more information visualization techniques are used (H2). Also, in alignment with most versions of TAM, and as outlined above, we claim that perceived ease of use positively impacts the perceived usefulness (H3) [67, 72]. Our model shows a simplified version of TAM, insofar as perceived ease of use and perceived usefulness directly impact the use of information visualization techniques and not the intention to use.

In TAM models, perceived ease of use is influenced by a set of external variables, and we consider three elements in particular: visual aptitude, familiarity with information visualization techniques, and support by the organization. *Visual aptitude* refers to a similar construct used in TAM called *self-efficacy* [73] and deals with the respondent's talent and inclination for visuals. The more someone feels like a visual person [74] and is gifted with visualizing, the more ready he or she will feel to deal with a visual (H4). Similarly, the more someone is familiar with dealing with information visualization techniques (i.e., someone who knows how to do visualizations and in which contexts to use them), the easier the use of the visual will seem (H5). When inquiring into the antecedents of ease of use, Venkatesh and Davis [75] suggested that direct experience and *familiarity* with a technology and tool impact the perceived ease of use (they suggested a moderation effect between objective usability and ease of use). Finally, we propose that *support* from an organization [76, 77] in the use of visual techniques, in terms of providing good resources or tools, positively relates to perceived ease of use (H6). Formulated negatively, when people have no easy tools to develop visualizations, they will feel that visualization is difficult and that it is very time intensive (i.e., low ease of use).

As for ease of use, we also suggest three antecedents for the second TAM variable, *perceived usefulness*: first, the organization's *acceptance* to use visual formats for communication; second, the *familiarity* of other organizational members with visualization; and third, the actual *use* of visualization techniques as an organization-wide practice. When proposing TAM2, Venkatesh and Davis [67] argued that social norm influences perceived usefulness so that the people strive to meet the expectations from their important referents in their technology use. Individuals aim at maintaining a favorable image within a community and if the use of a technology improves one's status, its perceived usefulness will be increased, as will one's intention to use it [67, 78]. Another aspect of social influence is that for a technology to be adopted, it needs to be compatible with the values, beliefs, past experiences, and current practices of a social system [79]. Moore and Benbasat [80] have shown that the compatibility of a technology with the way an organization works is central for its adoption.

Translating these insights to information visualization, we can propose that the more visualization is accepted in the organization as an effective way to communicate, the more useful it becomes to visualize information (H7). Also, only when other members of the organization are familiar with the conventions and uses of visualization can more sophisticated ways of visualizing be used without running the risk of being misunderstood. Thus, the more familiar others are with visualizing information, the more useful this form of communication becomes (H8). Finally, we suggest that the more visualization is used in an organization, the higher its perceived usefulness (H9).

Overall, the model we propose suggests that in order to adopt information visualization as a solution for information overload, a community not only has to perceive it as *useful* (i.e., effective in reducing overload) but also needs to deem it *easy to use*. Our model points out that these two aspects depend not only on the qualities of the solution but also on the characteristics of the user (i.e., familiarity, aptitude) and on a set of social dynamics within the organization (i.e., acceptance, support, part of organizational practice). In other words, we argue that if technical communicators want to successfully implement solutions to information overload, they have to think not only of an effective (technical) design of the solution but also of the current practices, perceptions, and the familiarity of the potential users and their organizations. The results from the study presented in the following section help us understand in more detail how these elements matter when introducing specific countermeasures to information overload.

10.5 Methods and Context of the Study

We draw on a study that we conducted in collaboration with the International Association for Business Communicators (IABC) in 2008 [81]. We carried out an online survey among a randomized sample of 5000 IABC members coming from the United States, Europe—all countries of the European Union plus Switzerland, Norway, Russia, and the Commonwealth of Independent States (CIS) countries—and the countries of the Middle East (members come from Egypt, Kuwait, Oman, Qatar, Saudi Arabia, and the United Arab Emirates). All respondents were communication professionals, i.e., communication specialists working in corporate communications, public affairs, investor relations, Web communication, video affairs, advertising, or other related fields. We asked them about their appreciation and use of information visualization techniques in their work as communication professionals. In particular, we were interested in understanding how they valued information visualization (in terms of ease of use and usefulness) and how they and their organizations used visualization formats. A total of 636 IABC members responded to the survey.

10.5.1 Measures

The survey began by exposing the respondents to three examples of information visualization, either conceptual diagrams or visual metaphors. We introduced a comparative element to the survey by asking half of our sample about their views and uses of conceptual diagrams and the other half about visual metaphors. (It is out of the scope of this chapter to elaborate on this comparison; for more details, see [81].)

In the second step, we asked the communication professionals about their own and their colleagues' *familiarity* with such visualizations, for example, "I know the conventions used when building such conceptual diagrams and can construct them myself" (both personal and organization-wide familiarity constructs were operationalized by three-item scales—see Table 10A.1). This was followed by a set of questions asking about the contexts and occasions in which the respondent *uses* or has seen the visual forms of communication to be used, for example, "Have you encountered such conceptual diagrams in your organization in internal reports, on the corporate website/intranet, or during presentations and workshops?"

Perceived usefulness was the third topic of the questionnaire. We measured usefulness along the qualitative attributes of information that relate to information overload (compare Table 10.1 for qualitative attributes of information with Figure 10.3 for items of

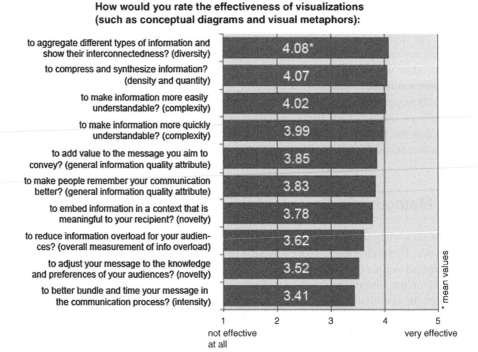

Figure 10.3 Perceived usefulness of information visualization. The mean values for the perceived usefulness of visualization clearly show that visuals are highly valued by communication professionals.

measurement). For example, we asked questions such as "How would you rate the effectiveness of conceptual diagrams to aggregate different types of information and show their interconnections?" (The question relates to the diversity of information—see Table 10.2.) Table 10.2 shows how the respondents could express their opinions regarding the effectiveness of information visualization for addressing the various qualitative drivers of information overload. In addition, we asked a direct question about how participants evaluated the effectiveness of information visualization for reducing information overload. The set of 12 questions shown in Table 10.2 allowed us to operationalize the perceived usefulness construct (of which two fell out in the factorial analysis).

Finally, a fourth set of questions asked the professionals about the challenges when using visual formats in their communication. For example, "If you occasionally find it problematic to use visual metaphors, why may this be the case?"—Answer: "They are often too complex to be understood quickly."

The questions aimed both at the respondent's individual perception (i.e., the respondent's own familiarity, use, appreciation, and perceived challenges) and at his or her assessment of organization-wide practices, values, or judgments regarding visuals and their capacity to reduce information overload. Questions addressing such organization-wide preferences and practices allowed us to measure the social influence constructs we defined as explaining variables for perceived usefulness. Table 10A.1 provides an overview on the items we used to operationalize and measure the constructs of the model summarized in Figure 10.2. All items were measured on a 5-point Likert scale.

T a b l e 10.2 Measuring Qualitative Drivers of Information Overload Through Perceived Usefulness Items

Attributes of Information Driving Information Overload	Items Measuring Perceived Usefulness
	How would you rate the effectiveness of visualizations to:
Intensity of information	– better bundle and time your messages in the communication process?
Density of information	– compress and synthesize information?
Quantity of information	
Diversity of information	– to aggregate different types of information and show their interconnections?
	– to standardize the way with which you convey information?
Novelty of information	– to embed information in a context that is meaningful to your recipient?
Uncertainty of information	
Ambiguity of information	– to adjust your message to the knowledge and preferences of your audience?
Complexity of information	– to make information more easily understandable?
	– to make information more quickly understandable?
General attributes of information quality	– to reduce information overload for the people receiving it?
	– to add value to the message you aim to convey?
	– to make people remember your communications better?
	– to also involve emotions in your communication?

10.5.2 Procedure and Analysis

The randomized group of IABC members in the United States was invited to participate in the online survey through an e-mail sent out by the IABC Research Foundation in spring 2008. Three weeks later, we sent out a reminder to those who had not yet taken part in the survey. After this first round of data collection was successfully completed, we applied the same procedure in early fall 2008 to invite IABC members from Europe and the Middle East to participate in the study. We tried to enhance participation by emphasizing that participants would learn about visualization techniques through the questionnaire and by announcing that we would provide results and feedback in the upcoming yearly conference of the association.

After a first descriptive analysis aimed at gaining a sense of the level of appreciation and use of visualization techniques, we conducted a confirmatory factor analysis combined with a reliability analysis to validate the scales of the presented model (see Table 10A.1). We then tested the hypotheses of the model by conducting a linear multiple regression analysis. Given the space limitations and the focus of this chapter, we did not perform systematic comparisons across the geographical regions of respondents and are not able to make a cultural argument around the use of visualization.

In the following discussion of our findings, we first present a descriptive analysis that shows that information visualization is appreciated by communication professionals to reduce information overload, but is not yet widely used. We then embed the high perceived usefulness but low use in the TAM model to better understand what is responsible for this knowing–doing gap. Finally, we discuss what communication professionals need to do in order to effectively implement solutions for information overload, such as information visualization.

10.6 Indications of the Knowing–Doing Gap: Visuals Are Valued, but Poorly Used

The descriptive analysis of the survey shows that visuals are highly valued by communication professionals (see Figure 10.3 showing the mean values for the perceived usefulness of visualization). They are seen to positively act on a multitude of information attributes that are central for reducing information overload. In fact, diagrams and visual metaphors are judged to be particularly beneficial for aggregating different types of information and showing their interconnections (thus reducing information diversity). This, of course, bears the risk of increasing information density in the visuals themselves, particularly if the employed summary diagrams are overloaded themselves or suffer from an unclear structure. They are also valuable for compressing and synthesizing information (thus reducing information density and quantity) and for making information more easily and more quickly understandable (thus reducing information overload more generally).

For other aspects relevant to reducing information overload, such as standardizing (reducing information diversity) or bundling information (reducing the intensity of the information flow), visualization does not seem to be the most effective method in the view of the surveyed communicators. Yet, the general tenor is that visualization is very beneficial for effectively communicating within an organization and positively shapes various qualities of information that contribute to information overload. In fact, when business communicators were directly asked whether visualization is an effective means to reduce information overload, they valued it as rather effective (with a mean value of 3.62 on the 5-point Likert scale).

In spite of this positive appraisal of visualization, visualization formats such as visual metaphors and conceptual diagrams are only sometimes used by communication professionals and by other members of the organizations for which they work. When asked about their personal use, communication professionals responded with an average value of 3.01 (on a scale from 1 = never to 5 = extremely often) and indicated a similar value of 3.04 for their colleagues (see Table 10A.1). These first descriptive findings seem to confirm the frequently discussed gap between knowing and doing: communication professionals know of the multiple benefits of visualization, but they are rather reluctant to engage in visualization in their everyday practices. What prevents them from doing so? In the following section, we will show how this gap between knowing and doing can be explained.

10.7 Understanding the Knowing–Doing Gap with TAM

Before presenting the results for the model that we propose, we briefly discuss our findings on a measurement level. The measurement scales exhibited strong reliability with all Cronbach alpha coefficients showing values above 0.77. Construct validity was strongly supported by the principal components analysis with varimax rotation (see Table 10A.1 for the detailed psychometric properties of the scales: factor loadings, explained variance, Cronbach alphas, mean values, and standard deviations).

Table 10.3 shows Pearson's interconstruct correlations (the strongest significant correlations with coefficients above 0.35 are shown in bold). It can be seen that the communication professionals' self-reported use of visualization is most strongly correlated with their perceived usefulness of visuals (0.503**) and with their use by other members of

T a b l e 10.3   Correlation Matrix of Construct Variables

	1	2	3	4	5	6	7	8	9
1. Perceived ease of use									
2. Perceived usefulness	**0.415****								
3. Self-reported use	**0.374****	**0.503****							
4. Visual aptitude	0.340**	0.208**	**0.393****						
5. Familiarity with visual	0.101*	−0.058	0.066	0.130**					
6. Support by organization to use visual	**0.373****	0.060	0.279**	**0.356****	0.043				
7. Acceptance of visual formats by organization	0.341**	0.217**	0.239**	0.111**	0.073	0.202**			
8. Familiarity of organization with visuals	0.231**	0.211	0.330**	0.230**	0.178**	0.243**	0.329**		
9. Use by other members of organization	0.183**	0.267**	**0.551****	**0.187****	0.144**	0.168**	*0.392***	**0. 475****	

*Significance at 0.05 level.
**Significance at 0.01 level.

the organization (0.551^{**}). There is further a rather strong correlation between the perceived ease of use and the support provided by the organization to engage in visualization (0.373^{**}). Finally, the self-reported use shows a rather strong, significant correlation with visual aptitude (0.393^{**}).

Figure 10.4 provides the results of hypothesis testing and regression analysis. We find that both "perceived usefulness" and "ease of use" significantly explain the use of information visualization techniques (hypotheses H1 and H2 are confirmed). In particular, and like in prior studies on TAM in information systems, we find that the perceived usefulness ($\beta = 0.55, p < 0.001$) more strongly predicts the use of visualization techniques than ease of use does ($\beta = 0.24, p < 0.001$) [67]. Together, perceived usefulness and ease of use explain a moderate 28% of the variance of the use of visualization ($R^2 = 0.28$). In addition, the model shows that 20% of the variance of "perceived ease of use" can be explained by the person's visual aptitude ($\beta = 0.24, p < 0.001$) and by the support provided by the organization to visualize information (e.g., through providing time, tools) ($\beta = 0.29, p < 0.001$) (H4 and H6 are confirmed). Instead, the familiarity with visualization techniques does not significantly explain the perceived ease of use (H5 is not confirmed). Finally, the "perceived usefulness" of engaging in visualization depends mainly on the "perceived ease of use" ($\beta = 0.33, p < 0.001$) and on whether visualization is commonly practiced by other members of the organization ($\beta = 0.14, p < 0.001$) (H3 and H9 are confirmed). However, we do not find support for the other two elements that we have suggested would further predict the perceived usefulness of visuals, i.e., the acceptance of other members of the organization to use visuals and their familiarity to deal with visualization (H7 and H8 are not confirmed).

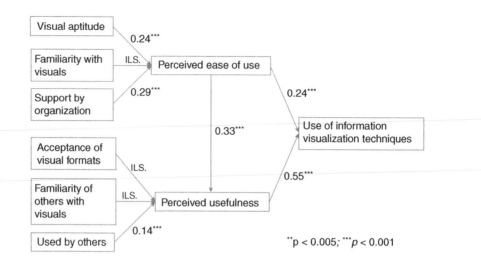

R^2 for use of visuals = 0.28; R^2 for perceived ease of use = 0.20; R^2 for perceived usefulness = 0.20

Figure 10.4 Results of adapted TAM model for information visualization. Both "perceived usefulness" and "ease of use" significantly explain the use of information visualization techniques, but perceived usefulness more strongly predicts the use of visualization techniques than ease of use does. Together, perceived usefulness and ease of use explain a moderate 28% of the variance of the use of visualization.

10.8 Discussion

Our analysis shows that information visualization is more likely to be perceived as highly valuable if visualization techniques and tools are easy to use and widely used by one's peers. Organizations that want to profit from the merits of visualization in order to reduce information overload are thus well advised to start with relatively simple visual communication formats that are easy to scale and deploy without excessive need for training or documentation. Examples of such straightforward and scalable visualization techniques are hand-drawn diagrams, as advocated in Dan Roam's recent best seller, *The Back of the Napkin* [82].

Our analysis also suggests that the perceived ease of visualization not only depends on the simplicity of visualizations themselves but also on the support provided by the organization. If organizations want to reduce information overload by using more effective means of communication, such as visuals, they must also provide the resources, such as time or simple tools, to make use of such means.

Our data reveals another insight, speculated upon by Dan Roam in his book [82], namely, that familiarity with visualization techniques does not significantly explain the perceived ease of use. This can be interpreted in the sense that visual communication should not be treated as a secret trade to be learned but rather as a natural skill that anyone can employ to make communication more effective. This interpretation is also supported by our finding that one's perceived visual aptitude can explain an important part of the perceived ease of use. For visualization methods to be introduced successfully into the everyday practice of organizations, it is important to convey to everybody the sense that we are all visual thinkers. As our study has shown, providing simple visualization methods and tools is one way to provide technical communicators a first positive experience with visualization. A main challenge with these simple visualization methods, however, is that they need to provide an adequate complexity to represent the often complex information that has to be communicated, but not themselves contribute to additional information overload by confusing recipients instead of informing them.

We did not find that social influence factors are particularly central for the perceived usefulness of visuals. In fact, our data do not suggest that the acceptance of visualization by other members of the organization or their familiarity with visualization techniques significantly impacts communicators' appreciation or use of visual forms of communication. This finding might be explained by the type of respondents we had in the survey. These were all communication professionals and, as such, they might be used to being ahead of the game in their communication behavior. Future research could address this limitation of our study by surveying a much broader group of practicing engineers and other professionals on their appreciation and use of visualization.

Finally, we set out by claiming that effective means to counter information overload should not be limited to outlining "technical" solutions but have to address how the knowing–doing gap can be closed. Our study only goes some way in achieving this objective. In fact, the perceived usefulness of visualization and its ease of use can only explain 30% of the variance of the use of visualization. We can only speculate about additional factors that further explain what defines the use of visuals in practice. Venkatesh [73] showed, for example, that extrinsic motivation and job fit are important additional predictors for use. Although such elements might complete the picture of why visualization is valued but only sometimes used in practice, we suggest that in-depth insight into the everyday communication practices can best be gained by observational studies. Such studies would allow us to develop a more precise understanding of why people engage in forms of communication that add to information overload although they are aware of alternative modes of communication that could contain the felt information load.

10.9 Conclusion

Information overload is a complex phenomenon that is not caused simply by "too much" information. Other characteristics of information make us feel overloaded as well, such as ambiguity, diversity, novelty, or complexity. In this chapter, we have focused exclusively on these qualitative aspects of information, and we have discussed visualization as one way to enhance the quality of information and reduce overload.

We have further argued that solutions to overload can only work in practice if specific social (e.g., organizational and interactional) contexts [29, 83] are taken into account. In view of this argument, we have investigated the knowing–doing gap regarding visual communication and we have gone some way in explaining what drives the use of information visualization in practice. Using the Technology Acceptance Model to frame our argument and translate it to the domain of information visualization, we have been able to show that the perceived usefulness and ease of use can explain a good part of whether or not visualization is used in internal communication practice. Major impediments to using visualization are a perceived lack of visualization skills, a need for simple tools to use visualizations effectively, and an insufficient amount of time to develop visuals. Overall, our study suggests that information visualization can become a feasible solution to information overload only if organizations promote simple forms of visualization and provide active support by granting good resources and tools for easier visualization.

10.10   Future Research Directions

In terms of future studies on information overload and information visualization, we see different possible directions. Further analysis is needed to better understand *which kind* of visualizations can help to contain and reduce information overload in specific *contextual circumstances*. For example, we have to better understand for which types of audiences (e.g., differing in their levels of expertise or visual aptitude), in which working situations (e.g., under large external pressure, high ambiguity), and in which (organizational) cultures the use of visual communication can address overload. In terms of the kinds of visualization, research should investigate how different processes of visualizing, for example, emerging visualizations with gradually growing complexity versus static visualizations, act on information overload. In this regard, media also matter, and future research should examine how (interactive) *digital* visualizations differ from *analogous* (i.e., hand drawn) ones in terms of their effects on information overload. Closely related to this issue is the question of how pictures can complement text, for example, in technical communication: When do pictures reduce overload in technical communication because they provide a synopsis, overview, or orientation, and when do they help to prevent information underload by providing additional perspectives or insights not conveyed in the text?

A related research direction consists of comparing different *genres* of information visualizations and their respective overload implications. One could, for example, experimentally evaluate the respective effects of visual metaphors versus conceptual diagrams on overload (as distinguished in our survey). Not only the form of visualization but also the *type of information* could be researched by investigating which kind of information is best represented graphically so that overload can be reduced. In this context, one should examine how much graphic representation of procedural, declarative, or experiential information is reducing or contributing to information overload.

Research in these areas will contribute to an understanding of information overload that is not static, but that takes into account the dynamic and contextualized qualities of the phenomena. In this way, information visualization is not a panacea against information overload, but needs to be carefully aligned to the specific organizational context, the characteristic of its audiences, and the existing communication practices. By addressing the knowing–doing gap of visualization, we hope that this chapter has gone some way to help the reader gain a handle on these contextual aspects of a potentially very promising solution to information overload, which is information visualization. We thus hope that our contribution encourages other researchers and practitioners to actively explore this fascinating communication practice and its potential for reducing information overload.

10A.1 Appendix

Table 10A.1 Measurement Analysis for the Knowing–Doing Gap Model of Information Visualization (Factor Loadings, Percentages of Variance Explained, Cronbach Alphas, Mean Values, and Standard Deviations of Constructs)

Factor	Item	Factor Loading	% of Variance Explained	Cronbach Alpha	Mean (S.D.)
1. Perceived ease of use (inversion of perceived difficulty to use)	1. They are NOT difficult to use (inv.)	0.693	54.46	0.77	2.85 (0.96)
	2. They are NOT too complex to be understood quickly (inv.)	0.600			
	3. They are NOT difficult to use (orga) (inv.)	0.824			
	4. They are NOT too complex to be understood quickly (orga) (inv.)	0.818			
2. Perceived usefulness	1. to compress and synthesize information?	0.810	64.94	0.939	3.81 (0.83)
	2. to aggregate different types of information and show their interconnections?	0.733			
	3. to adjust your message to the knowledge and preferences of your audience?	0.772			
	4. to add value to the message you aim to convey?	0.828			
	5. to embed information in a context that is meaningful to your recipient?	0.836			
	6. to better bundle and time your messages in the communication process?	0.713			
	7. to reduce information overload for the people receiving it?	0.796			
	8. to make information more easily understandable?	0.867			
	9. to make information more quickly understandable?	0.863			
	10. to make people remember your communications better?	0.823			
3. Reported use	1. Do you personally use visual metaphors in your communications with your colleagues?	0.915	86.6	0.922	3.01 (1.13)
	2. When you prepare workshops, meetings, presentations, reports, or other documents, do you use visual metaphors?	0.937			
	3. Are visual metaphors a part of your normal communication toolbox?	0.940			

(continued)

T a b l e 10A.1 (*Continued*)

Factor	Item	Factor Loading	% of Variance Explained	Cronbach Alpha	Mean (S.D.)
4. Visual aptitude	1. I consider myself a visual thinker	0.829	63.58	0.786	3.39 (0.88)
	2. I generally like working with images	0.860			
	3. I feel rather gifted at working with images	0.868			
	4. I DO NOT lack design and visualization know how (Inv.)	0.602			
5. Familiarity with visualization	1. I am familiar with visual metaphors such as the ones shown before (although I might not know the specific ones shown here)	0.791	71.24	0.797	2.78 (1.06)
	2. I know in which contexts it is suitable to use such visual metaphors	0.899			
	3. I know the conventions used when building such visual metaphors and can construct them myself	0.840			
6. Support by organization to use visualizations	1. I DO have enough time to develop them (inv.)	0.777	63.53	0.809	2.19 (0.90)
	2. There is NO lack of simple tools to quickly develop or adapt the diagram (inv.)	0.792			
	3. The employees in our organization DO have enough time to develop them (orga) (Inv.)	0.800			
	4. There is NO lack of simple tools to quickly develop or adapt the diagram (orga) (inv.)	0.819			
7. Acceptance of visual formats by organization	1. There is NOT a lacking acceptance by other members of the organization	0.754	68.53	0.881	3.57 (0.91)
	2. It IS compatible with our existing practices, values, and/or routines	0.817			
	3. In my organization, there is NO lacking acceptance by other members of the organization when someone uses visualization techniques	0.778			
	4. In my organization, it IS compatible with our existing practices and/or routines to use visualization techniques	0.892			
	5. In my organization, it IS compatible with our existing values to use visualization techniques	0.888			

8. Familiarity of organization with visuals	1. These visual metaphors are well-known in my organization	0.837	78.98	0.863	2.78 (0.158)
	2. People in my organization are well aware of the contexts in which it is suitable to use such visual metaphors	0.925			
	3. The members of my organizations know how to use and construct such visual diagrams	0.902			
9. Use by other members of organization	1. Are visual metaphors used in your organization for internal communications?	0.894	78.47	0.863	3.04 (0.24)
	2. Have you encountered such visual metaphors in internal reports in your organization or during presentations and workshops, or in websites?	0.879			
	3. Are visual metaphors considered a part of the communication toolbox that is used in your organization?	0.884			

REFERENCES

[1] G. F. Thomas and C. L. King, "Reconceptualizing e-mail overload," *J. Business Tech. Commun.*, vol. 20, pp. 252–287, Jul. 2006.

[2] Netcraft. (2011, Jun.). *Web Server Survey* [Online]. Available: www.netcraft.com Aug. 29, 2011 (last date accessed).

[3] D. Bawden, "Information overload," in *Library and Information Briefings*, London, U.K.: South Bank Univ., 2001, pp. 1–15.

[4] A. G. Schick, *et al.*, "Information overload: A temporal approach," *Account. Organiz. Soc.*, vol. 15, no. 3, pp. 199–220, 1990.

[5] K. L. Keller and R. Staelin, "Effects of quality and quantity of information on decision effectiveness," *J. Consumer Res. Policy*, vol. 14, no. 2, pp. 200–213, 1987.

[6] R. S. Owen, "Clarifying the simple assumption of the information load paradigm," *Adv. Consumer Res.*, vol. 19, pp. 770–776, 1992.

[7] S. C. Schneider, "Information overload: Causes and consequences," *Human Syst. Manag.*, vol. 7, no. 2, pp. 143–153, 1987.

[8] C. W. Simpson and L. Prusak, "Troubles with information overload. Moving from quantity to quality in information provision," *Int. J. Inform. Manag.*, vol. 15, no. 6, pp. 413–425, 1995.

[9] C. Klausegger, *et al.*, "Information overload: A cross-national investigation of influence factors and effects," *Market. Intell. Plan.*, vol. 25, no. 7, pp. 691–718, 2007.

[10] R. S. Wurman, *Information Anxiety*. Indianapolis, IN: Macmillan, 2001.

[11] E. M. Hallowell, "Overloaded circuits: Why smart people underperform," *Harvard Business Rev.*, vol. 83, no. 1, pp. 54–62, 2005.

[12] I. Basex. (2005). *The Cost of Not Paying Attention: How Interruptions Impact Knowledge Worker Productivity* [Online]. Available: http://bsx.stores.yahoo.net/coofnotpaat.html Aug. 29, 2011 (last date accessed).

[13] M. J. Albers and B. Mazur, *Content and Complexity. Information Design and Technical Communication*. Mahwah, NJ: Lawrence Erlbaum Associates, 2003.

[14] S. K. Card, *et al.*, *Readings in Information Visualization: Using Vision to Think*. Los Altos, CA: Morgan Kaufmann, 1999.

[15] J. Pfeffer and R. I. Sutton, *The Knowing-Doing Gap. How Smart Companies Turn Knowledge into Action*. Boston, MA: Harvard Business School Press, 2000.

[16] D. K. Allen and M. Shoard, "Spreading the load: Mobile information and communications technologies and their effect on information overload," *Inform. Res.*, vol. 10, no. 2, Jan. 2005.

[17] K. Koroleva, *et al.* (2010). "'Stop spamming me!' Exploring information overload on Facebook," in *Proc. AMICIS* [Online]. Available: http://aisel.aisnet.org/amcis2010/447/ Aug. 29, 2011 (last date accessed).

[18] M. J. Eppler and J. M. Mengis, "The concept of information overload: A review of literature from organization science, accounting, marketing, MIS, and related disciplines," *Inform. Soc.: An Intern. J.*, vol. 20, no. 5, pp. 325–344, 2004.

[19] D. Kirsch, "A few thoughts on cognitive load," *Intellectica*, vol. 1, no. 30, pp. 19–51, 2000.

[20] N. Kock, *et al.*, "The information overload paradox: A structural equation modeling analysis of data from New Zealand, Spain, and the USA," *J. Global Inform. Manag.*, vol. 17, no. 3, pp. 1–19, 2009.

[21] C. Speier, *et al.*, "The influence of task interruption on individual decision making: An information overload perspective," *Decision Sci.* vol. 30, no. 2, pp. 337–359, 1999.

[22] M. L. Tushman and D. A. Nadler, "Information processing as an integrating concept in organizational design," *Acad. Manag. Rev.*, vol. 3, no. 3, pp. 613–625, 1978.

[23] M. Grise and R. B. Gallupe, "Information overload: Addressing the productivity paradox in face-to-face electronic meetings," *J. Manag. Inform. Syst.*, vol. 16, no. 3, pp. 157–185, 1999 –2000.

[24] T. Klingberg, *The Overflowing Brain: Information Overload and the Limits of Working Memory.* New York: Oxford Univ. Press, 2009.

[25] J. Agnew and L. R. Szykman. (2004, Sep. 22). *Asset Allocation and Information Overload: The Influence of Information Display, Asset Choice and Investor Experience* [Online]. Available: http://ssrn.com/abstract=1142932 Aug. 29, 2011 (last date accessed).

[26] M. R. Swain and S. F. Haka, "Effects of information load on capital budgeting decisions," *Behavioral Res. Account.*, vol. 12, pp. 171–199, 2000.

[27] T. E. Muller, "Buyer response to variations in product information load," *Psychol. Rev.* vol. 63, no. 2, pp. 81–97, 1984.

[28] K. A. Braun-LaTour, *et al.*, "Mood, information congruency, and overload," *J. Business Res.*, vol. 60, no. 11, pp. 1109–1116, Nov. 2007.

[29] K. Weick and K. M. Sutcliffe, "Information overload revisited," in *The Oxford Handbook of Organizational Decision Making*, G. P. Hodgkinson and W. H. Starbuck, Eds. Oxford, U.K.: Oxford Univ. Press, 2008, pp. 56–75.

[30] D. Bawden and L. Robinson, "The dark side of information: Overload, anxiety and other paradoxes and pathologies," *J. Inform. Sci.*, vol. 35, no. 2, pp. 180–191, 2009.

[31] P. A. Herbig and H. Kramer, "The effect of information overload on the innovation choice process," *J. Consumer Market.*, vol. 11, no. 2, pp. 45–54, 1994.

[32] J. Jacoby, "Information load and decision quality: Some contested issues," *J. Market. Res.*, vol. 14, pp. 569–573, 1977.

[33] J. Jacoby, "Perspectives on information overload," *J. Consumer Res. Policy*, vol. 10, no. 4, pp. 432–436, 1984.

[34] N. K. Malhotra, *et al.*, "The information overload controversy: An alternative viewpoint," *J. Market.*, vol. 46, no. 2, pp. 27–37, 1982.

[35] P. R. Sparrow, "Strategy and cognition: Understanding the role of management knowledge structures, organizational memory and information overload," *Creativity Innovat. Manag.*, vol. 8, C no. 2, pp. 140–149, Jun. 1999.

[36] G. Wiederhold and M. Genesereth, "The conceptual basis for mediation services," *IEEE Expert*, vol. 12, no. 5, pp. 38–47, Sep.–Oct. 1997.

[37] A. D. Songer, *et al.*, "Multidimensional visualization of project control data," *Construct. Innovat.*, vol. 4, no. 3, pp. 173–190, 2004.

[38] K. M. Sutcliffe, "Information handling challenges in complex systems," *Int. Public Manag. J.*, vol. 8, no. 3, pp. 417–424, 2005.

[39] H. Lesca and E. Lesca, *Gestion de L'Information, Qualite de L'Information et Performances de L'Entreprise.* Paris, France: Litec, 1995.

[40] I. Mulder, *et al.*, "An information overload study: Using design methods for understanding," in *Proc. 18th Australia Conf. Comput.-Human Interact. Des.: Activities, Artefacts Environ.*, 2006, pp. 245–252.

[41] E. R. Iselin, "The effects of the information and data properties of financial ratios and statements on managerial decision quality," *J. Business Finance Account.*, vol. 20, no. 2, pp. 249–267, 1993.

[42] H. M. Schroder, *et al.*, *Human Information Processing. Individuals and Groups Functioning in Complex Social Situations.* New York: Holt, Rinehart, & Winston, 1967.

[43] E. R. Iselin, "The impact of information diversity on information overload effects in unstructured managerial decision making," *J. Inform. Sci.*, vol. 15, pp. 163–173, 1989.

[44] R. Axelrod, "Schema theory: An information processing model of perception and cognition," *Am. Political Sci. Rev.*, vol. 67, no. 4, pp. 1248–1266, 1973.

[45] C. Chen and Y. Yu, "Empirical studies of information visualization: A meta-analysis," *Int. J. Human–Comput. Stud.*, vol. 53, no. 5, pp. 851–866, 2000.

[46] C. S. Stewart and D. D. Stewart, "Group recall: the picture-superiority effect with shared and unshared information," *Group Dynamics: Theory, Res. Practice*, vol. 5, no. 1, pp. 48–56, 2001.

[47] A. M. Glenberg and W. E. Langston, "Comprehension of illustrated text: Pictures help to build mental models," *J. Memory Language*, vol. 31, no. 2, pp. 129–151, 1992.

[48] B. B. Bederson and B. Shneiderman, *The Craft of Information Visualization. Readings and Reflections*. San Francisco, CA: Morgan Kaufmann, 2003.

[49] J. H. Larkin and H. A. Simon, "Why a diagram is (sometimes) worth ten thousand words," *Cognitive Science*, vol. 11, no. 1, pp. 65–99, 1987.

[50] O. Turetken and R. Sharda, "Visualization support for managing information overload in the web environment," presented at the 22nd International Conference on Information System, New Orleans, LA, 2001, pp. 221–232.

[51] W. Chung, *et al.*, "A visual framework for knowledge discovery on the web: An empirical study of business intelligence exploration," *J. Manag. Inform. Syst.*, vol. 21, no. 4, pp. 57–84, 2005.

[52] J. Agnew and L. R. Szykman, "Information overload and information presentation in financial decision making," in *Handbook of Behavioral Finance*, B. Bruce, Ed. Cheltenham, U.K.: Edward Elgar Publishing, 2010, pp. 25–44.

[53] R. A. Burkhard, "Learning from architects: The difference between knowledge visualization and information visualization," presented at the 8th International Conference Inform. Visualization (IV), London, U.K., 2004.

[54] R. Boardman, "Bubble trees: The visualization of hierarchical information structures," presented at the Conference Human Factors in Computing Systems, The Hague, The Netherlands, 2000.

[55] G. Conti, *et al.*, "Countering security information overload through alert and packet visualization," *Comp. Graphics Applicat.*, vol. 26, no. 2, pp. 60–70, 2006.

[56] O. Turekten and R. Sharda, "Development of a fisheye-based information search processing aid (FISPA) for managing information overload in the web environment," *Decision Support Syst.*, vol. 37, no. 3, pp. 415–434, Jun. 2004.

[57] B. Shneiderman, "The eyes have it: A task by data type taxonomy for information visualization," presented at the IEEE Symposium on Visual Languages, Boulder, CO, 1996, pp. 336–343.

[58] R. A. Burkhard and M. Meier, "Tube map visualization: Evaluation of a novel knowledge visualization application for the transfer of knowledge in long-term projects," *J. Universal Comput. Sci.*, vol. 11, no. 4, pp. 473–494, 2005.

[59] N. Carroll, "Visual metaphor," in *Aspects of Metaphor*, J. Hintikka, Ed. Dordrecht, The Netherlands: Kluwer, 1994, pp. 189–218.

[60] M. J. Eppler, "The image of insight: The use of visual metaphors in the communication of knowledge," in *Proc. I-KNOW*, Graz, Austria, 2003, pp. 81–88.

[61] M. J. Eppler, *Managing Information Quality. Increasing the Value of Information in Knowledge-Intensive Products and Processes*, 2nd ed. Berlin, Germany: Springer, 2006.

[62] S. Few. (2011, Jul.). *Data Visualization for Human Perception*. [Online]. Available: http://www.interaction-design.org/encyclopedia/data_visualization_for_human_perception.html Aug. 29, 2011 (last date accessed).

[63] H. Wainer, "How to display data badly," *Am. Statist.*, vol. 38, no. 2, pp. 137–147, 1984.

[64] E. R. Tufte, *Envisioning Information*. Cheshire, CT: Graphic Press, 1990.

[65] S. Bresciani and M. J. Eppler, "The risks of visualization: A classification of disadvantages associated with graphic representations of information," in *Identität und Vielfalt der*

Kommunikationswissenschaft, P. J. Schulz, *et al.*, Eds. Konstanz, Germany: UVK Verlagsgesellschaft mbH, 2009, pp. 165–178.

[66] J. Pfeffer and R. I. Sutton, "Knowing 'what' to do is not enough: Turning knowledge into action," *California Manag. Rev.*, vol. 42, no. 1, pp. 83–108, 1999.

[67] V. Venkatesh and F. D. Davis, "The theoretical extension of the technology acceptance model: Four longitudinal field studies," *Manag. Sci.*, vol. 46, no. 2, pp. 186–204, 2000.

[68] P. Legris, *et al.*, "Why do people use information technology? A critical review of the technology acceptance model," *Inform. Manag.*, vol. 40, no. 3, pp. 191–204, 2003.

[69] M. Igbaria, *et al.*, "Personal computing acceptance factors in small firms: A structural equation model," *MIS Quar.*, vol. 21, no. 3, pp. 279–302, 1997.

[70] D. J. Jackson, "Revisiting sample size and number of parameter estimates: Some support for the N:q hypothesis," *Structural Equation Model.*, vol. 10, no. 1, pp. 128–141, 2003.

[71] R. Agarwal and J. Prasad, "The role of innovation characteristics and perceived voluntariness in the acceptance of information technologies," *Decision Sci.*, vol. 28, no. 3, pp. 557–582, 1997.

[72] F. D. Davis, *et al.*, "User acceptance of computer technology: A comparison of two theoretical models," *Manag. Sci.*, vol. 35, no. 8, pp. 982–1003, 1989.

[73] V. Venkatesh, "Determinants of perceived ease of use: Integrating perceived behavioral control, computer anxiety and enjoyment into the technology acceptance model," *Inform. Syst. Res.*, vol. 11, no. 4, pp. 342–365, 2000.

[74] R. E. Mayer and L. J. Massa, "Three facets of visual and verbal learners: Cognitive ability, cognitive style, and learning preferences," *J. Educ. Psychol.*, vol. 95, no. 4, pp. 833–841, 2003.

[75] V. Venkatesh and F. D. Davis, "A model of the antecedents of perceived ease of use: Development and test," *Decision Sci.*, vol. 27, no. 3, pp. 451–481, 1996.

[76] F. D. Davis, "User acceptance of information technology: System characteristics, user perceptions and behavioral impacts," *Int. J. Man-Machine Studies*, vol. 38, no. 3, pp. 475–487, 1993.

[77] R. L. Thompson, *et al.*, "Personal computing: Toward a conceptual model of utilization," *MIS Quar.*, vol. 15, no. 1, pp. 124–143, Mar. 1991.

[78] T. F. Stafford, *et al.*, "Determining uses and gratifications for the Internet," *Decision Sci.*, vol. 35, no. 2, pp. 259–288, 2004.

[79] E. Rogers, *Diffusion of Innovations*. 3rd ed. New York: Free Press, 1988.

[80] G. C. Moore and I. Benbasat, "Development of an instrument to measure the perceptions of adopting an information technology innovation," *Inform. Syst. Res.*, vol. 2, no. 3, pp. 192–222, 1991.

[81] M. J. Eppler and J. Mengis,"Preparing messages for information overload environments. What business communicators should know about information overload and what they can do about it," presented at the International Association Business Communicators, Seattle, WA, 2008.

[82] D. Roam, *The Back of the Napkin*. London, U.K.: Portfolio, 2008.

[83] L. Becker,"The impact of organizational information overload on leaders: Making knowledge work productive in the 21st century," Ph.D. dissertation, Univ. Idaho, Moscow, ID, 2009.

PRACTICAL INSIGHTS FROM ALVOGEN

Alvogen is a multinational pharmaceuticals company focused on complex generic products. Building on the 120 years of success and experience from Norwich Pharmaceuticals, Alvogen is determined to make its mark on the industry. The company has more than 200 pharmaceutical products in development and registration with over 100 product and market launches expected in 2012. The Alvogen product portfolio consists of a broad range of leading molecules for the treatment of conditions in the areas of oncology, cardiology, respiratory, neurology, and gastroenterology. Alvogen has access to first-class regulatory, development, and manufacturing capabilities in Asia, Europe, and North America.

The Challenges of Information Overload

It comes as no surprise then to learn that Alvogen executives are faced with the challenges of information overload. The fact that their executives are often on the road all over the globe just adds to the complicated process of handling communication with all their stakeholders, both internal and external. As in most companies, coping with e-mail is high on the list of contributors to their information overload issues. As one manager said, "Heavy meeting booking compounded with 200+ e-mails per day make time management quite a challenge." Another observed, "I work 20 hours a day, sleep with my I-phone and answer e-mails in the middle of the night. . . . The thing I miss the most is 'thinking time.'"

Both the writing style and the length of e-mails can contribute to information overload as much as their content. "Another issue is e-mails from people who don't know how to

Information Overload: An International Challenge for Professional Engineers and Technical Communicators,
First Edition. Edited by Judith B. Strother, Jan Ulijn, Zohra Fazal.
© 2012 Institute of Electrical and Electronics Engineers. Published 2012 by John Wiley & Sons, Inc.

write e-mails! My personal opinion is that an e-mail should always fit to the screen. If it is longer, it is not a good e-mail. . . . "

Too often, executives and other professionals have to figure out how to accomplish multiple tasks at the same time. One Alvogen executive shared how his attempts to manage his time often complicate the information overload problem.

> A really bad practice that I and many others do is 'multi-tasking' in meetings or on conference calls on my computer/blackberry. I try to avoid this because I think it is rude and leads to inefficient meetings because I am only listening 'part-time.' However, when I don't do this, usually after a week or so of good, attentive meeting behavior, I am so behind on e-mail that I struggle to get caught back up.

Strategies for Dealing with Information Overload

Considering Alvogen's success, it is clear that Alvogen executives employ various effective techniques to deal with information overload. Here are some of their tried and true strategies in their own words.

- "I filter my e-mails very carefully according to urgency. I think that a lot of people cannot judge properly what is of importance and what is not."
- "I screen e-mails very quickly as they come in and act on the ones that need action right away—no procrastinating."
- "Develop a quiet time. I use Saturday as the day to greatly reduce all types of information processing for work. Only critical items are processed. The balance waits till Sunday mid-afternoon."
- "I keep a notebook with me at all times that contains only actions—list of rolling actions, dates, responsible person—I go through this daily and follow up on open action items with responsible persons, so I am always on top of the key business drivers and hold people accountable."
- "I keep my iPhone away from the bed—in another room. I switch off the data package of my iPhone at least for few days while I am on a vacation."
- "I do not take calls/answer e-mails when I am doing something that requires more attention, e.g., reviewing/drafting contracts or reviewing commercial offers."
- "For e-mails, I use a personal 'subject line ROI' when I have limited time and a ton of things to get through. If it's from a high value external source (for example, a customer), I answer it first. I keep the preview pane open and try to scan other questionable e-mails to see if they need immediate attention."
- "I use rules to get e-mails structured, make task lists, and flag mail. I also dedicate times to work on projects and to check new e-mails/information."
- "In my view, [the problem of information overload] cannot be solved by the individual person alone; the whole organization has to help here:
  - Inform/train people about proper e-mail use, e.g., one should not always copy all colleagues in an answer, i.e., copy only those who really need to know.
  - Empower staff and actively delegate. I try to avoid interfering in e-mail dialogues, to keep the issue with the person in charge.
  - Set priorities: Do the important and time critical things first; the rest has to wait.

- Be organized: manage your desk, filing system, inbox, hard disc . . . .When people are not well organized, it creates additional workload for themselves and others.
- Last but not least: managers, reserve time for mental and physical exercise. As a manager, we should encourage this and set an example. Show staff that it is possible to reserve some hours per week for exercise. It will reduce the stress level and increase alertness."

Contributors to Practical Insights from Alvogen

- Doug Drysdale, CEO, Alvogen America
- Martina Feitzinger, Director, Regulatory Affairs
- Elin Gabriel, Chief Operation Officer
- Bill Hill, Vice President, Sales and Marketing America
- Dr. Georg Ingram, Executive Vice President, Business Development
- Peter Keil, Regional Director, Sales and Marketing Asia
- Joe Rutz, Chief Information Officer
- Petar Vazharov, Director, Business Development

11

DROWNING IN DATA: A REVIEW OF INFORMATION OVERLOAD WITHIN ORGANIZATIONS AND THE VIABILITY OF STRATEGIC COMMUNICATION PRINCIPLES

DAVID REMUND and DEBASHIS "DEB" AIKAT

> *The sheer availability of information . . . has launched a tsunami of seeking . . . at the same time, the information glut has contributed to pervasive cynicism, fragmentation, and a sense of helplessness.*
>
> Michael J. Gelb
> Author and Creativity Expert

ABSTRACT

Information overload has traditionally been viewed as a phenomenon created by, and affecting, individuals. This chapter reflects a contemporary shift in thinking and analyzing the information overload phenomenon from an organizational perspective. Indeed, organizations tend to act as incubators for information overload. As people in organizations try to cope with increasing amounts of information, they become less capable of identifying the most critical bits of information, distinguishing accurate information from that which is not accurate, and making decisions that are sensible for organizations. Based on literature review and meta-analysis of contemporary research, this chapter examines the primary effects of information overload in organizations, including implications on individual work performance, knowledge transfer, decision making, and, ultimately, organizational effectiveness. A practical approach to minimizing information overload within organizations is explored within this chapter. Specifically, the principles of strategic communication practiced by public relations specialists and other professional communicators may

Information Overload: An International Challenge for Professional Engineers and Technical Communicators, First Edition. Edited by Judith B. Strother, Jan Ulijn, Zohra Fazal.
© 2012 Institute of Electrical and Electronics Engineers. Published 2012 by John Wiley & Sons, Inc.

help keep organizations from drowning in data. Traditionally, disseminating information within an organization had been the domain of those in charge of public relations, employee communications, human resources, and/or operations. However, nearly anyone within an organization now has that kind of power, via e-mail and other technologies. Training and coaching individuals within the organization to use strategic communication principles may help people throughout the organization focus on meaningful data, make sound decisions, and improve performance.

11.1 Introduction

The concept of information overload has influenced the intellectual context of communication since the 1960s and the publication of John Naisbitt's thought-provoking book, *Megatrends*, which suggested that society is drowning in information yet starved for knowledge [1]. In the twenty-first century, concerns over information overload continue to escalate as computers become faster, more powerful, and more prevalent. People seem to be drowning in data.

Thanks to technological innovations and social developments, organizations emerged during the twentieth century as a primary force of power and action within society. Today, organizations play a pivotal role in economic growth and individual livelihood. Organizations also flood the world with information. Unfortunately, having more information does not necessarily help individuals within organizations share knowledge or make decisions. In fact, the opposite is often the case; individuals who receive more information than they can process in a given timeframe may struggle with decision making and other responsibilities—or withdraw from these responsibilities altogether.

Based on a meta-analysis of scholarly research and professional insights from a number of disciplines and industries, this chapter addresses the effects of information overload within organizations and on the individuals who comprise organizations. Traditional strategies for addressing information overload are highlighted and explained. Emphasis is placed on an underutilized approach—training people in the art of strategic communication—as a viable model for reducing information overload. Principles and standards of strategic communication are outlined and applied to the dynamic of information overload within organizations.

The ultimate lesson from this chapter is that organizations and their members need to focus on information sharing and decision making that is actionable and situation specific. In order to reduce individuals' stress, improve productivity, and enhance overall effectiveness, the flood of seemingly endless data and largely irrelevant information within organizations must end. Strategic communication could be the key.

11.2 Defining Information Overload within Organizations

There may well be no universal definition of information overload. Table 11.1 highlights several important variations in the definition of information overload. However, as it relates to individuals within organizations, information overload may be defined as the self-perception of having too much information for the amount of time available to process the information, causing a person to feel stressed, which, in turn, affects decision making [2, 3]. At the organizational level, these self-perceptions are spread to such a degree among members of the organization that they reduce the overall effectiveness of the organization [3, 4].

Other researchers, including Gordon, Haka, O'Reilly, and Schick, have defined information overload as an imbalance within an organization [4, 5]. In this view, organizations that fail to balance their information processing capabilities with the load of information encountered,

T a b l e 11.1   Definitions of Information Overload

Definition	Published Study (in Chronological Order by Initial Publication Date)
The point at which information inputs cannot be processed and utilized by an individual, resulting in cognitive ineffectiveness or terminated information processing	Rogers and Agarwala-Rogers (1975) [6]
A condition of confusion, cognitive strain, and other negative effects occurring when a person tries to process too much information in too little time	Maholtra (1984) [9]
The result of uncertain, ambiguous, novel, complex, or intense information	Schneider (1987) [16]
An imbalance (within organizations) between information processing capabilities and the load of information encountered, generated, or distributed	O'Reilly (1980) [5] Schick *et al.* (1990) [4]
A situation in which information inputs exceed the individual's ability to process such information	Chewning and Harrell (1990) [7]
Self-perception of having too much information for the amount of time available to process the information	Edmunds and Morris (2000) [2] Allen and Wilson (2003) [3]
The situation in which the quantity of information one requires for processing may advance more rapidly than physical capability to process that information depending, perhaps at least in part, upon an individual's decision-making efficacy	Bock *et al.* (2010) [27]

generated, or distributed will, in simple terms, experience information overload. Individual capacity for information consumption and processing may vary based on role, experience, and other factors. Perceived information overload in individuals, then, can be defined as a breakdown—i.e., a point at which information inputs cannot be processed and utilized by an individual, resulting in cognitive ineffectiveness or terminated information processing [6–8].

Information overload essentially describes a condition of confusion, cognitive strain, and other negative effects occurring when a person tries to process too much information in too little time [9]. These themes apply to many types of organizations as well, from corporations to hospitals to libraries [10, 11]. Information overload may perhaps be even more problematic for virtual organizations, such as online communities, e-markets, and distributed sales teams [12–14].

In a groundbreaking review of information overload research across multiple management disciplines, Mengis and Eppler [15] (see Chapter 10) noted that information overload—as defined by business scholars and experts—often reflects performance. An individual's performance will improve, to a certain point, with the amount of information received. However, if information inputs exceed the individual's ability to process such information, performance will rapidly decline. Drops in individual performance may affect department or team performance, which, in turn, may negatively impact overall organization performance. From this foundational study, it is possible to envision how information overload might ultimately drown an entire organization. (See Chapter 7 for a further discussion of the organizational costs of information overload.)

Information overload often results from information that is uncertain, ambiguous, novel, complex, or intense—a well-established definition that has not changed significantly over the past few decades despite advances in technology [16]. Indeed, the causes of information overload are similar for individuals and organizations, though impacts and solutions are not necessarily so.

Research about heightened awareness of information overload has drawn attention to the need for greater information literacy or what researchers define as the ability to discern between accurate and inaccurate information, as well as relevant and irrelevant data [17]. As technological innovation continues and data further flood the world, further studies will certainly refine the definition of information overload and its relationship to individuals and organizations.

11.3 Evolution of the Information Overload Concept in Organizations

Drowning in data is not a new phenomenon. Indeed, cultures thrive and evolve through information sharing. Developed societies have been particularly dependent upon a knowledge-driven economy since the rise of industry and technology, and they continue to become increasingly dependent on data and information.

Information overload has traditionally been viewed as a phenomenon created by, and affecting, individuals. However, given the prominence of organizations in modern society, there may be significant merit in analyzing information overload from an organizational perspective. Industrialization and urbanization in the twentieth century have given rise to organizations, many of which are large corporations, and made them an integral part of society and the economy in the United States. Understanding the communication practices and patterns of organizations is critical to understanding how work gets accomplished.

For the greater part of the past 40 years, scholars have argued that the process of communication itself is what constitutes an organization [18, 19]. Without communication, they argue, there is no such thing as an organization. This concept is known as *organizing theory*.

How then can information overload be viewed and analyzed within the context of organizations? To answer this question, it is helpful to reflect on the concept of information overload as articulated by French philosopher Paul Virilio. He believed that the development and adoption of communication based on electromagnetic media (e.g., radio, television, the Internet) ushered in an entirely new world within which people had to learn how to cope and function [20]. Virilio posited that because electromagnetic technology enables the transmission of data at the speed of light, space and time no longer have meaning. Messages can be sent and retrieved anytime and anywhere. On the surface, this would seem to aid the flow of information within organizations. However, human perception is rooted in the concepts of time and space, so this "anytime anywhere" model of communication inevitably overwhelms individuals, and, ultimately, organizations [20, 21]. This conundrum applies not only to organizational communication but also to other aspects of modern life, as well, including advertising exposure, consumer behavior, and purchasing decisions [9, 22, 23]. Individuals and organizations simply have a finite capacity for absorbing and understanding information that is being shared in real-time across boundless distances.

Organizational dynamics have been at play in terms of how information overload happens. Over a major part of the twentieth century, data and information were disseminated in a top–down model of leaders sharing information as they deemed appropriate [24]. With such controlled flow, the issue of information overload within organizations was not as pressing as it has become today. In the 24/7 digital age, information flows continually in a multilateral trajectory. Most professionals today are knowledge workers valued for their specialized expertise and possessing the ability to create, publish, and distribute information as frequently and freely as they wish [25,26]. This free flow of data can cloud the decision-making process for knowledge workers in organizations, literally paralyzing individuals and workgroups to varying degrees, depending upon how they use or intend to use the information.

The traditional approach to researching information overload has been to focus on the impact on individuals. Some contemporary studies have begun to examine the phenomenon of information overload from more of an organizational perspective [16,20,27,28]. Conceptually, an argument can be made that organizations concentrate and catalyze the information overload effect, with multiple individuals participating in and contributing to the communication and information-sharing processes.

For decades, scholars have claimed that organizations should function as information processing systems, helping, not hurting, individuals in their attempt to obtain important data, analyze options, and make prudent decisions [29–31]. Indeed, with the concept of an information processing system as the model, it is important to better understand information overload within organizations.

11.4   Implications of Information Overload within Organizations

As data proliferates and cognitive capacities become strained, the stress-related effects for individuals and organizations multiply exponentially. Table 11.2 highlights some of the implications of information overload within organizations.

11.4.1   Organizational Implications

Sharing information with members is important for organizational performance. For example, studies have shown that 90% of employees who are kept fully informed are willing to work harder while nearly 80% of those who are not kept informed are not willing to take on extra work or put in extra hours [32]. Effective sharing of information can be a positive and powerful influence on performance.

What, then, is the ideal amount of information an organization can handle? Each organization is different, and each member within that organization is also different, so there is no single answer to this question. However, research indicates that individuals suffer when their ability to absorb and understand information has been breached. For example, a survey of more than 2000 insurance agents suggested that increases in the amount of information shared with agents can ultimately have an indirect and inverse influence on sales performance [14]. A separate but related study of marketers found that as more information was shared with employees, the more structure and formality employees needed in order to make sense of the wealth of information and data [33]. Gauging and identifying the right amount of information to share with individuals in an organization, and designing systems to realize that quota most efficiently, are paramount. (See Chapter 8.)

Managing the information environment is difficult not only because of sheer quantity and volume, but also because much of the information that organizations deal with is unpredictable [18]. As noted earlier in this chapter, organizing theory contends that people

Table 11.2   Implications of Information Overload

Key Findings	Published Study (in Chronological Order by Initial Publication Date)
Implications on Organizations	
Information overload within organizations results in inefficient work styles, decreased job satisfaction, loss of creativity and clear thinking, stress, illness, and burnout.	O'Reilly (1980) [5] Janssen and de Poot (2006) [34]
Increases in amount of information shared with insurance agents can have an indirect and inverse influence on sales performance.	Hunter (2004) [14]
Employees need more structure and formality to make sense of information as more information is shared.	Souchon *et al.* (2004) [33]
• Nine of ten employees who feel informed are willing to work harder as a result • Eight of ten employees who do not feel informed are not willing to take on extra work or put in extra hours	Hoen (2006) [32]
Implications on Individual Members of Organizations	
"Mental fog" is experienced most often when information does not relate to a specific situation or does not help one along the decision-making path.	O'Reilly (1980) [5]
As people in organizations try to cope with increasing amounts of information, they become less capable of identifying the most critical bits of information, distinguishing accurate information from that which is not accurate, and making decisions that are sensible for the organization.	Sparrow (1999) [37] Edmunds and Morris (2000) [2] Kock (2000) [36] Baldacchino *et al.* (2002) [35]
Information overload is most often related to e-mail, in the form of message ambiguity, cascades/avalanches, sheer volume, or poor netiquette.	Janssen and de Poot (2006) [34]
Nearly 85% of professional communicators believe e-mail overload negatively impacts their productivity at least part of the time.	Williams and Williams (2006) [39]
Information overload is identified as just one of several contributing factors to the negative impacts of pervasive technology on knowledge workers; other contributing factors are communication overload and system feature overload.	Karr-Wisniewski and Ying (2010) [26]

within organizations may need to gather more information to understand a situation or decision [18,19]. However, if the information already received is ambiguous, people do not need more information; they need to simply make a decision based on the interpretation they think is the best [19]. Regardless of the cause, information overload within organizations has been shown to result in inefficient work styles, decreased job satisfaction, loss of creativity and clear thinking, stress, illness, and burnout [5,34]. Helping individuals find the right amount and balance of information seems critical for organizational effectiveness. In order to achieve such balance, it is first important to more fully examine and understand the implications of information overload on individual employees.

11.4.2 Employee Implications

Organizations are ultimately nothing more than structured networks of individuals working collaboratively to achieve shared goals. Using this foundation and applying the concept of organizing theory, which espouses that communication is the heart of an organization, it becomes apparent that combating information overload must happen on an individual-by-individual basis.

Research in cognitive processing has shown that the brain responds to the environment when trying to absorb and analyze information. As people in organizations try to cope with increasing amounts of information, they become less capable of identifying the most critical bits of information, distinguishing accurate information from that which is not accurate, and making decisions that are sensible for the organization [2, 35–37]. In short, as the working environment becomes more clouded, so too does the individual employee's brain, creating a kind of "mental fog" in day-to-day performance. This has significant implications on information exchange, knowledge transfer, and organizational effectiveness [38]. Another issue is that individuals experience information overload most often when the information provided does not relate to a specific situation or does not help one along the decision-making path [5].

Information overload may result not only from too much information for an individual to handle but also from work that has grown increasingly complex [34]. Complexity of job responsibilities is evidenced in the information age by increasingly fragmented work and more frequent interruptions than workers have known in the past. Therefore, people often spend just a few minutes on a task before moving on to the next thing that has to be accomplished. Individuals can delegate or relinquish some tasks and duties or lower their ambitions. But, these strategies are not solutions for information overload; they are simply remedies that may help individuals cope with the symptoms of a data-flooded organization.

One study identified the root causes leading to critical incidents of information overload—incidents that have significant negative consequences—as experienced by senior managers within various divisions and departments of a company [34]. The managers believed that the factors most often contributing to information overload were ambiguous e-mail (21% of all critical incidents) and inaccessibility of necessary information (17%). A full 60% of all critical incidents of information overload, however, somehow related to e-mail, whether due to message ambiguity, cascades/avalanches, sheer volume, or poor netiquette [34]. Clearly, misuse and overuse of e-mail is seen as significantly contributing to information overload.

The International Association of Business Communicators (IABC), composed of people who work in public relations and other communication-related disciplines, reports e-mail as a growing problem [39]. This is particularly notable given that the members of this professional organization are people trained to communicate effectively. Given that they cite e-mail as a problem, it stands to reason that nonprofessional communicators would have similar, if not greater, concerns. In a survey of more than 13,000 members, IABC found that 85% of its members believe e-mail overload negatively impacts productivity at least part of the time, while 40% of members are regularly or frequently affected. In companies with more than 5000 employees, workers between the ages of 45 and 55 are most likely to feel the negative effects of e-mail. Employees under 30 who work in an organization with fewer than 50 employees felt the least overloaded by e-mail [39].

However, people typically spend no more than 20% of each workday reading and responding to e-mail [32], and e-mail volume tends to be perceived more as an irritation than an overload [40]. So, while e-mail appears to be a notable contributor to information overload for individuals, it should not be the sole focus of mitigation. Through participant observation and interviews, researchers found that information workers for a private company in the United States worked an average of 3 minutes on each specific task, investing 12 minutes on a collection of related tasks before moving on to the next project or group of related tasks, often requiring a switch to different documents, tools, and reference materials [41]. This same study showed that up to 70% of an information worker's day is spent doing desk work or being involved in meetings—significantly more time than what is spent managing e-mail. These findings were generally consistent with prior related studies. Notably, this study found that individuals interrupt their work themselves about as often as they are interrupted by external influences.

11.5 Traditional Strategies for Addressing Information Overload

Organizations and individuals do not have to be passive victims to the growing onslaught of data and information. Several successful strategies have been identified.

11.5.1 Organizational Strategies

Over the years, organizations have tried to mitigate the effects of information overload by modifying job responsibilities, providing more flexibility in work schedules, flattening organizational charts, and empowering individuals to take more control of decision making [42]. They have also implemented technical solutions such as decision support and recommender systems to help manage and streamline the flow of data and information, and software interface agents to assist in information filtering, information retrieval, information customization, mail management, and related tasks [43–46]. At the same time, employees have been encouraged to manage personal stress and maintain positive attitudes [47]. None of these interventions has targeted the source of the problem, which is people's inability to quickly and easily create, reproduce, and distribute seemingly unlimited amounts of information.

At the organizational level, the most likely strategies for dealing with information overload include the following:

- establishing and enforcing e-mail rules that go beyond merely appropriate usage and company monitoring (e.g., using the "to" line only for those who need to take action, and listing all others only on the "cc" line simply to keep them informed);
- heightening awareness of information overload across the organization (e.g., self-assessments so that people understand the problem and how they contribute to it);
- developing or implementing more helpful technologies (e.g., Intranet, wikis, knowledge collaboration tools, and file and task management software);
- providing training or coaching to help individuals better manage information and reduce stress [34, 48–51].

Organizations may also push their members to reduce complexity by seeing what can be learned from information that is already available rather than pushing for the collection and distribution of even more information related to the same issue or topic [52]. Likewise, some organizations maintain balanced scorecards, which are performance measurement processes that anchor information sharing and decision-making processes [53, 54]. The measures are established by senior management and serve as the overarching organizational strategy. A survey of more than 1500 accountants showed that those who work in organizations with balanced scorecards more often have the information necessary to make decisions [25]. In a balanced scorecard environment, more information may be flowing, but that information is more targeted than in nonscorecard environments, again hinting that it is not necessarily the volume of information that causes overload, but rather the relevance and quality of the information.

11.5.2 Individual Strategies

Steps that organizations take in an attempt to address information overload are only part of the puzzle. Research indicates that the degree to which individuals suffer from information overload seems to relate to the individual strategies used in trying to deal with the problem [34]. Perhaps, what the individual can do to tame overload is more important than, or at least as equally important as, measures adopted by the organization.

As individuals strive to reduce information overload, they can try to use their time more efficiently or reduce the number of tasks they perform [4]. Time management steps may include following the same procedures for routine activities, using computer-based systems, and prioritizing daily work. With regard to reducing the number of tasks, individuals can collect fewer data, produce more simplified reports, reduce their self-expectations, and delegate or share responsibilities.

Researchers in academia and business have identified tangible steps individuals can take to manage personal levels of information overload [11, 41, 47, 55, 56]. These strategies include such actions as focusing only on the information most critical to the problem at hand; establishing self-imposed time constraints for research; delegating or sharing the task of research; single-tasking instead of multitasking; keeping critical documents handy; and intermittently disconnecting from technologies such as the telephone, e-mail, and Internet (i. e., media diet). Likewise, David Allen's popular "Getting Things Done" model emphasizes making decisions and completing tasks efficiently by collecting, processing, organizing, and reviewing data in a structured, systematic fashion [57].

Professional communicators, such as the members of the International Association of Business Communicators, are people who manage employee communication or other communications within an organization. Like researchers, these professional communicators also tout the importance of setting priorities for each workday; dealing with a message only once; scheduling regular times to manage e-mail; bundling bits of related information into a single message and linking to an Intranet, if possible, for more details; taking advantage of any available collaboration tools such as wikis, blogs, and portals; and using real simple syndication to control inflow [39]. These corporate communication practitioners also believe in proper netiquette when it comes to e-mail (e.g., eliminating distribution lists of more than five names and using succinct and clear subject lines). Other professional communicators offer similar advice for streamlining cybercommunication, such as restricting each e-mail to a single request or topic and supplying all relevant details in a message rather than assuming that recipients have the information they need readily at hand [58].

In the digital age, it is important to understand communication nuances within virtual organizations and working relationships. In virtual organizations, individuals typically use one of two strategies to deal with information overload: leaving the community, or filtering and ignoring messages. Individuals who communicate mostly or exclusively in a virtual world will withdraw their active participation as information overload increases. Short, simple messages are what get through to those feeling the threat of overload [59]. In Germany, a study of 33,000 members of an online community over a 3-year period validated this point. Messages with the key point expressed in the subject line were more than twice as likely to generate a response from individual members as messages that did not have a specific subject line [12].

Information overload remains a challenge despite the many organizational and individual strategies taken to address the problem. A more holistic approach may be needed.

11.6   Strategic Communication Principles: A Viable Solution?

The twentieth century gave rise to organizations and technology, as well as to the modern practice of public relations, including the management of organizational communications. Professional communicators, of course, contribute to the deluge of information flowing within organizations. However, unlike their colleagues trained in other disciplines, formal communicators have been specifically trained to pinpoint key facts, construct cogent arguments, and write and report information in a succinct, easy-to-understand fashion. Drawing upon such principles, organizations and their members can employ strategies to help reduce information overload within the organization [60–63].

The strategic communication process (see Figure 11.1) typically involves five steps [64] :

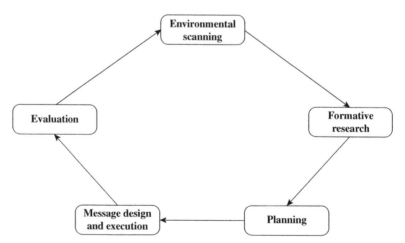

Figure 11.1 Strategic communication process. Managing communication in a strategic fashion involves five distinct, yet interdependent, steps. This process typically starts by scanning the organizational environment for emerging issues and conducting research to better understand the issue and related perceptions of that issue. Only then are communications efforts planned, implemented, and evaluated [64].

- *Environmental Scanning* (monitoring issues and trends that may affect the organization);
- *Formative Research* (including reviewing available data as well as conducting new research);
- *Planning* (identifying the stakeholders who need to be informed and the information they need to take action or make a decision);
- *Message Design and Execution* (drawing upon each of the prior steps to summarize key points in a succinct, straightforward manner) (see Chapters 6 and 7);
- *Evaluation* (taking the pulse of stakeholder groups to see if they are able to absorb, understand, and use the information being shared through the communication process).

When effectively adopted, the five-step process of strategic communication is continuously repeated. Emerging issues spark new research, which influences plans and messages, both of which become further refined upon evaluation and reevaluation. Communication becomes a fluid, interactive, evolving process.

Practicing engineers and other nonprofessional communicators can help reduce information overload within organizations by respecting and following the tenets of effective strategic communication to which their professional communication counterparts adhere. For example, engineers, architects, and contractors perpetually pursue new projects in order to stay challenged, innovative, and perhaps most importantly, profitable. Winning a bid, though, is just the first step in the complicated challenge of moving a project from vision to reality. An engineering firm committed to clear communication within its project team and with the client would employ the principles of strategic communication. A senior leader would have ultimate authority for project-related communication and, ideally, some prior experience managing organizational, operational, and/or technical communication.

Step one would involve monitoring the current situation for project-related issues that have emerged or are emerging. Next, some research would be done regarding these issues and the stakeholders would be involved in these issues. Only then would communication processes be designed and initiated. The specific communication method would be tailored to the project and the groups of people involved. Regardless of the method used, all communication would be treated as two-way dialogue, not simply one-way information transfer; listening and being responsive would be considered just as important as relaying information. A formal method of evaluating communication effectiveness would also be developed and implemented. This could be as simple as an informal series of conversations with key project leaders, or a more formal periodic survey of everyone involved in the project, perhaps via Web-based software that would automatically capture responses and calculate results. A project may require communication monitoring as frequently as weekly, daily, or even hourly, depending upon the project's scope and visibility, as well as where the project stands relative to its targeted completion.

The strategic communication process would not stop there, though. It would be a continuous cycle, in which every bit of evaluative feedback would not only assess communications that have happened to date but also help identify any newly emerging issues that must then be researched and addressed. A savvy firm would have the right kind of leadership in place to ensure project-related communications are approached in just such a systematic, strategic fashion.

11.7   Putting Strategic Communication into Practice

Nonprofessional communicators may put principles of strategic communication into practice by focusing on the following:

- *Solutions.* Colleagues seek information from one another in order to know what to do and/or how to do something. Concentrating too heavily on the details or reasons behind a problem distracts from the often urgent need to resolve a problem or answer a nagging question. By focusing on solutions when communicating with others, nonprofessional communicators can help their colleagues get more quickly to the information, answers, and advice they need.
- *Alternative Sources.* If a solution or answer does not immediately come to mind, the engineer or other nonprofessional communicator should point the way to alternative sources of information, such as other colleagues with specialized expertise or information repositories such as databases or archival records.
- *Conceptualization.* There are times when it is appropriate to help a colleague better conceptualize or define a challenge. Perhaps, the problem could be viewed more broadly or maybe a particular aspect needs greater attention. The focus should stay on solving the problem, though, not rehashing in detail the circumstances or events that lead up to the current problem.

Additionally, nonprofessional communicators may help colleagues by simply giving them the opportunity to present ideas and concepts, share feedback, or provide approval or validation. All these practices help create and foster a culture of strategic communication.

Training and coaching people within organizations to adopt and adhere to the strategic communication process should also help alleviate the amount of irrelevant information hitting colleagues' desks. The training program should emphasize the following:

- being mindful of the key issues or topics;
- researching what information is already available to help aid decision making;
- giving thought as to who actually needs the information;
- carefully designing communications such as e-mails, presentations, and reports for maximum readability by using aids such as bullet points and subheadlines;
- evaluating the effectiveness of such messages.

Such training could be provided during the new employee orientation process. The professional communicators within the organization could be available on a consultative basis to various departments and project teams, in a continuing effort to improve information quality and reduce information overload across the organization. A communication strategist could be specifically assigned to each major new initiative, much like a design engineer or project manager would be. It would be the responsibility of this professional communicator to ensure that the principles of strategic communication are employed throughout the duration of the project.

All these measures can help individuals dramatically streamline and improve the information they share with others in the organization. Ultimately, strategic communication should help prevent the breakdown associated with information overload—i.e., helping ensure that information is shared in a way that falls within the cognitive processing capacity of those intended to receive the information.

11.8 Further Research

Much remains to be learned about the role strategic communication plays in mitigating information overload within organizations. Empirical research is necessary to test hypotheses that will help inform our understanding of the relationship between these two concepts.

Experimental studies by researchers in organizational communication or technical communication, for example, could examine the effects of tailored, or strategic, communication versus nonstrategic communication on groups of workers with similar responsibilities. Likewise, experimental research could test messages that have been refined to varying degrees, employing strategic communication principles, to better understand the point at which a common degree of understanding and processing is achieved across groups of similar research subjects. While results from experimental studies cannot be generalized to all organizations across all industries, the empirical data from these studies would be valuable in designing field research or surveys focused on specific types of organizations or work groups.

11.9 Conclusion

This chapter reviews scholarly research and professional insights into the issue of information overload with organizations. Traditionally, the strategies for addressing information overload have focused on treating the symptoms rather than the causes. In this information age, work is complex and individuals' workdays are fractured. It may be possible for individuals to keep up with the increasing flow of information, but only if the information is relevant and highly applicable to decision making. Information must be actionable in order to be impactful.

Adding more systems and software will do little to help keep organizations from drowning in data. Training and coaching individuals within the organization to use strategic communication principles may help people focus on the most critical information, weed out meaningless data and unnecessary reports, and ensure that information is tailored to the varying needs others might have, such as information gathering, decision making, or problem solving. Moreover, following a strategic communication model will help people share situation-specific information in a way that will be more quickly absorbed and more readily understood by others, in turn improving decision-making and knowledge transfer processes within the organization.

REFERENCES

[1] J. Naisbitt, *Megatrends: Ten New Directions Transforming Our Lives*. New York: Grand Central Publishing, 1988.

[2] A. Edmunds and A. Morris, "The problem of information overload in business organizations," *Int. J. Inform. Manag.*, vol. 20, no. 1, pp. 17–28, Feb. 2000.

[3] D. Allen and T. D. Wilson, "Information overload: Context and causes," *New Rev. Inform. Behaviour Res.*, vol. 4, no. 1, pp. 31–44, 2003.

[4] A. Schick, *et al.*, "Information overload: A temporal approach," *Account., Organiz. Soc.*, vol. 15, no. 3, pp. 199–220, 1990.

[5] C. O'Reilly, "Individuals and information overload in organizations: Is more necessarily better?" *Acad. Manag. J.*, vol. 23, no. 4, pp. 684–696, 1980.

[6] E. Rogers and R. Agarwala-Rogers, "Organizational communication," in *Communication Behavior*, G. L. Hanneman and W. J. McEwen, Eds. Reading, MA: Addison-Wesley, 1975.

[7] E. Chewning and A. Harrell, "The effect of information load on decision makers' cue utilization levels and decision quality in a financial distress decision task," *Account., Organiz. Soc.*, vol. 15, no. 6, pp. 527–542, 1990.

[8] C. Beaudoin, "Explaining the relationship between Internet use and interpersonal trust: Taking into account motivation and information overload," *J. Comput.-Mediated Commun.*, vol. 13, no. 3, pp. 550–568, 2008.

[9] N. Malhotra, "Reflections on the information overload paradigm in consumer decision making," *J. Consumer Res.*, vol. 10, no. 4, pp. 436–440, 1984.

[10] A. Hall and G. Walton, "Information overload within the health care system: A literature review," *Health Inform. Libraries J.*, vol. 21, no. 2, pp. 102–108, 2004.

[11] M. J. Rudd and J. Rudd, "The impact of the information explosion on library users: Overload or opportunity?" *J. Acad. Librarianship*, vol. 12, no. 5, pp. 304–306, 1986.

[12] T. Schoberth, *et al.*, Exploring communication activities in online communities: A longitudinal analysis in the financial services industry," *J. Organiz. Comput. Electron. Commerce*, vol. 16, nos. 3–4, pp. 247–265, 2006.

[13] V. Grover, J. Lim, and R. Ayyagari, "The dark side of information and market efficiency in e-markets," *Decision Sci.*, vol. 37, no. 3, pp. 297–324, 2006.

[14] G. Hunter, "Information overload: Guidance for identifying when information becomes detrimental to sales force performance," *J. Personal Selling Sales Manag.*, vol. 24, no. 2, pp. 91–100, 2004.

[15] M. Eppler and J. Mengis, "The concept of information overload: A review of literature from organization science, accounting, marketing, MIS, and related disciplines," *Inform. Soc.*, vol. 20, no. 5, pp. 325–344, 2004.

[16] S. C. Schneider, "Information overload: Causes and consequences," *Human Syst. Manag.*, vol. 7, no. 2, pp. 143–153, 1987.

[17] M. Aro and E. Olkinuora, "Riding the information highway: Towards a new kind of learning," *Int. J. Lifelong Edu.*, vol. 26, no. 4, pp. 385–398, 2007.

[18] K. Weick, *The Social Psychology of Organizing*. Reading, MA: Addison-Wesley, 1969.

[19] M. Dainton and E. Zelley, *Applying Communication Theory for Professional Life: A Practical Introduction*. Thousand Oaks, CA: Sage, 2005.

[20] D. Kloock, "Aesthetics of speed," in *Media Theory: An Introduction*, D. Kloock and A. Spaar. Eds., Munich, Germany: Wilhelm Fink, 1997, pp. 134–164.

[21] B. Jungwirth and B. Bruce, "Information overload: Threat or opportunity?" *J. Adolescent Adult Literacy*, vol. 45, no. 5, pp. 400–406, 2002.

[22] B. Southwell, "Information overload? Advertisement editing and memory hindrance," *Atlantic J. Commun.*, vol. 13, no. 1, pp. 26–40, 2005.

[23] N. Lurie, "Decision making in information-rich environments: The role of information structure," *J. Consumer Res.*, vol. 30, no. 4, pp. 473–486, 2004.

[24] S. deBakker, "Information overload within organizational settings: Exploring the causes of overload," in Int. Commun. Assoc. Annu. Meeting Conf. Papers, 2007, pp. 1–25.

[25] L. McWhorter, "Does the balanced scorecard reduce information overload?" *Manag. Account. Quar.*, vol. 4, no. 4, pp. 23–27, 2003.

[26] P. Karr-Wisniewski, and L. Ying, "When more is too much: Operationalizing technology overload and exploring its impact on knowledge worker productivity," *Comput. Human Behavior*, vol. 26, no. 5, pp. 1061–1072, 2010.

[27] G. Bock, *et al.*, "The impact of information overload and contribution overload on continued usage of electronic knowledge repositories," *J. Organiz. Comput. Electron. Commerce*, vol. 20, no. 3, pp. 257–278, 2010.

[28] J. Cho, *et al.*, "The rate and delay in overload: An investigation of communication overload and channel synchronicity on identification and job satisfaction," *J. Appl. Commun. Res.*, vol. 39, no. 1, pp. 38–54, 2011.

[29] J. Gailbraith, *Organizational Design*. Reading, MA: Addison-Wesley, 1977.

[30] C. A. O'Reilly, III, and L. R. Pondy, "Organizational communication," in *Organizational Behavior*, S. Kerr, Ed. Columbus, OH: Grid, 1979, pp. 119–150.

[31] M. L. Tushman and D. A. Nadler, "Information processing as an integration concept in organizational design," *Acad. Manag. Rev.*, vol. 3, no. 3, pp. 82–98, 1978.

[32] F. Hoen, "Make yourself heard: Are there more effective ways of getting your employees' attention than e-mail?" *Commun. World*, 23, no. 2, pp. 35–37, 2006.

[33] A. Souchon, *et al.*, "Marketing information use and organizational performance: The mediating role of responsiveness," *J. Strategic Marketing*, vol. 12, no. 4, pp. 231–242, 2004.

[34] R. Janssen and H. de Poot,"Information overload: Why some people seem to suffer more than others," in *Proc. NordiCHI 2006: Changing Roles*, Oslo, Norway, Oct. 2006, pp. 397–400.

[35] C. Baldacchino, *et al.*, "Information overload: It's time to face the problem," *Manag. Services*, vol. 46, C no. 4, pp. 18–19, 2002.

[36] N. Kock, "Information overload and worker performance: A process-centered view," *Know. Process Manag.*, vol. 7, no. 4, pp. 256–264, 2000.

[37] P. Sparrow, "Strategy and cognition: Understanding the role of management knowledge structures, organizational memory, and information overload," *Creativity Innovat. Manag.*, vol. 8, no. 2, pp. 140–149, 1999.

[38] C. Cramton, "The mutual knowledge problem and its consequences for dispersed collaboration," *Organiz. Sci.*, vol. 12, no. 3, pp. 346–371, 2001.

[39] T. Williams, and R. Williams, "Too much e-mail!" *Commun. World*, vol. 23, nos. 11–12, pp. 38–41, 2006.

[40] C. Kimble, *et al.*, "The role of contextual clues in the creation of information overload: Matching technology with organisational needs," *Proc. 3rd UKAIS Conf.*, University of Lincoln, U.K., Apr. 1998, pp. 405–412.

[41] V. Gonzalez and G. Mark,"Constant, constant, multi-tasking craziness: Managing multiple work spheres," in *Proc. CHI*, Vienna, Austria, Apr. 2004, pp. 113–120.

[42] L. Hudson, "Lessons from Wal-mart and the *Wehrmacht*: Team Wolfowitz on administration in the information age," *Middle East Policy*, vol. 9, no. 2, pp. 25–38, 2004.

[43] L. Orman, "Fighting information pollution with decision support systems," *J. Manag. Inform. Syst.*, vol. 1, no. 2, pp. 64–71, 1984.

[44] P. Maes, "Agents that reduce work and information overload," *Commun. ACM*, vol. 37, no. 7, pp. 31–40, 1994.

[45] H. Berghel, "Cyberspace 2000: Dealing with information overload," *Commun. ACM*, vol. 40, no. 2, pp. 19–24, 1997.

[46] Y. Wei, *et al.*, "A market-based approach to recommender systems," *ACM Trans. Inform. Syst.*, vol. 23, no. 3, pp. 227–266, 2005.

[47] S. Sidle, "Workplace stress management interventions: What works best?" *Acad. Manag. Perspectives*, vol. 22, no. 3, pp. 111–112, 2008.

[48] D. Delen, and S. Al-hawamdeh, "Framework for knowledge discovery and management," *Commun. Assoc. Comput. Machinery*, vol. 52, no. 6, pp. 141–145, 2009.

[49] D. Arnesen, and W. Weis, "Developing an effective company policy for employee Internet and email use," *J. Organiz. Culture, Commun. Conflict*, vol. 11, no. 2, pp. 53–65, 2007.

[50] S. Nishida, *et al.*, "Information filtering for emergency management," *Cybern. Syst.: Int. J.*, vol. 34, no. 3, pp. 193–206, 2003.

[51] J. Zhao, *et al.*, "Workflow-centric information distribution through e-mail," *J. Manag. Inform. Syst.*, vol. 17, no. 3, pp. 45–72, 2001.

[52] A. Perry-Kessaris, "Recycle, reduce, and reflect: Information overload and knowledge deficit in the field of foreign investment and law," *J. Law Soc.*, vol. 35, no. 1, pp. 67–75, 2008.

[53] R. S. Kaplan and D. P. Norton, *The Strategy-Focused Organization: How Balanced Scorecard Companies Thrive in the New Business Environment*. Boston, MA: Harvard Business School Press, 2001.

[54] Neumann, *et al.*, "Information search using the balanced scorecard: What matters?" *J. Corporate Account. Finance*, vol. 21, no. 3, pp. 61–66, 2010.

[55] F. Heylighen, and C. Vidal, "Getting things done: The science behind stress-free productivity," *Long Range Plan.*, vol. 41, no. 6, pp. 585–605, 2008.

[56] T. Barkow, "Information overload: Workers awash in a sea of information ask if they're really better off," *PR Tactics*, vol. 11, no. 11, p. 12, 2004.

[57] D. Allen, *Getting Things Done: The Art of Stress-Free Productivity*. New York: Penguin, 2001.

[58] R. Goldsborough, "A new year's resolution: Battling information overload," *PR Tactics*, vol. 16, no. 1, p. 15, 2009.

[59] Q. Jones, *et al.*, "Information overload and the message dynamics of online interaction spaces: A theoretical model and empirical exploration," *Inform. Syst. Res.*, vol. 15, no. 2, pp. 194–210, 2004.

[60] R. Cross and L. Sproull, "More than an answer: Information relationship for actionable knowledge," *Organiz. Sci.*, vol. 15, no. 4, pp. 446–462, 2004.

[61] H. McNeely, "Effective business communication," *Int. J. Commerce Manag.*, vol. 3, pp. 95–96, 2004.

[62] D. Denton, "Better decisions with less information," *Industr. Manag.*, vol. 43, no. 4, pp. 21–26, 2001.

[63] M. Hegarty, "Individual differences in use of diagrams as external memory in mechanical reasoning," *Learn. Individual Differences*, vol. 9, no. 1, pp. 19–42, 1997.

[64] W. T. Coombs and S. Holladay, *PR Strategy and Application: Managing Influence*. Malden, MA: Wiley-Blackwell, 2010.

PRACTICAL INSIGHTS FROM THE DUTCH EMPLOYERS' ASSOCIATION

The Dutch Employers' Association (AWVN) was founded in 1919 to promote its members' interest and provide services in the social and economic fields. With about 850 large companies and over 70 sectors, AWVN is the largest employers' organization in The Netherlands. AWVN is a professional service organization acting for and on behalf of its members, providing a broad range of employment related services for companies and institutions small and large across many industries, the services sector, building sector, the world of harbors, transportation, logistics, knowledge institutions, and the like. AWVN plays an advisory role in, for instance, over 500 collective labor agreements, representing the majority of the collective labor agreements in The Netherlands. In the fields of job evaluation and remuneration management, AWVN has already been a market leader for many decades in The Netherlands. In all fields connected to terms of employment and labor relations, such as reorganizations, labor law, pensions, employee participation, and international labor mobility, AWVN provides high quality services, and its ambition is to grow further.

Operating on the crossroads between market-oriented companies and socio-economic institutions of various kinds, AWVN has always been amidst dense flows of information

Information Overload: An International Challenge for Professional Engineers and Technical Communicators,
First Edition. Edited by Judith B. Strother, Jan Ulijn, Zohra Fazal.
© 2012 Institute of Electrical and Electronics Engineers. Published 2012 by John Wiley & Sons, Inc.

from a variety of sources. From a historical perspective, three distinct phases in information management can be distinguished:

1. acting as an information resource,
2. focusing on the added value of information, and
3. co-creating added value in interaction with companies.

Although the major part of the assignment of AWVN requires tacit knowledge and an "implicit awareness" of what is relevant in labor relations, only some of the explicit elements of information are highlighted here.

Acting as an Information Resource

Up to the 1990s, at AWVN, information processing was based on producing and distributing information on paper in many different appearances. The main objective of the organization and its employees was "to know everything" about industrial relations, and its social economic role was to act as an "invisible hand" in industrial relations. Back then, from a distance, the AWVN building must have looked like a library, and inside, the procedures had, indeed, a far reaching similarity with the ones applied in a library. The documents regarding the collective labor agreements or the job descriptions of all member-organizations had to be available any moment in case of questions or strikes, or for updates. AWVN literally held the key to industrial labor relations. However, unlike the goal of a library, the goal of AWVN was to keep social peace and minimize the risk of strikes. In this regard, its goal was similar to the role of an insurance company. A collective labor agreement could be seen as a document of "frozen trust," because an agreement had to be written out in detail to prevent possible differences in interpretation. More overload of such explicitly categorized information was hardly imaginable, and in a way, overload was a "built-in" function of the system. In its existence since 1919, AWVN has proven to be an acknowledged and appreciated partner in knowledge building and distributing.

Focusing on the Added Value of Information

For different reasons, during the 1990s, the static position of the first phase changed in a fundamental way. The main reason was the decentralization of the negotiation process between employers and trade unions. In the changing socio-economic context, AWVN no longer acted as an invisible hand in the negotiations with the trade unions, as the companies and branches had started to negotiate themselves.

The decentralization of the negotiation process required the development of negotiating skills of the representatives of the companies, and AWVN developed a training method to transfer negotiating skills. Initially, the training was based on the so called Harvard principles. This Anglo-Saxon approach was soon developed further into the context of the Rhineland vision on labor relations as described by Joosse [1]. Moreover, the emerging global economy challenged the existing industrial relations with the need for more flexibility. In a fast pace, industrial relations transformed from a "stand alone" process in human resources towards an integration with other business processes. Labor relations became a tool of general management. Furthermore, the introduction of Information and

Communication Technology (ICT) offered possibilities to process information in a new way and the digital office was emerging.

These developments imposed a set of new requirements in the way AWVN had to manage information. Instead of "knowing everything," the role of AWVN focused on coordinating the decentralized processes of negotiation in companies and branches. The role of AWVN changed from that of an expert to that of an advisor, although on specialized subjects, the expert role still stood. This resulted in the vision that information itself now could be distributed among members for free, contrary to the previous phase where AWVN and its advisors were "sitting on information."

To develop a focus on an advisory role, several major changes were implemented. For instance, in order to support the movement towards the use of a collective labor agreement as a tool of management, AWVN introduced a management cycle of social policy for companies, with the following key steps: development of vision, exploring of interests, negotiation, implementation, and evaluation. Another major adaption was the restructuring of the job evaluation system that was applied in member-companies as a deduction of management goals. Both examples illustrate that the information process had to be reorganized not only within AWVN, but also within the associated market oriented companies.

The context for information had changed drastically, and with it the meaning changed of what was seen as information overload. While in the first phase, information overload was a way of life, now, information was mainly relevant within the context of coordination of the processes of negotiation in companies and branches [2]. In this phase, an awareness of overload of information was dawning.

Co-Creating Added Value in Interaction with Companies

In the first decade of this new century, the third phase in information management at AWVN was born out of the growing pluriformity among its member-companies with respect to their information need. In terms of information management, this required less information that contained a common content for all and more focus on differentiating between groups of members. With fewer standard questions than ever, and the need for a faster response, the answers had to be organized in a new way. This information process was, among others, tackled by an "employer's helpdesk" where short answers are given either by phone or email, in no time amounting to over 40,000 interactions a year. The ICT offered possibilities for queries in the database of questions and answers to identify patterns.

Another trend was that the structure of the Dutch economy was heading towards a knowledge economy, which went along with the renewal of the chains of added value between companies. Not only did the content of added value in the chain change, but the associated networks were also gradually becoming more global. Today, about 80 percent of the AWVN members have so-called "mother-daughter" relationships, i.e., they have headquarters in The Netherlands and daughter companies abroad, or visa versa. This required intensive interaction between the Dutch company sites and their overseas head-quarters. AWVN has developed, for instance, a training program on "Dutch Labor Relations" to transfer relevant knowledge to foreign managers.

For AWVN, this trend also implied a widening of the advisory role on labor relations towards consultancy in a wider sense of business administration. AWVN believes in competitive advantage due to good employership. In this perspective, AWVN wants to be a leading center of knowledge, co-creating with her member-companies. Being a consultancy now, AWVN is operational in the whole field of employership, employment conditions, and employment relationships [3]. This required a standardization of the advisory process, in order to speak the same internal and external language and to deliver even more predictability in deliverables for companies.

The next issue in information overload was knocking on the door: a dilemma between standardization of working processes versus an outreach for new challenges. Further standardization of working processes will invoke a specific information overload (all information that would not fit the standard), and therefore a certain discarding of weak signals. On the contrary, an outreach for new challenges is necessary to cope with the dynamics and risks in the economy and will lead to co creation of new added value, requiring mavericks who cross borders.

A Final Observation

Where does the description of the three phases lead to in anno 2012? Given the turbulence in the Dutch economy as well as in the social system of the Dutch society—a mirror of what is happening around the world—critical reflection on the current mechanism of coordinating information is needed. What information is required in a system of constructive labor relations, or in terms of business administration in an entrepreneurial climate? If interaction between stakeholders is a key issue, what information process is facilitating a dialogue? How can we balance the supply and demand of information? Who will suffer from information overload? A review of such questions might yield some answers for an outlook on the future.

Contributor to Practical Insights from The Dutch Employers' Association

• Dr. Arjen Verhoeff, Senior Advisor

REFERENCES

[1] D. J. B. Joosse, *Negotiation in Industrial Relations*. Den Haag, The Netherlands: AWVN, 2010.

[2] W. Buitelaar and P. Van den Toren, *Moving Time. AWVN 1919-1999. An Employers' Organization in the Concerted Economy* (in Dutch). Den Haag, The Netherlands: AWVN, 1999.

[3] AWVN. *The Van Walraven Testament* (in Dutch). Den Haag, The Netherlands: AWVN, 2009.

12

BLINDFOLDED THROUGH THE INFORMATION HURRICANE?

A REVIEW OF A MANAGER'S STRATEGY TO COPE WITH THE INFORMATION PARADOX

ARJEN VERHOEFF

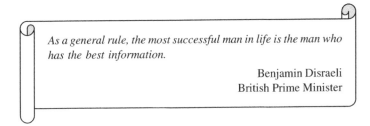

As a general rule, the most successful man in life is the man who has the best information.

Benjamin Disraeli
British Prime Minister

ABSTRACT

Dealing with information might be seen as a basic competence of managers, but today, information is produced in such quantities and is available from so many sources that dealing with it might not be straightforward anymore. This study departs from the theoretical perspective of Simon [1], who identified asymmetry in information between stakeholders. The current research explores, in a European context, the research question is whether a manager is aware of the need for an information strategy within and between companies and if he or she needs a new or revised strategy. A decomposition of the information paradox indicates that more symmetry in information might offer a direction for solutions. The results of a preliminary survey among Dutch managers are interpreted

---

Information Overload: An International Challenge for Professional Engineers and Technical Communicators, First Edition. Edited by Judith B. Strother, Jan Ulijn, Zohra Fazal.
© 2012 Institute of Electrical and Electronics Engineers. Published 2012 by John Wiley & Sons, Inc.

using the backbone model for analyzing the information flows between internal and external stakeholders. Inspired by the various chapters of this book, learning points are discussed. The general conclusion is that the information paradox is an important issue that needs to be addressed. Companies need managers who are capable of achieving a more symmetrical helix of information between the internal and external stakeholders. However, this chapter illustrates that the orientations of managers in daily life are not quite consistent with the new information reality; therefore, there is a need to reflect on integrated relational and technical solutions for the information paradox.

12.1 Introduction

For a manager, dealing with information is a basic competence in controlling the various business processes. Of course, an average MBA-educated manager is supposed to understand what it takes to apply, for instance, the insights of Simon's bounded rationality [1]. Bounded rationality refers to the limits of human capability, for instance, to select information, to take into account the consequences of choice, or to adjudicate among multiple goals. Gathering, selecting, and sharing information seem to be simple tasks among the many complex ones that a manager is required to fulfill today. Still, more than ever, information is produced in continuously increasing quantities and is available nowadays from a great variety of (digital) sources. Is managing information really as straightforward as one is inclined to believe on face value? This contribution explores whether a manager needs an information strategy, or whether he or she can leave it to, for example, the forces of supply and demand on information in the daily life within a firm. The context for an empirical grounding is The Netherlands in Europe. This quest results in a research agenda on managing information and information overload.

First, the reason that the amount of and diversity in information has become a management issue needs to be established. *Information* is based on accurate and timely verified *data*, and is specific and organized for a purpose. Information is presented within a context that gives it meaning and relevance and can lead to an increase in understanding and decrease in uncertainty. In an organizational context, *knowledge* is the sum of what is known and resides in the intelligence and the competence of people. Information *overload* is stress induced by reception of more information than is necessary to make a decision (for additional definitions, see [2], and Chapters 1 and 2).

In the past decades, the average level of education and emancipation has increased, contributing to the accumulation of knowledge in society at large and in market-oriented companies. However, in Europe, the accumulated amount of knowledge of employees in companies and knowledge institutions has not sufficiently led to innovation, a situation that has been identified as the *innovation paradox* [3–5]. The innovation paradox is the inability or reluctance of (manufacturing) firms to pursue the strategies that build the operational capabilities necessary for innovation to provide both profitability and growth [6]. The innovation paradox is also known as the knowledge paradox in a knowledge economy. Although the knowledge paradox is often seen as a policy issue of Europe as a whole or of individual nations, it is by definition mirrored in individual companies. While managers may think dealing with information is a minor task, the knowledge paradox suggests a fundamental issue in the processing of business information when related to achieving added value. Instead of moving forward to develop *knowledge* further, it might be relevant to revisit how managing *information* may contribute to solving the knowledge paradox or, perhaps more appropriately, the information paradox [7].

Paraphrasing the above definition, the information paradox is the inability of managers in an organization to develop and pursue an information strategy that builds a mindset among stakeholders [8] and associated technical auxiliaries necessary for innovation to achieve goals. This chapter's research question addresses: (1) whether a manager is aware of the need for an information strategy within and between companies, and (2) if he or she needs a new or renewed strategy.

12.2 Decomposing the Information Paradox

The assignment of a top manager consists of, among other elements, controlling the information regarding the exploitation process in order to ensure short-term profit as well as exploring what information is needed for the continuity of the firm. Of course, information can be viewed primarily from a technical perspective. The various chapters in this book show that there is a vast domain of technical issues to be researched regarding information and information overload, such as human-centered approaches to Internet searches (Chapter 8), storage and query techniques (Chapter 9), or visualization (Chapters 4, 6, and 10).

The various chapters also maintain that information is a matter of mindset. As Vossen states in Chapter 3, people use their emotional, perceptual, and cognitive functions to cope with information, while this capability is extended by auxiliary facilities for information processing in various appearances. Vossen puts forward that current approaches to managing information overload are doomed to fail, because they do not offer a solution to the imbalance between information production and the limited capacity of stakeholders to attend to and process so much information. His study suggests that analyzing only technical aspects of the information paradox will not be sufficient to expose the fundamentals of the information paradox. Ulijn and Strother broaden this concept in Chapter 5 by linking the individual to various aspects of the culture of a nation, company, or profession. These are all strong pointers that relations matter in analyzing the information paradox.

When decomposing the information paradox, the interaction between stakeholders [8] is taken as a point of departure. In executing an assignment, a manager needs to address various stakeholders, such as employees, customers, suppliers, financing institutions, or environmental organizations. In order to illustrate the first part of the research question—Is a manager aware of changes in information flows within and between companies?—the information paradox can now be decomposed to focus on

1. the control of the internal information process;
2. the control of the external information process.

Although the context for the following analysis is a European setting of labor relations with its Rheinlandian history, the key issues might—*mutatis mutandis*—also apply in an Anglo Saxon culture. This could be a subject of further study.

12.2.1 The Control of the Internal Information Process

The control of the internal information process starts with the relationship between the manager and the employee. By definition, in a hierarchical relationship, a manager is in control and decides what information is passed on to the employee and what information is extracted from the input of an employee. On the other hand, given the hierarchy, an employee has to set and follow his or her compass on the information received. In other words, a traditional hierarchical relationship [8, 9] implies an asymmetrical way of organizing information. Asymmetry in information occurs in a situation that favors the more knowledgeable party in a transaction [2] and is a condition in which at least some relevant information is known to some but not all parties involved. In, for instance, traditional craftwork, information is organized by apprenticeship, with a strong asymmetry in information as indicated in the expression "master apprentice." In this kind of relationship, one might argue that the master (= sender) steadily produces information overload in the apprentices (= receivers) to keep them in the learning mode. This also refers to the development of tacit knowledge, which has primarily been organized in traditional industrial firms by operational routines [10, 11]. In service-oriented organizations, information is partly organized as knowledge in terms of relevant (sales or support) scripts.

Essentially, there is nothing wrong with asymmetry in an organization as it contributes to efficiency. However, side effects may occur, such as a lack of motivation of an employee to report information relevant for the achievement of business goals or other system goals [12–14]. For example, when an organizational system regularly blames an employee for mistakes while not taking action to adapt the system or develop the competences of the employee, the employee may use an individual defense mechanism and limit his or her actions to tasks that cannot go wrong. Such a case represents a mirrored asymmetry as the employee—probably quite unconsciously—does not tell the boss (again) what is really going on. In my consultancy practice, over the years, I have collected hundreds of what Scott-Morgan calls "causal maps" of individuals, mostly born from a negative motivation on how to survive the organizational system. Speaking of information, such causal maps usually illustrate a risky type of information *underload* and might indicate insufficient leadership [15, 16].

The asymmetry in the information flow in market-oriented companies has been challenged in various ways. In the past decade, for instance, in most companies in The Netherlands, delayering and decentralization of tasks have led to a situation where the manager no longer is solely responsible for control. Nowadays, a machine operator is not only restricted to the mere execution of tasks but also manages control tasks. Instead of only cost as a business criterion, now, quality and flexibility are also relevant [5]. Today, managers and employees are required to organize and develop their own knowledge based on competences and results instead of the static job descriptions of the old days. These developments, together with technology-enhanced learning, have opened a perspective on a more symmetrical relationship in information between a manager and an employee. Even more so, when quality and flexibility are at stake, it is important to organize the information flow as a two-way street.

12.2.2 The Control of the External Information Process

Managers and employees of today might have the same limitations in their "administrative behavior" as Simon already described in 1976 [17], but they can build their decisions on a quite different external information process. Back then, for instance, organizations were mostly structured according to the principles of scientific management [18], with only a very limited number of windows on the external world by a limited number of persons in the organization. This resulted in an asymmetry in the external information process for most employees. Since the globalizing of the economy at the end of the last century and the possibilities of the Internet and other ICT, market-oriented firms have organized their external information process in a fundamentally new way. The concept of open technical innovation has emerged in the past decade at a fast pace and on a large scale [19–21].

The innovative force induced by the proximity of companies in Silicon Valley is well known. Other regions have developed in a similar way. A notable region is the Brainport in the Dutch region of Eindhoven (www.brainport.nl), which, in June 2011, was declared to be the most "Intelligent Community of the Year 2011" [22]. The magic of the Brainport region is a strongly coordinated information flow between firms, knowledge institutions of all kinds, and governmental institutions, a "triple helix." However, open innovation is not only based on personal contacts in a specific regional setting but also stretches globally by digital links and vast external networks of personal ties. In fact, in open innovation, the external chain is at least as important as the internal relations. So, in addition to maintaining quality and flexibility, in innovation, it is important to organize the information flow as a two-way street. Thus, open innovation requires the renewal of relational patterns of stakeholders, not just internally but also externally, with external stakeholders in the production chain or with external knowledge institutions. Although open innovation is still primarily perceived as a technical domain, it cannot be denied that interests and roles of related stakeholders are shifting. Instead of an asymmetry in information, one might maintain that today nothing prevents a manager from organizing symmetry in information flows between internal and external stakeholders, except for maybe the limitations of PR-rules of publicly held companies.

The importance of symmetry in information in an organization has gained even more importance by the phenomenon of the new world of work [23]. In this new world, enabled by the individualized ICT facilities such as e-mail, Internet, and mobile phones, there are new working patterns and a more symmetrical, reciprocal flow of information. In this environment, the employee is supposed to be working *independent of time and place*, which requires a manager oriented toward trust instead of control. Here, in theory, the psychological contract [9, 20] between a manager and an employee is based more on partnership than hierarchy.

In the past decade in The Netherlands, a new phenomenon—social innovation in market-oriented firms—has emerged [24], stating that information should be seen in the context of a cultural-economic system representing an organization. Social innovation is the exploring and exploiting of new or renewed ideas, rules, routines, or learning processes in a market-oriented network or organization, initiated by stakeholders and leading to added value in the interest of stakeholders [24]. In this perspective, symmetry of information between stakeholders might be a necessary condition for sustainable success.

In this section, the information paradox is decomposed into a discussion of the control of the internal and the external information processes. In both processes, a shift in interests from asymmetry toward symmetry in information could be noted. This might indicate that symmetry in information flows is important for solving the information paradox, and this is worth further theoretical grounding. Still, the internal information process with its hierarchical elements is fundamentally different from the external process based on co-creation, and this is the departure for the description of a framework to analyze the information paradox.

12.3 A Framework to Analyze the Information Paradox

The main guideline of the framework used to analyze the information paradox is the information strategy regarding stakeholders. On all levels of an organization, a manager

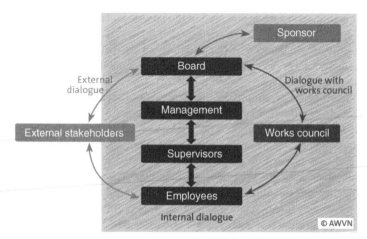

Figure 12.1 Backbone model. The backbone model is used to describe communication flow among both internal and external stakeholders. It may contribute to breaking down the information paradox as it distinguishes between if and how a strategy is used (vertical axis) and the interaction with external stakeholders (horizontal axis) [26].

needs to develop the relations with stakeholders [8, 20, 25, 26] and facilitate the necessary conditions to achieve sustainable partnership. First, this requires committed leadership between managers and employees [27]. This "vertical" relationship is a dissatisfier for building relations with other stakeholders. That is, without such a trustful hierarchical relationship, the quality of the flow of internal and external information is at risk. Second, a "horizontal" relationship between internal and external stakeholders is distinguished. Here, not only a manager but also other representatives, for example, a work or business council, meet external stakeholders—customers or knowledge partners in the chain of production or service—in order to co-create added value. The panorama of stakeholders is expressed by the vertical and horizontal axes of the backbone model in Figure 12.1 [26].

For over 20 years, the model of Figure 12.1 has proved its internal validity and eclectic value in industrial labor relations in The Netherlands [26]. Moreover, the model has illustrated multicultural qualities in that it has successfully and repeatedly been applied in other national cultures, such as in various countries in Eastern Europe, Africa, and the Far East (e.g., see www.decp.nl). When systematically applying the model in, for instance, training situations, the open exchange of information appears to be a necessary condition for successful negotiation. Given such experiences, the model has sufficient external and ecological validity to analyze information strategies.

As a general proposition, the model states that a manager needs a strategy to control information. The model may contribute to breaking down the information paradox as it distinguishes between if and how a strategy is used (vertical axis) and the interaction with external stakeholders (horizontal axis). The interaction between the vertical and horizontal axis should ignite this "helix of innovation." This decomposition of the information paradox leads to the following three questions.

1. Do managers experience issues regarding information?
2. Do managers use an information strategy?
3. Do managers use a strategy to transform information into added value?

12.3.1   Do Managers Experience Issues Regarding Information?

The very premise of this book, the existence of information overload, demonstrates the relevance of this question, and many chapters in this book address it. In Chapter 2, for example, the history of the issue of information overload is extensively addressed. Aikat and Remund identify information overload as a twentieth century phenomenon by the example of *Time Magazine*, which set a standard in 1923 for simplifying the supply of information. Other chapters identify factors that lead to the perception of information overload. Will a manager recognize such factors as important in the information process in his or her organization in this modern era?

12.3.2   Do Managers Use an Information Strategy?

The second question about whether a manager uses an information strategy is derived from the "vertical" axis of the backbone model, shown in Figure 12.1. Without such a strategy, supervisors and employees will have no sense of direction and will not be able to set the right time frame to collect and digest information in an effective way. This requirement has been illustrated well in Chapter 4, where Caborn and Cooper study the focus of technical communicators. Interestingly, they maintain that the use of metaphors is a powerful means to convey information. One of their examples is the role of the doctor in the twenty-first century (sender), which is not only to know what is best for the patient (receiver) but also to communicate this to the patient in such a way that the patient concurs with the doctor's decision. This is exactly the kind of leadership element that a manager is supposed to master [28], even though in the case of a manager and an employee, the roles of the sender and receiver may also, and simultaneously, be reversed. In other words, an information strategy is not a collective monologue, but a dialogue in a constructive entrepreneurial culture, as Ulijn and Strother show in Chapter 5.

Ulijn and Strother build on Hall's research [29] about communicating in a high or low context of information, and their insights might also apply to managers. In a high context (company) culture, the participants of the system know a lot from sources within the culture itself. Managers in such a company culture tend to prefer indirect patterns of communication as well as higher levels of politeness and ambiguity. Managers from low context cultures, which are quite heterogeneous, share much less knowledge, so everything must be explained clearly and in much more detail. Managers from low context cultures usually prefer messages that are direct and concise. One might maintain that the information on the "vertical" axis of the backbone model is to be associated with a high context culture. This would be worthwhile to study further.

12.3.3   Do Managers Use a Strategy to Transform Information into Added Value?

The third question, whether a manager uses an information strategy, is derived from the "horizontal" axis of the backbone model. Stakeholders need entrepreneurial space to explore new ways of organizing added value [24]. In terms of Hall's high and low context of information, the horizontal axis could be associated with a low context (company) culture, as the stakeholders are usually quite heterogeneous in their interests and share much less knowledge. In the cooperation in the external chain, it might be relevant for a manager to apply an information strategy with the aim that the low context culture will turn into a high context culture. In such a reframing, the information strategy element can be incorporated as brought forward by Alton and Manning in Chapter 6 about the effect of color, visual form, and textual information. Strategic alliances require a balance in cost and revenues, and Bidkar analyzes an essential cost element in Chapter 7, by putting forward a framework for calculating the cost of information overload. Another theoretical element that Ulijn and Strother bring forward in Chapter 5, the differences in the cultural patterns of discourse, might be relevant here. It might be interesting to verify whether the vertical axis of the backbone model is linear and direct, handling one thing at a time (monochronic), and whether the horizontal axis allows for digressions, discussing several things at the same time (polychronic).

With this elaboration of the backbone model of Figure 12.1, an attempt can be made to analyze the ability of managers to develop and pursue an information strategy aimed at internal and external stakeholders. The three questions are operationalized and applied on a dataset as described in the next section.

12.4 Illustrating the Framework with Some Dutch Empirical Evidence

After a general impression on how information and information overload fits in the Dutch business case, the outcome of a survey among Dutch managers is presented. The backbone model of the previous section has been operationalized and applied to firms in The Netherlands.

12.4.1 The Importance of an Information Strategy

This section presents some figures on the effects of the new world of work in order to demonstrate the extent to which the structure of the information flows in firms in The Netherlands has already been changed. In The Netherlands, teleworking—when employees have access to ICT from a location outside of a company—doubled between 2003 and 2009. In 2003, 28% of the companies offered teleworking to their employees, but by 2009, this percentage had increased to 56% [30]. Furthermore, a survey on the new world of work among employees of 30 firms with an average of 520 employees yielded the top 10 potential improvements, shown in Table 12.1 [31].

Table 12.1 illustrates several task-oriented points of improvement, such as information concentrated in one spot (1), clear status of actions and project tasks (2), or access to

T a b l e 12.1   Overview of Potential Improvement in the New World of Work

1. Information concentrated in one spot
2. Clear status of actions and (project) tasks
3. Access to information of external stakeholders
4. Ability to work efficiently with systems
5. Open communication
6. One time entry of information in systems
7. Clear vision of management
8. Common goal setting
9. Transparency of the organization and its processes
10. Discovery of knowledge/expertise of colleagues

information of external stakeholders (3). It also shows stakeholder-related points, such as communicating openly (5), common goalsetting (8), or finding knowledge/expertise of colleagues (10). The adoption of the new world of work by companies is an evolutionary process, and such working patterns are to be disseminated on a large scale in the years to come. In some sectors of an economy, such as many jobs in industry, health care, or law enforcement, it would simply be impossible to work independently of time and space. For many managers, this way of organizing information is a guideline for development of their own companies. It requires a careful balance between technical information and information focused on developing a partnership with stakeholders. Table 12.1 illustrates that a manager cannot achieve a symmetrical information flow or prevent information overload without an information strategy.

12.4.2 Preliminary Survey Among Dutch Managers

In July 2011, a preliminary survey was conducted among CEOs and HR managers of market-oriented companies in The Netherlands. The survey was intended to map the current state of information management. The three questions of the previous section were included, each with a corresponding four-point Likert scale with precoded answers, along with an open question at the end. The question on whether a manager experiences any issues regarding information was precoded with an option for an open answer, while the questions about whether a manager uses an information strategy and the one about a strategy to transform information into added value were simply precoded. The open question solicited for suggestions on how to deal with information overload. The number valid responses was $N = 72$, and respondents represented small, medium, and large companies in industry and service.

Do Managers Experience Any Issues Regarding Information? Almost all the managers were experiencing some kind of issue with information. One out of three mentioned the large variety of sources of information. One out of five was experiencing a problem with setting priorities in what type of information is needed. Also, almost one out of five saw the amount of information as well as the speed with which information decays as problematic. Evaluating where information has led was an issue for one out of six managers. One out of ten was experiencing a problem with selecting information or coordinating information with other organizations. Also, one out of ten managers estimated that the current knowledge is not compliant with the business strategy of the company for the next years.

Do Managers Use an Information Strategy? All managers use some kind of information strategy, and most use more than one. Three out of four managers apply a generic strategy to increase the overall level of information and knowledge in the organization. Also, three out of four are oriented toward information enhancing the craftsmanship or professionalism in the organization. Three out of five managers prefer a more specific strategy, for instance, to improve technical/administrative routines or to introduce new techniques or systems. With fewer than one out of two, a strategy stimulating an entrepreneurial company culture to absorb and share information is the least popular.

Do Managers Use a Strategy to Transform Information into Added Value?
Managers seemed inclined to prefer a task-oriented strategy. For instance, four out of five

managers focused on increasing the cost-effectiveness of the organization or developing (new) products or services. This underlines the importance of the contributions Bidkar offers in Chapter 7 with the framework for calculating the cost of information overload. On the contrary, only one out of two managers was oriented toward a stakeholder-related approach, for instance, to intensify interaction with customers or challenge employees to take the initiative in transforming information into added value. This indicates a weakly developed capability in managing stakeholders' relations, which are so important in preventing or resolving the information paradox. If a manager is not at the steering wheel, he or she is not able to achieve a balance of knowledge.

12.5 Discussion and Conclusion: Lessons in Information Strategy

The research questions were: (1) whether a manager is aware of the need for an information strategy within and between companies, and (2) if he or she needs a (renewed) strategy. After a discussion of the results and the methodological grounding, some lessons are drawn in the context of the other chapters of this book. Together with the applied implications, a program for further research is presented.

12.5.1 Discussion

The first observation is that the greatest issue managers experience regarding information overload is the multiplicity of sources of information. This needs interpretation. If managers do not prioritize the sources of information, is it because they have an effective personal way of dealing with huge flows of information or do they focus only on what is needed to achieve their assignment? Do they speak only for themselves or do managers have a clear view on what others in their organizations are experiencing? The preliminary questionnaire was not equipped to give us a window on such important issues; perhaps, in-depth interviews could reveal more. Furthermore, managers indicate applying various information strategies. In a way, they might have been biased by the precoded answers of the second question. Managers might pretend to have an information strategy, but the analysis suggests that they have a long way to go. This should be verified later, to include an analysis of the symmetry/asymmetry in information flows on the vertical axis of the backbone model. The third question concerned the horizontal axis on co-creation between internal and external stakeholders. Here, a task orientation prevails over a stakeholder orientation. This is an intriguing result in an era dominated by emancipated customers and empowered employees. Such a mindset certainly does not favor symmetry in information flows and might lead to information overload of the wrong kind. Are managers aware of this orientation?

When the insights about the vertical and horizontal axes of the backbone model are combined, the result might help us better understand the information paradox. An important theoretical insight in the beginning of this chapter was that symmetry in information flows might contribute to solving the information paradox. A strong symmetry on the vertical axis would result in a better focus of all internal stakeholders, while a strong symmetry on the horizontal would contribute to co-creation. The interaction between the vertical and horizontal axes should ignite the "helix of innovation." However, the preliminary survey did not yield strong evidence that this is a common managers' concern or practice.

12.5.2 Methodological Grounding

The preliminary survey did not fully answer the question of whether managers are confronted with information overload, as this would require a more elaborate study. It illustrates that more solid research ground is needed in order to investigate what strategies are useful against information overload. The expression *overload* contains a dimension of time in that the amount of information apparently exhibits a certain density at a given time frame. In addition to time, overload also expresses a dimension of (mental) space, where it concerns interaction between specific stakeholders or pluriformity where uniformity is expected. The analysis of the research question revealed links between several applied disciplines, such as information/ICT, communication, and knowledge, as well as psychology, economics, sociology, and business administration. Further research will benefit from a multidisciplinary approach, as elaborated in Chapter 3 by Vossen. Furthermore, the issue of information overload can be qualified as a multilevel domain.

Finally and almost by definition, information flows are multiregional. By taking such properties into account, validated instruments might be developed to offer managers and their related stakeholders a level playing field for their decision space. The preliminary questionnaire used in this chapter will be developed further in this direction. In order to investigate symmetry in information flows, it is important to take not only managers into account but also other stakeholders. Furthermore, analyzing the information paradox not only requires quantitative research but also requires qualitative insights in order to identify what organizations and their stakeholders need in order to develop the ability to operate in multiple frames of time and space, and enhance learning modes [31].

12.5.3 Learning Points

That the greatest issue managers experienced is the multiplicity of sources of information is not a big surprise in an era of multimedia. The exploding variety of ICT-driven sources of information, such as Internet, e-mail, Twitter, and other kinds of community-like social media, is discussed in this book in various chapters. The development of ICT-applications is outlined in Chapter 2 by Aikat and Remund; search strategies are analyzed in Chapter 8 by Hoenkamp; storage and query techniques are discussed in Chapter 9 by Calders *et al.*; and visualization is explored in Chapter 4 by Caborn and Cooper, in Chapter 6 by Alton and Manning, and in Chapter 10 by Mengis and Eppler. For instance, in Chapter 8, Hoenkamp focuses on three ways to mitigate information overload related to search engines: being specific in what we ask, amending our requests when we do not find what we need, and making our retrieval techniques more human centered. These are directions for solutions that managers can and should underline. This might also require a *new standard*, following the standard of *Time Magazine* dating from 1923 as Aikat and Remund describe in Chapter 2. Contrary to the standard of *Time Magazine* today, visualization also counts. In this perspective, in Chapter 10, Mengis and Eppler discuss why visualization is still not frequently used by communication professionals, even though its potential for reducing information overload is explicitly acknowledged. Here, there is still leadway for improvement.

The second subquestion revealed that managers have indicated applying various information strategies. Two of the strategies mentioned by managers can be interpreted as task or technology oriented: improve technical and administrative routines and introduce new techniques and systems. On the other hand, the other two mentioned strategies can be seen as stakeholder oriented, as is inherent in social innovation. The most popular, enhancing craftsmanship or professionalism, is oriented toward individual workers, which relates to professional culture. Here, information load is dependent on their professional culture (see Chapter 5 by Ulijn and Strother).

Stimulating an entrepreneurial company culture to absorb and share information is the least popular strategy. Why is this? In Chapter 11, Remund and Aikat examine the effects of information overload within organizations and present strategies for improving knowledge transfer and decision making. They conclude that managers focus their information strategy primarily on individuals, while the aggregate effect of information overload within organizations does not have priority. Their finding seems to support the findings in our preliminary study even further, for both the vertical axis and the horizontal axis. Are managers more oriented toward their own professional culture instead of the company culture? If a strategy to prevent information overload would be dependent only on individual managers instead of a strategy embedded in an organizational vision, no longer-term information policy would be effective. Remund and Aikat stress the need for organizations and the relevant stakeholders in organizations to focus on information sharing that is situation specific and decision making that is actionable. It would be interesting to explore if their direction for a solution is more appropriate on the vertical axis than on the horizontal axis of the backbone model.

These insights culminate in the question of what company culture is needed to ensure the development and operationalization of an information strategy that aims to prevent information overload. As Ulijn and Strother show in Chapter 5, this requires a thorough analysis of mindset, attitudes, and actual behavior, taking into account theoretical approaches such as monochronic versus polychronic time perspective or a (mental) space perspective in a high or a low context company culture. The third subquestion illustrates that in creating added value from information, managers in The Netherlands preferred a task-oriented strategy above a stakeholder-related strategy. This outcome supports the idea that managers are more oriented toward their own profession than toward the desire to create a company culture inviting stakeholders for an innovative dialogue. These learning points may lead to the insight that the current way of maintaining information flow does not contribute to solving the information paradox.

The most important learning point might be that a manager is mainly aware of the need for an information strategy in the context of his or her own professional culture. Given that few managers are aware that information overload needs a company-level strategy, managers are not likely to leave the comfort zone of their own culture and venture out to solve the company's information problems.

12.5.4 Applied Innovative Directions

This contribution takes a stakeholder approach to analyze the information paradox. This does not mean that the technical solutions presented elsewhere in this book are not relevant for everyday solutions; in fact, those solutions regarding information function and form need to be integrated into corporate strategy.

In order to guide managers in spotting situations or sources of information overload, one can distinguish between the dimensions of (entrepreneurial) space and time (or timing). More specifically, in innovation, two states are usually defined, exploration and exploitation. These two "axes" are combined to produce Table 12.2, with examples of four typical situations of information overload (derived from Verhoeff [24]).

The four examples of situations of information overload in innovation, as expressed in Table 12.2, can be elaborated in the following way. The first situation expresses a pluriformity of sources and targets in information (Cell A). In innovation, it is important

Table 12.2   Examples of Situations of Information Overload in Innovation

Cells: Types of Information Overload Between Stakeholders		Conditions for Innovation	
		Entrepreneurial Space	Time and Timing
Innovation (social and technical)	Explore	A. Pluriformity of sources and targets	B. Ambidexteriosity in information function
	Exploit	C. Adapting routines by focus in information	D. Effectivity of information forms

to achieve pluriformity of information, such as by ideation, networking, routining, or learning strategies. Examples of this perspective can be found in *Practical Insights* from Media Group Limburg, Alvogen, and The Dutch Employers' Association (AWVN).

The second situation is ambidexteriosity in information (Cell B). Ambidexteriosity occurs when information from two extremes meets, for instance, a historical perspective and a future scenario [33]. An example of this perspective can be found in *Practical Insights* from Laboratory for Quality Software (LaQuSo). The pathways to cope with the different paces of the interests of internal (vertical axis of the backbone model) and external stakeholders (horizontal axis of the backbone model) must be considered. Examples of this perspective can be found in "Practical Insights" from IBM and Xerox.

The third situation is adapting routines in information (Cell C). Continuous improvement of routines may be conflicting with the natural habitat or the inclination of the persons involved to stay in their comfort zone and be less receptive to new information. Still, successful organizations have discovered ways to overcome organizational defenses and adapt [34]. Examples of this perspective can be found in "Practical Insights" from Applied Global Technologies (AGT) and Harris.

The fourth situation concerns the effectivity of information (Cell D). This seems straightforward, but timing, or just in time stock management, holds true not only for physical logistics but also for information flow. Too many companies, for instance, still work with a multitude of databases that are not compatible and therefore do not allow streaming information in a timely manner. If one ignores making information uniform in relation to the innovation process, the result is overload for internal or external stakeholders. Examples of this perspective are found in "Practical Insights" from IBM and Xerox.

Paul-Peter Feld of Xerox made the following insightful observation about the importance of the need to examine the issues of the digital age with multiple perspectives, especially focused on shared leadership and responsibility of all stakeholders:

The last years it turned out that conventional thinking reached its boundaries. In a rapidly changing, connected, and complex world, linear thinking will not lead us to "next practices." We have to create resilience through new ways of collaborating, organising, (un)learning and leadership. This can only be done together. Leadership becomes a social process filled in by individuals with specific qualities. Leadership is the result of co-creation and the quality will be improved/enhanced by "just" sharing it. As a consequence of this, everybody can (and must) be a leader but even more important: stakeholders have to take that responsibility together. Shared leadership will strengthen and accelerate our social innovation, the foundation for every future success. The Twenty-first Century is about collaboration, connections, ideas, and people. Welcome to the human age [36].

12.5.5   Toward an Innovative Research Agenda

Several domains can be identified for further innovative research on the information paradox.

a. This study illustrates that information is directly related to stakeholders. A solution for information overload might not be simply within reach by appointing a Chief Information Officer (CIO). It requires a conceptual framework as well as a consistent information strategy [35]. Qualitative research is needed to explore how information flows are related to stakeholders, which is an important methodological issue of the research agenda on social innovation in market-oriented companies. One specific research field can be found in the symmetry/asymmetry of information between stakeholders, and the here-assumed importance of symmetry in information flows. It is not clear why managers underestimate the importance of company culture in information strategy. This stresses the importance of analyzing differences in information patterns between company and professional cultures as discussed in Chapter 5 by Ulijn and Strother. A more applied field of research is the role of a manager and the meaning of leadership amidst the information flows in the new world of work, as a manager is a key player in innovation of relations between stakeholders.

b. A more quantitative route for research on the information paradox is the assessment of existing information patterns, comprising the analysis of interactions between stakeholders in terms of reciprocal meaning, modes of representing information, or connectedness of databases. This contributes to solving the information paradox in that it may identify what kind of information is necessary to achieve a sense of symmetry in exchanging information.

c. An exploratory research path would focus on the rules and regulations to process information between stakeholders. This could depart from a change of paradigm in information management. Further theoretical grounding of the information paradox might reveal the conditions of (mental) space to develop the interaction between, for instance, the vertical and horizontal axes of the backbone model, and their helix-like interaction. In this perspective, an appreciative inquiry into the differences in the meaning of information exchange between the European Rheinland and the Anglo Saxon and Asian national cultures is encouraged.

d. A last research path, but not the least important, is how information and information overload are related to technical and social innovation in market-oriented firms. Here, the link can be seen within the scope of Chapter 1 and this concluding chapter on how to integrate technical solutions with interests of stakeholders when resolving the information paradox.

12.5.6   General Conclusion

The information paradox is an important issue that needs to be addressed, as is illustrated by elements in the various chapters in this book. Companies need managers who are capable of making the information helix between the internal and external stakeholders work. However, this chapter illustrates that the orientations of managers in daily life are not quite consistent with the new information reality. In this perspective, managers and other readers of this book are encouraged to reflect on integrated relational and technical solutions for the information paradox.

REFERENCES

[1] W. Callebout, "Herbert Simon's silent revolution," *Biologic. Theory*, vol. 2, no. 1, pp. 76–86, 2007.

[2] *BusinessDictionary.com* [Online]. Available: http://www.businessdictionary.com/definition/information-overload.html Jul. 21, 2011 (last date accessed).

[3] R. Farson and R. Keyes, *The Innovation Paradox: The Success of Failure, The Failure of Success*. New York: Free Press, 2003.

[4] M. Knockaert, *et al.*, "The knowledge paradox explored: What is impeding the creation of ICT spin-offs?" *Technol. Anal. Strategic Manag.*, vol. 22, no. 4, pp. 479–493, 2010.

[5] M. Porter, "Innovation and competitiveness: Findings on The Netherlands," *Innovation Lecture for the Dutch Ministry of Economic Affairs. Den Haag: Ministry of Economic Affairs*, 2001.

[6] V. Baard and T. Watts, "Breaking the paradox," *Australasian Account. Business Finance J.*, vol. 1, no. 2, pp. 25–33, 2007.

[7] J. Thorp, *The Information Paradox*. Whitby, ON, Canada: McGraw-Hill Ryerson, 2003.

[8] E. Freeman, *Strategic Management: A Stakeholder Approach*. Boston, MA: Pitman, 1984.

[9] D. M. Rousseau, *I-deals: Idiosyncratic Deals Workers Bargain for Themselves*. New York: M. E. Sharpe, 2005.

[10] A. Hoeve and L. F. M. Nieuwenhuis, "Learning routines in innovation processes," *J. Workplace Learn.*, vol. 18, no. 3, pp. 171–185, 2006.

[11] I. Nonaka and H. Takeuchi, *The Knowledge Creating Company: How Japanese Companies Create the Dynamics of Innovation*. New York: Oxford Univ. Press, 1995.

[12] S. Beer, *The Heart of the Enterprise*. New York: Wiley, 1995.

[13] B. Oshry, *Seeing Systems*. San Francisco, CA: Berrett-Koehler, 1996.

[14] P. Scott-Morgan, *The Unwritten Rules of the Game*. New York: McGraw-Hill, 1994.

[15] P. Senge, *The Fifth Discipline*. New York: Doubleday, 1990.

[16] J. F. Manzoni and J. L. Barsoux, "The set-up-to-fail syndrome," *Harvard Business Rev.*, vol. 76, no. 2, pp. 101–113, Mar.–Apr. 1998.

[17] H. A. Simon, *Administrative Behavior. A Study of Decision-Making Processes in Administrative Organization*, 3rd ed. London, U.K.: Free Press, Collier Macmillan Publishers, 1996.

[18] F. W. Taylor, *The Principles of Scientific Management*. New York: Norton, 1911.

[19] H. Chesbrough, *et al.*, Eds., *Open Innovation: Researching a New Paradigm*. London, U.K.: Oxford Univ. Press, 2005.

[20] V. Gilsing, *et al.*, "Network embeddedness and the exploration of novel technologies: Technological distance, betweenness centrality and density," *Res. Policy*, vol. 37, no. 10, pp. 1717–1731, 2008.

[21] J. Ulijn, *et al.*, Eds. *Strategic Alliances, Mergers and Acquisitions*. Cheltenham, U.K.: Edward Elgar, 2010.

[22] Intelligent Community Forum, *Intelligent Community of the Year* [Online]. Available: https://www.intelligentcommunity.org/index.php?submenu=Awards&src=gendocs&ref=ICF_Awards&category=Events&link=ICF_Awards Sep. 4, 2011 (last date accessed).

[23] B. Gates, "The New World of Work," *Microsoft Executive E-mail*, May 19, 2005 [Online]. Available: http://www.microsoft.com/mscorp/execmail/2005/05-19newworldofwork.mspx Sep. 4, 2011 (last date accessed).

[24] A. A. Verhoeff, "No technical innovation without social innovation," Ph.D. dissertation, Open Universiteit Nederland, Heerlen, The Netherlands, 2011.

[25] C. W. L. Hill and T. M. Jones, "Stakeholder-agency theory," *J. Manag. Studies*, vol. 29, no. 2, pp. 131–154, 1992.

[26] D. J. B. Joosse, *Negotiation in Industrial Relations*. Den Haag, The Netherlands: AWVN, 2010.

[27] S. Grönfeldt and J. B. Strother, *Service Leadership: The Quest for Competitive Advantage*. Thousand Oaks, CA: Sage, 2005.

[28] R. Quinn, "Building the bridge when you walk on it," *Leader Leader*, vol. 2004, no. 34, pp. 21–26, 2004.

[29] E. T. Hall, *Beyond Culture*. New York: Doubleday, 1976.

[30] Central Bureau for Statistics, *ICT, kennis en economie* [ICT, Knowledge and Economy], The Hague, The Netherlands, Rep., 2001.

[31] Microsoft Nederland and Boer & Croon, *Het Nieuwe Werken: Resultaten van een kwalitatief onderzoek* [The New World of Work: Results of a Qualitative Study], Rep., 2007.

[32] R. E. Quinn and K. S. Cameron, *Paradox and Transformation: Toward a Theory of Change in Organization and Management*. Cambridge, MA: Ballinger Publishing Co., 1988.

[33] C. A. O'Reilly, III, and M. I. Tushman, "The ambidextrous organization," *Harvard Business Rev.*, vol. 82, no. 4, pp. 74–81, 2004.

[34] C. Argyris, *Overcoming Organizational Defenses*. Boston, MA: Allyn & Bacon, 1990.

[35] M. Broadbent and E. Kitzis, *The New CIO Leader: Setting the Agenda and Delivering Results*. Stamford, CT: Gartner, 2005.

[36] Private communication with Paul-Peter Feld, Director, Human Resources and Organization, Xerox, The Netherlands.

LIST OF REFERENCES FOR BOXED QUOTATIONS

Chapter 1

J. Spira, *Overload! How Too Much Information Is Hazardous to Your Organization.* Hoboken, NJ: Wiley, 2011.

Chapter 2

H. A. Simon, "Designing organizations for an information-rich world," *Computers, Communication, and the Public Interest.* Baltimore, MD: John Hopkins Press, 1971.

Chapter 3

M. Madsen. Tony Buzan's mindmapping, *Personal Entry* [Online]. Available: http://clicks.robertgenn.com/html/archive.php?clickback=develop_ideas.html&id=478 Jan. 23, 2012 (last date accessed).

Chapter 4

B. Gamarekian,"Working profile: Daniel J. Boorstin. Helping the Library of Congress fulfill its mission," *The New York Times, B6,* Jul. 8, 1983.

Information Overload: An International Challenge for Professional Engineers and Technical Communicators, First Edition. Edited by Judith B. Strother, Jan Ulijn, Zohra Fazal.
© 2012 Institute of Electrical and Electronics Engineers. Published 2012 by John Wiley & Sons, Inc.

Chapter 5

G. Hofstede, *et al.*, *Cultures and Organizations: Software of the Mind*, 3rd ed. New York: McGraw-Hill, 2010.

Chapter 6

M. Hattingh. Family of blue giraffes, *Personal Entry* [Online]. Available: http://clicks. robertgenn.com/html/archive.php?clickback=moose.htm&id=251 Jan. 23, 2012 (last date accessed).

Chapter 7

S. Mills. Epicurean simplicity, *Entry* [Online]. Available: http://quote.robertgenn.com/ auth_search.php?authid=5657 Jan. 23, 2012 (last date accessed).

Chapter 8

M. Twain. *Letter to George Bainton October 15, 1888* [Online]. Available: http://www. twainquotes.com/Lightning.html Jan. 23, 2012 (last date accessed).

Chapter 9

P. Schieber,"The wit and wisdom of Grace Hopper," *The OCLC Newsletter*, no. 167, Mar.– Apr., 1987.

Chapter 10

R. McDaniel. Information, *Entry* [Online]. Available: http://quote.robertgenn.com/auth_ search.php?authid=534 Jan. 23, 2012 (last date accessed).

Chapter 11

M. Gelb. Information, *Entry* [Online]. Available: http://quote.robertgenn.com/auth_ search.php?authid=2877 Jan. 23, 2012 (last date accessed).

Chapter 12

B. Disraeli, *Endymoin*. London, U.K.: British Library, 2011.

AUTHOR INDEX

Information Overload: An International Challenge for Professional Engineers and Technical Communicators,
First Edition. Edited by Judith B. Strother, Jan Ulijn, Zohra Fazal.
© 2012 Institute of Electrical and Electronics Engineers. Published 2012 by John Wiley & Sons, Inc.

SUBJECT INDEX

Information Overload: An International Challenge for Professional Engineers and Technical Communicators,
First Edition. Edited by Judith B. Strother, Jan Ulijn, Zohra Fazal.
© 2012 Institute of Electrical and Electronics Engineers. Published 2012 by John Wiley & Sons, Inc.

Books in the IEEE PCS
PROFESSIONAL ENGINEERING
COMMUNICATION SERIES

Sponsored by IEEE Professional Communication Society

Series Editor: Traci Nathans-Kelly

This series from IEEE's Professional Communication Society addresses professional communication elements, techniques, concerns, and issues. Created for engineers, technicians, academic administration/faculty, students, and technical communicators in related industries, this series meets a need for a targeted set of materials that focus on very real, daily, on-site communication needs. Using examples and expertise gleaned from engineers and their colleagues, this series aims to produce practical resources for today's professionals and pre-professionals.

Information Overload: An International Challenge for Professional Engineers and Technical Communicators · Judith B. Strother, Jan M. Ulijn, and Zohra Fazal (editors and authors)

Negotiating Cultural Encounters: Case Studies in Intercultural Engineering and Technical Communication · Han Yu and Gerald Savage (editors and authors)

Forthcoming:

Slide Rules: Design, Build, and Archive Presentations in the Engineering and Technical Fields · Traci Nathans-Kelly and Christine G. Nicometo (authors)

Communication Practices in Engineering, Manufacturing, and Research for Food, Drug, and Water Safety · David Wright (editor and author)

Decisions: An Engineering and Executive Perspective · Gerard "Gus" Gaynor (author)

Teaching and Training for Global Engineering: Perspectives on Culture and Communication Practices · Kirk St. Amant and Madelyn Flammia (editors and authors)

International Virtual Teams: Engineering Successful Global Communication · Pam Estes Brewer (author)

Printed and bound by CPI Group (UK) Ltd, Croydon, CR0 4YY

27/10/2024

14580280-0002